D0938860

Web and Network Data Science

Modeling Techniques in Predictive Analytics

THOMAS W. MILLER

Editor-in-Chief: Amy Neidlinger
Executive Editor: Jeanne Glasser
Operations Specialist: Jodi Kemper
Cover Designer: Alan Clements
Managing Editor: Kristy Hart
Project Editor: Andy Beaster
Senior Compositor: Gloria Schurick
Manufacturing Buyer: Dan Uhrig

©2015 by Thomas W. Miller
Published by Pearson Education, Inc.
Upper Saddle River, New Jersey 07458

Pearson offers excellent discounts on this book when ordered in quantity for bulk purchases or special sales. For more information, please contact U.S. Corporate and Government Sales, 1-800-382-3419, corpsales@pearsontechgroup.com. For sales outside the U.S., please contact International Sales at international@pearsoned.com.

Company and product names mentioned herein are the trademarks or registered trademarks of their respective owners.

All rights reserved. No part of this book may be reproduced, in any form or by any means, without permission in writing from the publisher.

Printed in the United States of America

First Printing December 2014

ISBN-10: 0-13-388644-1
ISBN-13: 978-0-13-388644-3

Pearson Education LTD.
Pearson Education Australia PTY, Limited.
Pearson Education Singapore, Pte. Ltd.
Pearson Education Asia, Ltd.
Pearson Education Canada, Ltd.
Pearson Educacin de Mexico, S.A. de C.V.
Pearson Education—Japan
Pearson Education Malaysia, Pte. Ltd.
Library of Congress Control Number: 2014956958

Contents

Preface

"Scotty, beam me up."

—William Shatner as Captain Kirk in
Star Trek IV: The Voyage Home (1986)

The web is a network of linked pages. The web is a communication medium. The web is the locus of the world's information. We spend much of our time searching the web, extracting relevant data, and analyzing those data. Our lives are easier when we can work efficiently on the web. This book shows how.

The book emerged from a course I teach at Northwestern University. The course started as an introduction to website analytics, looking at usage statistics and performance in search. Then I added concepts from network science and social media. After teaching the course for two years, I realized that gathering information from the web provided a unifying theme. There is much to learn about web and network data science. This book, like the course, provides a guide.

Web and network data science is data science and network science combined, focusing on the web as an information resource. And the best way to learn about it is to work through examples. We include many examples in this book. We help researchers and analysts by providing a ready resource and reference guide for modeling techniques. We show programmers how to build on a foundation of code that works to solve real business problems.

v

The truth about what we do is in the programs we write. It is there for everyone to see and for some to debug. To promote student learning, each program includes step-by-step comments and suggestions for taking the analysis further. Data sets and computer programs are available from the book's website at `http://www.ftpress.com/miller/`.

Python gets its name from Monty Python. We see packages with devious names such as Twisted and Scrapy. R has its lubridate and zoo. Good things come from people who work and have fun at the same time. It is fun rather than profit or fame that motivates contributors to open source, and I am happy to be part of the Python and R communities. Let the fun begin.

When working on web and network problems, some things are more easily accomplished with Python, others with R. And there are times when it is good to offer solutions in both languages, checking one against the other. Together, Python and R are good at gathering web and network data and analyzing those data.

There is a long list of programming tools we mention only in passing. Web masters, charged with the task of making things happen on the web, rely on additional languages and technologies, including JavaScript, Apache and .Net web services, and database systems. We discuss these technologies but do not provide programming code.

Most of the data in the book were obtained from public domain data sources. Supporting data for the cases come from the University of California–Irvine Machine Learning Repository and the Stanford Large Network Dataset Collection. Movie information was obtained courtesy of The Internet Movie Database, used with permission. IMDb movie reviews data were organized by Andrew L. Mass and his colleagues at Stanford University. William W. Cohen of Carnegie Mellon University maintains the data for the Enron case. Maksim Tsvetovat maintains the data for the Quake Talk case. We are most thankful to these scholars for providing access to rich data sets for research.

Many have influenced my intellectual development over the years. There were those good thinkers and good people, teachers and mentors for whom I will be forever grateful. Sadly, no longer with us are Gerald Hahn Hinkle in philosophy and Allan Lake Rice in languages at Ursinus College, and Herbert Feigl in philosophy at the University of Minnesota. I am also most thankful to David J. Weiss in psychometrics at the University of Minnesota and Kelly Eakin in economics, formerly at the University of Oregon. Good teachers—yes, great teachers—are valued for a lifetime.

Thanks to Stan Narusiewcz who gave me my first job in business as a network engineer and to Tom Obinger who showed me how to be successful in selling computer systems as well as networks. Along with Bill JoBush and Brian Hill, they served as able managers and colleagues across various parts of my career as an information systems professional.

Thanks to Michael L. Rothschild, Neal M. Ford, Peter R. Dickson, and Janet Christopher who provided invaluable support during our years together at the University of Wisconsin–Madison. I am most grateful to the students and executive advisory board members of the A. C. Nielsen Center for Marketing Research and to Jeff Walkowski and Neli Esipova who worked with me in exploring online surveys and focus groups when those methods were just starting to be used for primary research.

I am fortunate to be involved with graduate distance education at Northwestern University's School of Professional Studies. Thanks to Glen Fogerty, who offered me the opportunity to teach and take a leadership role in the predictive analytics program at Northwestern University. Thanks to colleagues and staff who administer this exceptional graduate program. And thanks to the many students and fellow faculty from whom I have learned.

ToutBay is an emerging firm in the data science space. With co-founder Greg Blence, I have great hopes for growth in the coming years. Thanks to Greg for joining me in this effort and for keeping me grounded in the practical needs of business. Academics and data science models can take us only so far. Eventually, to make a difference, we must implement our ideas and models, sharing them with one another.

I live in California, four miles north of Dodger Stadium, teach for Northwestern University in Evanston, Illinois, and direct product development at ToutBay, a data science firm in Tampa, Florida. Such are the benefits of a good Internet connection.

Amy Hendrickson of TeXnology Inc. applied her craft, making words, tables, and figures look beautiful in print—another victory for open source. Thanks to Donald Knuth and the TeX / LaTeX community for their contributions to this wonderful system for typesetting and publication.

The book draws on materials developed for the web and network data science course at Northwestern University. Students from that course provided ideas and inspiration. Lorena Martin reviewed the book and provided much needed feedback. Candice Bradley served dual roles as a reviewer and copyeditor. I am most grateful for their help and encouragement. Thanks also to my editor, Jeanne Glasser Levine, and publisher, Pearson/FT Press, for making this book possible. Any writing issues, errors, or items of unfinished business, of course, are my responsibility alone.

My good friend Brittney and her daughter Janiya keep me company when time permits. And my son Daniel is there for me in good times and bad, a friend for life. My greatest debt is to them because they believe in me.

Thomas W. Miller
Glendale, California
November 2014

Figures

Tables

Exhibits

1

Being Technically Inclined

"Why don't you come up sometime and see me?"

—MAE WEST AS LADY LOU IN *She Done Him Wrong* (1933)

I began my business career working as a network engineer in Roseville, Minnesota. Just out of graduate training in statistics at the University of Minnesota, I was well schooled in math and models but lacking business understanding. It did not take long to learn that success in my job meant coming up with meaningful answers for management.

In the dial-up and leased-line world of the late 1970s, asynchronous, bisynchronous, and synchronous connections ruled the day. We translated network protocols into polling and message bits and noted the bits per second that each communication line could accommodate. Queuing theory and discrete event simulation guided the analysis.

A bank teller would make a request, hitting the return key at a terminal. The terminal was connected to a control unit, which in turn was connected to a remote concentrator processor. Leased lines went from remote concentrator processors to a front-end processor, providing a channel to the mainframe computer. These were the nodes and links of networks at the time. The queuing problem involved estimating how long the bank teller would have to wait to get a response from the mainframe.

Fast forward forty years. We have moved away from dial-up and leased lines. Protocols are packet-switched and mobile. Users of networks are everywhere, not just at banks, businesses, and research establishments. Most mainframes have been replaced by clusters of microcomputers. We carry the smallest of computers in our pockets. We wear computers if we like. Of course, when making requests of remote systems, we are still waiting for responses, although now we wait wherever we are and whatever we are doing.

With computer hardware looking more like a commodity and software going open-source, established technology firms seek out new opportunities in business intelligence and data science. IBM moves from hardware to software to consulting. HP splits into two firms, one focused on hardware, the other on business services and utilities. Meanwhile, Apple fights battles with Amazon and Google over the distribution of media, while suing Samsung for copyright violations.

The big battles of today concern information and its online distribution. Intellectual property, special knowledge, competitive intelligence, expertise, and art—these add value in an online world that otherwise appears to offer information for free.

It is hard to resist the allure of the web. She is the ultimate seductress, holding the promise of unlimited information and connection to all. The web is a huge data repository, a path to the world's knowledge, and the research medium through which we develop new knowledge.

Web and network data science is a collection of technologies and modeling techniques, some well understood, others emerging, that help us to understand the web and the networks in our lives. The technologies of the web are many, with current market shares tracked by Alexa Internet (2014) and W3Techs (2014), among others.

To work efficiently in web and network data science, it helps to be technically inclined, with some understanding of at least three languages: Python, R, and JavaScript. Python is the tool of choice for data preparation (or data munging, as it is sometimes called). R provides specialized tools for modeling and data visualization. And JavaScript is the client-side language of the web, available on every major web browser. When working on web and network problems, it also helps to know HTML5, CSS3, XPath, a variety of text and image file formats, Java, Linux, Apache, .Net web services, database systems, and server-side languages such as Perl and PHP. It helps to be technically inclined, but there is a limit to what we can cover in one book. We provide a glossary of terms as the final appendix in the book.

From its humble beginnings as a language that Brendan Eich developed in ten days in 1995 at the former company Netscape, JavaScript has emerged as the client-side language of the web, a browser-based engine for managing user interaction. JavaScript is dominant on the client side, with an estimated 88 percent of websites using the technology and with 11.8 percent of websites being pure/static HTML sites with no client-side programming (W3Techs 2014).

Crockford (2008) tells us what is right and wrong with JavaScript. Others tell us how to use it in practice (Stefanov 2010; Flanagan 2011; Resig and BearBibeault 2013). Recently, with the emergence of Node.js, JavaScript has taken on a role on the server side (Hughes-Croucher and Wilson 2012; Wanderschneider 2013; Cantelon, Harter, Holowaychuk, and Rajlich 2014). There are those who promote end-to-end JavaScript applications with client- and server-side programs and document databases (Mikowski and Powell 2014). JavaScript Object Notation (JSON), a data interchange format, is more readable than XML and easily integrated into a MongoDB document database (Chodorow 2013; Copeland 2013; Hoberman 2014), for example. JavaScript would certainly rule the web if it had sufficient capabilities as a modeling and analysis language. It does not.

Today's world of data science brings together statisticians fluent in R and information technology professionals fluent in Python. These communities have much to learn from each other. For the practicing data scientist, there are considerable advantages to being multilingual.

Designed by Ross Ihaka and Robert Gentleman, R first appeared in 1993. R represents an extensible, object-oriented, open-source scripting language for programming with data. It is well established in the statistical community and has syntax, data structures, and methods similar to its precursors, S and S-Plus. Contributors to the language have provided more than five thousand packages, most focused on traditional statistics, machine learning, and data visualization. R is the most widely used language in data science, but it is not a general-purpose programming language.

Guido van Rossum, a fan of Monty Python, released version 1.0 of Python in 1994. This general-purpose language has grown in popularity in the ensuing years. Many systems programmers have moved from Perl to Python, and Python has a strong following among mathematicians and scientists. Many universities use Python as a way to introduce basic concepts of object-oriented programming. An active open-source community has contributed more than fifteen thousand Python packages.

Sometimes referred to as a "glue language," Python provides a rich open-source environment for scientific programming and research. For computer-intensive applications, it gives us the ability to call on compiled routines from C, C++, and Fortran. We can also use Cython to convert Python code into optimized C. For modeling techniques or graphics not currently implemented in Python, we can execute R programs from Python.

Some problems are more easily solved with Python, others with R. We benefit from Python's capabilities as a general-purpose programming language. We draw on R packages for traditional statistics, time series analysis, multivariate methods, statistical graphics, and handling missing data. Accordingly, this book includes Python and R code examples and represents a dual-language guide to web and network data science.

Browser usage has changed dramatically over the years, with the rise of Google Chrome and the decline of Microsoft Internet Explorer (IE). Table 1.1 and figure 1.1 show worldwide browser usage statistics from October 2008 through October 2014. It is good to have some familiarity with browsers and the tools they provide for examining the text elements and structure of web pages.

The challenge of "big data," as they are sometimes called, is not so much the volume of data. It is that these data arise from sources poorly understood, in particular the web and social media. Data are everywhere on the web. We need to find our way to the relevant data and obtain those data in an efficient manner.

Application programming interfaces (APIs) are one way to gather data from the web, and Russell (2014) provides a useful review of social media APIs. Unfortunately, APIs have syntax, parameters, and authorization codes that can change at the whim of the data providers. We employ a different ap-

Table 1.1. *Worldwide Web Browser Usage Percentages (2008–2014)*

Year	IE	Chrome	Firefox	Safari	Other
2008	67.68	1.02	25.54	2.91	2.85
2009	57.96	4.17	31.82	3.47	2.58
2010	49.21	12.39	31.24	4.56	2.60
2011	40.18	25.00	26.39	5.93	2.50
2012	32.08	34.77	22.32	7.81	3.02
2013	28.96	40.44	18.11	8.54	3.95
2014	19.25	47.57	17.00	10.95	5.23

Data obtained from StatCounter (2014).

Figure 1.1. *Worldwide Web Browser Usage (July 2008 through October 2014)*

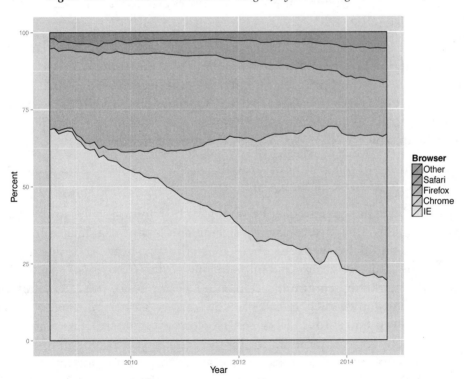

Data obtained from StatCounter (2014).

Figure 1.2. Web and Network Data Science: Online Research Process

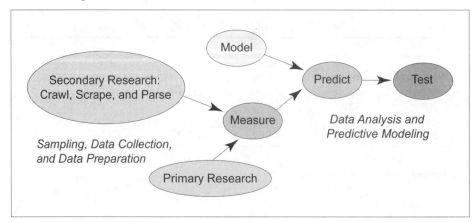

proach, focusing on general purpose technologies for automated data acquisition from the web.

Figure 1.2 summarizes the online research process. Sampling, data collection, and data preparation consume much of our time, with secondary research dominating primary research. Online secondary research draws from existing web data. We review secondary research methods in chapter three and use them in many subsequent chapters. Primary research online is facilitated by the web. We cover these methods in appendix B.

The domain of web and network data science is large. There are many questions to address, as shown in the list to follow.

- **Website design and user behavior.** Web analytics, as it is understood by many, involves collecting, storing, and analyzing data from users of a particular website. There are many questions to be addressed. How shall we design and implement websites (for ease of use, visibility, marketing communication, good performance in search, and/or conversion of visits to sales)? How can we gather information from the web efficiently? How can we convert semi-structured and unstructured text into data for input to analysis and modeling? What kinds of website and social media measures make the most sense? Who are the users of a website, and how do they use it? How well does a website do in serving user needs? How well does a website do compared with other websites?

- **Network paths and communication.** Web and network data science is much more than website analytics. We look at each website in the context of others on the web. We think in terms of networks—information nodes connected to one another, and users communicating with one another. What is the shortest, fastest, or lowest cost path between two locations? What is the fastest way to spread a message across a network? Which activities are on the critical path to completing a project? How long must we wait for a response from the server?

- **Communities and influence.** Social media provide a glimpse of electronic social networks in action. Here we have the questions of social network analysis. Are there identifiable groups of people in this community? Who are the key players, the most important people in a group? Who are the people with prestige, influence, or power? Who is best positioned to be the leader of a group?

- **Individual and group behavior.** As data scientists, we are often called on to go beyond description and provide predictions about future behavior or performance. So we have more questions to address. Will this person buy the product, given his/her connections with other buyers or non-buyers? Will this person vote for the candidate, given his/her connections with other voters? Given the motives of individuals, what can we predict for the group? Given growth in the network in the past, what can we expect for the future?

- **Information and networks.** As an information resource, the web is unparalleled. Additional questions arise about the nature of online information. Which are the best websites for getting information about a particular topic? Who are the most credible sources of information? How shall we characterize a domain of knowledge? How can we use the web to obtain competitive intelligence? How can we utilize web-based information as a database for answering questions (domain-specific and general questions)?

This book is designed to provide an overview of the domain of web and network data science. We illustrate measurement and modeling techniques for answering many questions, and we cite resources for additional learning. Some of the techniques may be regarded as basic, others advanced. All are important to the work of data science.

Some say that data science is the new statistics. And in a world dominated by data, data science is beginning to look like the new business and the new IT as well. Nowhere is this more apparent than when working on web and network problems. With unlimited data mediated and distributed through the web, there is certainly enough to keep us busy for a long time.

To begin the programming portion of the book, exhibit 1.1 lists a Python program for exploring web browser usage statistics. Exhibit 1.2 shows the corresponding R program and draws on graphics software from Wickham and Chang (2014).

Exhibit 1.1. *Analysis of Browser Usage (Python)*

```python
# Analysis of Browser Usage (Python)

# prepare for Python version 3x features and functions
from __future__ import division, print_function

# import packages for data analysis
import pandas as pd  # data structures for time series analysis
import datetime  # date manipulation
import matplotlib.pyplot as plt

# browser usage data from StatCounter Global Stats
# retrieved from the World Wide Web, October 21, 2014:
# \url{http://gs.statcounter.com/#browser-ww-monthly-200807-201410
# read in comma-delimited text file
browser_usage = pd.read_csv('browser_usage_2008_2014.csv')
# examine the data frame object
print(browser_usage.shape)
print(browser_usage.head())

# identify date fields as dates with apply and lambda function
browser_usage['Date'] = \
    browser_usage['Date']\
    .apply(lambda d: datetime.datetime.strptime(str(d), '%Y-%m'))
# define Other category
browser_usage['Other'] = 100 -\
    browser_usage['IE'] - browser_usage['Chrome'] -\
    browser_usage['Firefox'] - browser_usage['Safari']

# examine selected columns of the data frame object
selected_browser_usage = pd.DataFrame(browser_usage,\
    columns = ['Date', 'IE', 'Chrome', 'Firefox', 'Safari', 'Other'])
print(selected_browser_usage.shape)
print(selected_browser_usage.head())

# create multiple time series plot
selected_browser_usage.plot(subplots = True,  \
    sharex = True, sharey = True, style = 'k-')
plt.legend(loc = 'best')
plt.xlabel('')
plt.savefig('fig_browser_mts_Python.pdf',
    bbox_inches = 'tight', dpi=None, facecolor='w', edgecolor='b',
    orientation='portrait', papertype=None, format=None,
    transparent=True, pad_inches=0.25, frameon=None)

# Suggestions for the student:
# Explore alternative visualizations of these data.
# Try the Python package ggplot to reproduce R graphics.
# Explore time series for other software and systems.
```

Exhibit 1.2. *Analysis of Browser Usage (R)*

```
# Analysis of Browser Usage (R)

# begin by installing necessary package ggplot2

# load package into the workspace for this program
library(ggplot2)  # grammar of graphics plotting

# browser usage data from StatCounter Global Stats
# retrieved from the World Wide Web, October 21, 2014:
# \url{http://gs.statcounter.com/#browser-ww-monthly-200807-201410
# read in comma-delimited text file
browser_usage <- read.csv("browser_usage_2008_2014.csv")
# examine the data frame object
print(str(browser_usage))
# define Other category
browser_usage$Other <- 100 -
    browser_usage$IE - browser_usage$Chrome -
    browser_usage$Firefox - browser_usage$Safari

# define time series data objects
IE_ts <- ts(browser_usage$IE, start = c(2008, 7), frequency = 12)
Chrome_ts <- ts(browser_usage$Chrome, start = c(2008, 7), frequency = 12)
Firefox_ts <- ts(browser_usage$Firefox, start = c(2008, 7), frequency = 12)
Safari_ts <- ts(browser_usage$Safari, start = c(2008, 7), frequency = 12)
Other_ts <- ts(browser_usage$Other, start = c(2008, 7), frequency = 12)

# create a multiple time series object
browser_mts <- cbind(IE_ts, Chrome_ts, Firefox_ts, Safari_ts, Other_ts)
dimnames(browser_mts)[[2]] <- c("IE", "Chrome", "Firefox", "Safari", "Other")
# plot multiple time series object using standard R graphics
pdf(file="fig_browser_mts_R.pdf",width = 11,height = 8.5)
ts.plot(browser_mts, ylab = "Percent Usage", main="",
    plot.type = "single", col = 1:5)
legend("topright", colnames(browser_mts), col = 1:5,
    lty = 1, cex = 1)
dev.off()

# define Year as numeric with fractional values for months
browser_usage$Year <- as.numeric(time(IE_ts))

# build data frame for plotting a stacked area graph
Browser <- rep("IE", length = nrow(browser_usage))
Percent <- browser_usage$IE
Year <- browser_usage$Year
plotting_data_frame <- data.frame(Browser, Percent, Year)

Browser <- rep("Chrome", length = nrow(browser_usage))
Percent <- browser_usage$Chrome
Year <- browser_usage$Year
plotting_data_frame <- rbind(plotting_data_frame,
    data.frame(Browser, Percent, Year))
```

```
Browser <- rep("Firefox", length = nrow(browser_usage))
Percent <- browser_usage$Firefox
Year <- browser_usage$Year
plotting_data_frame <- rbind(plotting_data_frame,
    data.frame(Browser, Percent, Year))

Browser <- rep("Safari", length = nrow(browser_usage))
Percent <- browser_usage$Safari
Year <- browser_usage$Year
plotting_data_frame <- rbind(plotting_data_frame,
    data.frame(Browser, Percent, Year))

Browser <- rep("Other", length = nrow(browser_usage))
Percent <- browser_usage$Other
Year <- browser_usage$Year
plotting_data_frame <- rbind(plotting_data_frame,
    data.frame(Browser, Percent, Year))

# create ggplot plotting object and plot to external file
pdf(file = "fig_browser_usage_stacked_area_R.pdf", width = 11, height = 8.5)
area_plot <- ggplot(data = plotting_data_frame,
    aes(x = Year, y = Percent, fill = Browser)) +
    geom_area(colour = "black", size = 1, alpha = 0.4) +
    scale_fill_brewer(palette = "Blues",
        breaks = rev(levels(plotting_data_frame$Browser))) +
    theme(legend.text = element_text(size = 15))  +
    theme(legend.title = element_text(size = 15)) +
    theme(axis.title = element_text(size = 15))
print(area_plot)
dev.off()
```

2

Delivering a Message Online

"You had me at 'hello.' "

—RENÉE ZELLWEGER AS DOROTHY BOYD IN *Jerry Maguire* (1996)

My management style is simple: Hire good people and let them do their thing. By spending very little time telling others what to do, I have more time to do the things I like to do. My hands-off approach to management works as long as the general message—the vision for the organization—is clearly communicated and as long as others believe in the message.

Delivering a message is especially important when working online. Lacking face-to-face meetings, we rely on electronic communication. We put a message on the web and hope that people see and understand it. We wait for a reaction from an online community.

We can gauge success in delivering messages from the server side or the client side. On the server side there is the log of user requests. It is complete in showing where each request comes from (IP location and browser source), the web page being requested, and the status of the request. Timestamps reveal arrival and service times. The server log provides a server-centric view of website performance.

To understand a website from the user's point of view, we turn to service providers. The current market leader in client-side monitoring is Google.[1]

Client-side performance monitoring requires explicit compliance by website providers and implicit compliance by website visitors. An organization that elects to utilize Google client-side monitoring must place JavaScript code on each web page it wants to monitor.

Website managers can choose to use Google website monitoring or not. Most choose to use it because the base service is free. An estimated 49.8 percent of websites worldwide use Google web monitoring services, accounting for 81.5 percent of the traffic on the web (W3Techs 2014).

When a user/client receives a page from the website, the client's browser executes the tracking code, placing a cookie on the user's computer and sending that cookie with information (including the user's IP address) to Google. The user's computer retains the cookie text file for future reference. So when the user revisits the website, Google can identify the user as a repeat visitor to the website. The cookie remains on the user's computer until the user erases it.

For client-side web monitoring to work, users must allow cookies to be placed on their devices, be they computers, tablets, or smartphones. Most users of the web allow cookies. They may not understand cookies or how to disallow cookie collection. Or they may like the benefits that come from a personalized, cookie-mediated web experience.

For participating websites and users, Google tracks every request of every website. Extensive data are available to website managers who use Google's client-side monitoring services.

The *ToutBay Begins* case from appendix C (page 284) includes website performance data from Google. These data summarize website traffic from the first twenty-three weeks of operation of the ToutBay website (April 12 through September 19, 2014). By analyzing these data, we can learn about browser and operating system usage by website visitors, as shown in figures 2.1 and 2.2, respectively.

---

[1] In October 2014 Google web monitoring products included Google Analytics and Google Universal Analytics. Base versions of these products were free. Premium versions were also available. We restrict our discussion to the Google Analytics base product for client-side web monitoring.

Figure **2.1.** *Browsers Used by Website Visitors*

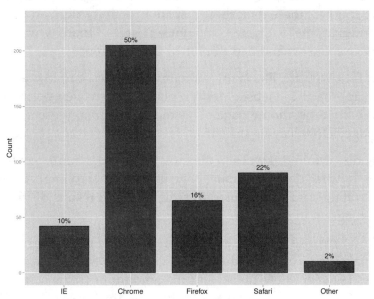

Figure **2.2.** *Operating Systems Used by Website Visitors*

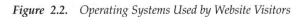
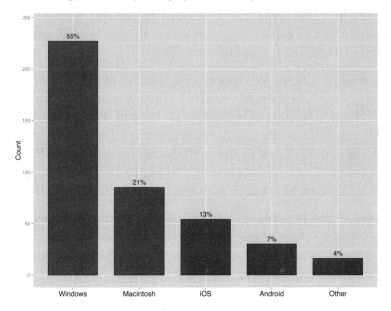

We can track website traffic by noting the numbers of sessions and page views and the average length of sessions (in seconds), as shown in figure 2.3. For ToutBay, these data show a spike in activity right around the time of the International R Users' Conference (UseR!). ToutBay was one of the conference sponsors and had a booth at the conference.

The ToutBay website, like many contemporary sites, employs a single-page website approach to website design, placing extensive information about the company on its home page. This means that many users see only the home page before leaving the site.

To provide meaningful website diagnostics, ToutBay implemented code to measure how far down the home page users were scrolling. An analysis of these scrolling data is presented in table 2.1, with results displayed in the Sankey diagram in figure 2.4.

To make a Sankey diagram, also called a process or river diagram, we think of web pages or locations within web pages as nodes in a network. We identify links between these nodes and the amount of traffic across the links. The amount of traffic is reflected by the line thickness of each link. The Sankey diagram provides a picture of user navigation across the website. It helps in evaluating website structure and performance.

Online retailers seek orders. They want to convert website visitors into customers. We can expect online retailers to pay special attention to the links between landing pages and order pages.

There is another side to web monitoring that is less frequently discussed— the privacy side. As long as we allow cookies on our computers, tablets, or smartphones, website service providers such as Google can see every request we make of every monitored website on the World Wide Web. And whether we allow cookies or not, with server logs, every website manager knows about every request we make to his or her website. Our privacy depends on those who control, manage, and monitor web activity.

One study suggested that 99 percent of web users allow JavaScript code execution and 85 percent allow JavaScript and both first-party and third-party cookies (Priebe 2009). These numbers may change as more users learn about cookies and IP addresses.

Figure 2.3. *Website Traffic Analysis*

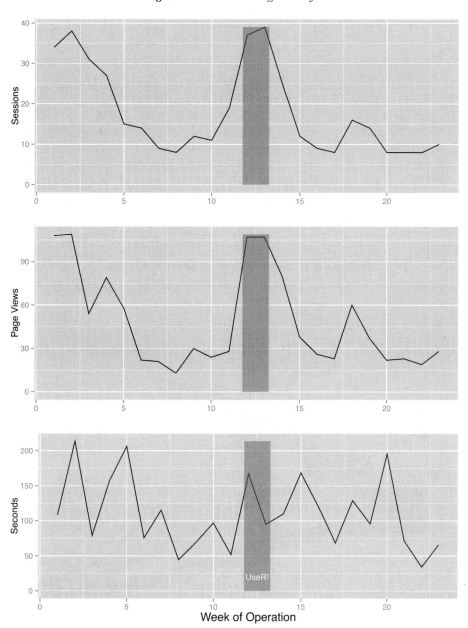

Table 2.1. *Website Home Page Scrolling*

Scroll Position	Enter Count	Exit Count	Percentage of Visitors from Previous Scroll Position	Percentage of Total Visitor Sessions
Top of Home Page	412			
ToutBay Video	296	116	72	72
What's ToutBay?	234	62	79	57
How does it work?	193	41	82	47
FAQ	174	19	90	42
News	125	49	72	30

Figure 2.4. *Sankey Diagram of Home Page Scrolling*

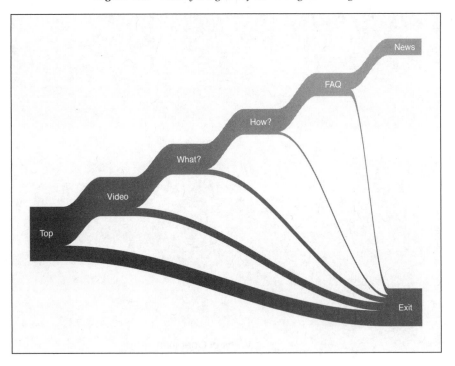

The European Union's e-Privacy Directive requires users' prior consent before cookies may be placed on their computers unless those cookies are necessary for the delivery of services (European Parliament 2002; Wikipedia 2014b). An IP address may be regarded as personally identifiable information. In some countries (including Germany) collection of IP addresses is illegal without the explicit prior consent of the user. When doing business in these countries, it is necessary to anonymize IP addresses (Clifton 2012).

Exhibit 2.1 shows the R program for analyzing client-side data for the Tout-Bay website. It draws on R packages from Auguie (2014), Grolemund and Wickham (2014), Neuwirth (2014), Weiner (2014), and Wickham and Chang (2014).

Exhibit 2.1. *Website Traffic Analysis (R)*

```
# Website Traffic Analysis (R)

# begin by installing necessary packages

# load packages into the workspace for this program
library(gridExtra)  # grid plotting utilities
library(ggplot2)  # grammar of graphics plotting
library(lubridate)  # date and time functions
library(riverplot)  # Sankey diagrams
library(RColorBrewer)  # colors for plots

# user-defined function to convert hh:mm:ss to seconds
make_seconds <- function(hhmmss) {
    hhmmss_list <- strsplit(hhmmss, split = ":")
    3600 * as.numeric(hhmmss_list[[1]][1]) +
        60 * as.numeric(hhmmss_list[[1]][2]) +
        as.numeric(hhmmss_list[[1]][3])
    }

# read in data from ToutBay Begins case
toutbay_begins <- read.csv("toutbay_begins.csv", stringsAsFactors = FALSE)

# examine the data frame object
print(str(toutbay_begins))

# set date as date object
toutbay_begins$date <- parse_date_time(toutbay_begins$date, "mdy")

# convert ave_session_duration to ave_session_seconds, total_session_seconds
toutbay_begins$ave_session_seconds <- numeric(nrow(toutbay_begins))
for (i in seq(along = toutbay_begins$ave_session_duration))
    toutbay_begins$ave_session_seconds[i] <-
        make_seconds(toutbay_begins$ave_session_duration[i])
# compute total seconds across all sessions in the day
toutbay_begins$total_session_seconds <-
    toutbay_begins$ave_session_seconds * toutbay_begins$sessions

# 161 days = 23 weeks so we can index by weeks
week <- NULL
for (i in 1:23) week <- c(week, rep(i, times = 7))
toutbay_begins$week <- week

# compute other_browser browser counts
toutbay_begins$other_browser <- toutbay_begins$sessions -
    toutbay_begins$chrome - toutbay_begins$safari -
    toutbay_begins$firefox - toutbay_begins$internet_explorer

# compute other_system operating system counts (Linux and others)
toutbay_begins$other_system <- toutbay_begins$sessions -
    toutbay_begins$windows - toutbay_begins$macintosh -
    toutbay_begins$ios - toutbay_begins$android
```

```
# extract daily counts and totals
toutbay_daily <- toutbay_begins[,
    c("date", "week", "sessions", "users", "pageviews", "scroll_videopromo",
      "scroll_whatstoutbay", "scroll_howitworks", "scroll_faq",
      "scroll_latestfeeds", "internet_explorer", "chrome", "firefox",
      "safari", "other_browser", "windows", "macintosh", "ios", "android",
      "other_system", "total_session_seconds")]
# examine the daily data frame
print(str(toutbay_daily))
print(head(toutbay_daily))

# aggregate by week using sum()
toutbay_weekly <-
    aggregate(toutbay_daily[, setdiff(names(toutbay_daily), c("date","week"))],
    by = list(toutbay_daily$week), FUN = sum)
names(toutbay_weekly)[1] <- "week"  # rename first column of data frame

# compute average session duration in seconds
toutbay_weekly$ave_session_seconds <-
    toutbay_weekly$total_session_seconds / toutbay_weekly$sessions

# examine the weekly data frame
print(str(toutbay_weekly))
print(head(toutbay_weekly))

# create browser data frame for plotting
Browser <- "IE"
Count <- sum(toutbay_weekly$internet_explorer)
browser_data_frame <- data.frame(Browser, Count)
Browser <- "Chrome"
Count <- sum(toutbay_weekly$chrome)
browser_data_frame <- rbind(browser_data_frame,
    data.frame(Browser, Count))
Browser <- "Firefox"
Count <- sum(toutbay_weekly$firefox)
browser_data_frame <- rbind(browser_data_frame,
    data.frame(Browser, Count))
Browser <- "Safari"
Count <- sum(toutbay_weekly$safari)
browser_data_frame <- rbind(browser_data_frame,
    data.frame(Browser, Count))
Browser <- "Other"
Count <- sum(toutbay_weekly$other_browser)
browser_data_frame <- rbind(browser_data_frame,
    data.frame(Browser, Count))

pdf(file = "fig_toutbay_begins_user_browsers_R.pdf", width = 11, height = 8.5)
browser_bar_plot <- ggplot(data = browser_data_frame,
    aes(x = Browser, y = Count)) +
    geom_bar(stat = "identity", width = 0.75,
        colour = "black", fill = "darkblue") +
    ylim(0, 225) +
    theme(axis.title.y = element_text(size = 15, colour = "black")) +
```

```
        theme(axis.title.x = element_blank()) +
        theme(axis.text.x = element_text(size = 15, colour = "black")) +
        annotate("text", x = 1, y = browser_data_frame$Count[1] + 5,
            label = paste(as.character(round(100 * browser_data_frame$Count[1]/
                sum(browser_data_frame$Count), digits = 0)), "%", sep = "")) +
        annotate("text", x = 2, y = browser_data_frame$Count[2] + 5,
            label = paste(as.character(round(100 * browser_data_frame$Count[2]/
                sum(browser_data_frame$Count), digits = 0)), "%", sep = "")) +
        annotate("text", x = 3, y = browser_data_frame$Count[3] + 5,
            label = paste(as.character(round(100 * browser_data_frame$Count[3]/
                sum(browser_data_frame$Count), digits = 0)), "%", sep = "")) +
        annotate("text", x = 4, y = browser_data_frame$Count[4] + 5,
            label = paste(as.character(round(100 * browser_data_frame$Count[4]/
                sum(browser_data_frame$Count), digits = 0)), "%", sep = "")) +
        annotate("text", x = 5, y = browser_data_frame$Count[5] + 5,
            label = paste(as.character(round(100 * browser_data_frame$Count[5]/
                sum(browser_data_frame$Count), digits = 0)), "%", sep = ""))
print(browser_bar_plot)
dev.off()


# create operating system data frame for plotting
System <- "Windows"
Count <- sum(toutbay_weekly$windows)
system_data_frame <- data.frame(System, Count)
System <- "Macintosh"
Count <- sum(toutbay_weekly$macintosh)
system_data_frame <- rbind(system_data_frame,
    data.frame(System, Count))
System <- "iOS"
Count <- sum(toutbay_weekly$ios)
system_data_frame <- rbind(system_data_frame,
    data.frame(System, Count))
System <- "Android"
Count <- sum(toutbay_weekly$android)
system_data_frame <- rbind(system_data_frame,
    data.frame(System, Count))
System <- "Other"
Count <- sum(toutbay_weekly$other_system)
system_data_frame <- rbind(system_data_frame,
    data.frame(System, Count))


pdf(file = "fig_toutbay_begins_user_systems_R.pdf", width = 11, height = 8.5)
system_bar_plot <- ggplot(data = system_data_frame,
    aes(x = System, y = Count)) +
    geom_bar(stat = "identity", width = 0.75,
        colour = "black", fill = "darkblue") +
    ylim(0, max(system_data_frame$Count + 15)) +
    theme(axis.title.y = element_text(size = 15, colour = "black")) +
    theme(axis.title.x = element_blank()) +
    theme(axis.text.x = element_text(size = 15, colour = "black")) +
    annotate("text", x = 1, y = system_data_frame$Count[1] + 5,
        label = paste(as.character(round(100 * system_data_frame$Count[1]/
            sum(system_data_frame$Count), digits = 0)), "%", sep = "")) +
```

```
      annotate("text", x = 2, y = system_data_frame$Count[2] + 5,
          label = paste(as.character(round(100 * system_data_frame$Count[2]/
              sum(system_data_frame$Count), digits = 0)), "%", sep = "")) +
      annotate("text", x = 3, y = system_data_frame$Count[3] + 5,
          label = paste(as.character(round(100 * system_data_frame$Count[3]/
              sum(system_data_frame$Count), digits = 0)), "%", sep = "")) +
      annotate("text", x = 4, y = system_data_frame$Count[4] + 5,
          label = paste(as.character(round(100 * system_data_frame$Count[4]/
              sum(system_data_frame$Count), digits = 0)), "%", sep = "")) +
      annotate("text", x = 5, y = system_data_frame$Count[5] + 5,
          label = paste(as.character(round(100 * system_data_frame$Count[5]/
              sum(system_data_frame$Count), digits = 0)), "%", sep = ""))
print(system_bar_plot)
dev.off()

# plot multiple time series for sessions, pageviews, and session duration
pdf(file = "fig_toutbay_begins_site_stats_R.pdf", width = 8.5, height = 11)
sessions_plot <- ggplot(data = toutbay_weekly,
    aes(x = week, y = sessions)) + geom_line()  +
    ylab("Sessions") +
    theme(axis.title.x = element_blank()) +
    annotate("rect", xmin = 11.75, xmax = 13.25,
        ymin = 0, ymax = max(toutbay_weekly$sessions),
        fill = "blue", alpha = 0.4)
pageviews_plot <- ggplot(data = toutbay_weekly,
    aes(x = week, y = pageviews)) + geom_line() +
    ylab("Page Views") +
    theme(axis.title.x = element_blank()) +
    annotate("rect", xmin = 11.75, xmax = 13.25,
        ymin = 0, ymax = max(toutbay_weekly$pageviews),
        fill = "blue", alpha = 0.4)

duration_plot <- ggplot(data = toutbay_weekly,
    aes(x = week, y = ave_session_seconds)) + geom_line() +
    xlab("Week of Operation") +
    ylab("Seconds") +
    theme(axis.title.x = element_text(size = 15, colour = "black")) +
    annotate("rect", xmin = 11.75, xmax = 13.25,
        ymin = 0, ymax = max(toutbay_weekly$ave_session_seconds),
        fill = "blue", alpha = 0.4) +
    annotate("text", x = 12.5, y = 20, size = 4, colour = "white",
        label = "UseR!")

mts_plot <- grid.arrange(sessions_plot, pageviews_plot,
    duration_plot, ncol = 1, nrow = 3)
print(mts_plot)
dev.off()

# construct Sankey diagram for home page scrolling
pdf(file = "fig_toutbay_begins_sankey_R.pdf", width = 8.5, height = 11)
nodes <- data.frame(ID = c("A","B","C","D","E","F","G"),
    x = c(1, 2, 3, 4, 5, 6, 6),
    y = c(7, 7.5, 8, 8.5, 9, 9.5, 6),
```

```
    labels = c("Top",
               "Video",
               "What?",
               "How?",
               "FAQ",
               "News",
               "Exit"),
               stringsAsFactors = FALSE,
               row.names = c("A","B","C","D","E","F","G"))
edges <- data.frame(N1 = c("A","B","C","D","E",
                           "A","B","C","D","E"),
                    N2 = c("G","G","G","G","G",
                           "B","C","D","E","F"),

   Value = c(
   sum(toutbay_weekly$sessions)- sum(toutbay_weekly$scroll_videopromo),
   sum(toutbay_weekly$scroll_videopromo) -
       sum(toutbay_weekly$scroll_whatstoutbay),
   sum(toutbay_weekly$scroll_whatstoutbay) -
       sum(toutbay_weekly$scroll_howitworks),
   sum(toutbay_weekly$scroll_howitworks) - sum(toutbay_weekly$scroll_faq),
   sum(toutbay_weekly$scroll_faq) - sum(toutbay_weekly$scroll_latestfeeds),
   sum(toutbay_weekly$scroll_videopromo),
   sum(toutbay_weekly$scroll_whatstoutbay),
   sum(toutbay_weekly$scroll_howitworks),
   sum(toutbay_weekly$scroll_faq),
   sum(toutbay_weekly$scroll_latestfeeds)), row.names = NULL)

selected_pallet <- brewer.pal(9, "Blues")
river_object <- makeRiver(nodes, edges,
   node_styles =
   list(A = list(col = selected_pallet[9], textcol = "white"),
    B = list(col = selected_pallet[8], textcol = "white"),
    C = list(col = selected_pallet[7], textcol = "white"),
    D = list(col = selected_pallet[6], textcol = "white"),
    E = list(col = selected_pallet[5], textcol = "white"),
    F = list(col = selected_pallet[5], textcol = "white"),
    G = list(col = "darkred", textcol = "white")))
plot(river_object, nodewidth = 4, srt = TRUE)
dev.off()
```

3
Crawling and Scraping the Web

Danny: "You gotta walk before you crawl."

Rusty: "Reverse that."

—GEORGE CLOONEY AS DANNY OCEAN AND BRAD PITT
AS RUSTY RYAN IN *Ocean's Eleven* (2001)

It may be my age. It may be the exercise regimen I employ as I try desperately to lose weight. It may be my refusal to use stimulant drugs of any kind, including those delivered through coffee and tea. Whatever the reason, I like to nap.

It is not unusual for me to lie down in the middle of the day and wake up two or three hours later, wondering where the time has gone. I suspect that my napping behavior is abnormal. But I am not sure if it is a sign of poor health. Should I be concerned about my napping? I do not know. I suppose I should learn more about the sleeping habits of men my age.

I could spend hours surfing the web, looking for information about napping behavior. I could go to health and fitness sites. I could look for medical advice and summaries of medical research. But to learn about napping in a systematic way—in a way I could do while napping—I will crawl and scrape the web. Then I will parse the data so they are easy to read.

Crawling is faster than surfing. It is automated. A crawler or spider does more than gather data from one web page. Like an actual spider in its web, a World Wide Web crawler or spider traverses links with ease. It follows web links from one web page to another gathering information from many sources.

Firms such as Google, Microsoft, and Yahoo! crawl the entire web to build indexes for their search engines. Our crawler represents a much more restricted effort, or what is known as a *focused crawl*.

A focused crawl has a starting point or points, usually a list of relevant web addresses. We surf a bit before we crawl in order to identify web addresses from which to start. Table 3.1 shows the ten website locations we selected for a focused crawl about napping behavior.

A focused crawl has a stopping point. We restrict the range of the crawl to named domains. We can stop the crawl of any given domain after a specified number of pages have been downloaded. We can set selection rules for pages being downloaded. And we can limit the total number of pages to be downloaded in a given crawl.

A focused crawl has a defined purpose. Therefore, we retain web pages that meet specific criteria. For the task at hand—gathering information about napping and health—we use keywords to guide page selection: "sleep," "napping," and "age."

With the crawl well in hand, the task turns to scraping or extracting the specific information we want from web pages. Each web page includes HTML tags, defining a hierarchy of nodes (or tree structure). We call this tree the *Document Object Model (DOM)*. And *XPath*, available through Python and R packages, is a specialized syntax for navigating across the nodes and attributes of the DOM, extracting relevant data. We gather data from the nodes, such as text within paragraph tags.

After extracting text data from a web page, we can parse those data using *regular expressions*. Page formatting codes, unnecessary spaces, and punctuation can be removed from each document. A successful crawl should end with a corpus of relevant documents for text analytics.

Table 3.1. *Web Addresses for a Focused Crawl*

Short Name	Description and Web Address

AMA	American Medical Association
	Journal of the American Medical Association (JAMA, Sleep)
	`http://jama.jamanetwork.com/solr/searchresults.aspx?q=sleep&fd_`
	`JournalID=67&f_JournalDisplayName=JAMA&SearchSourceType=3`

HARVARD1	Harvard Medical School (Press Releases)
	`http://www.health.harvard.edu/press_releases/snoozing-without-guilt--`
	`a-daytime-nap-can-be-good-for-health/`

HARVARD2	Harvard Medical School (Healthy Sleep)
	`http://healthysleep.med.harvard.edu/healthy/science/variations/`
	`changes-in-sleep-with-age`

MAYO1	Mayo Foundation for Medical Education and Research `http://www.mayoclinic.org/`
MAYO2	Mayo Foundation for Medical Education and Research (Napping) `http://www.mayoclinic.org/napping/ART-20048319?p=1`
NIH	National Institutes of Health `http://nih.gov/`
SLEEP	National Sleep Foundation `http://sleepfoundation.org/sleep-topics/napping`
HHS	U.S. Department of Health and Human Services (Sleep and Napping) `http://healthfinder.gov/search/?q=sleep+and+napping`
WEBMD	WebMD (Aging and Sleep) `http://www.webmd.com/sleep-disorders/guide/aging-affects-sleep`
WIKINAP	Wikipedia (Nap) `http://en.wikipedia.org/wiki/Nap`

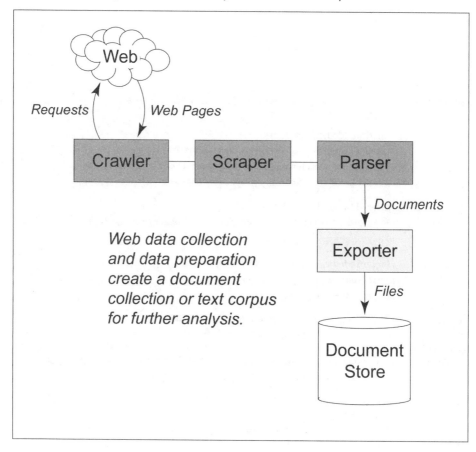

Figure 3.1. *Framework for Automated Data Acquisition*

Figure 3.1 shows a framework for automated data acquisition on the web that is consistent with Python web crawling and scraping software developed by Hoffman et al. (2014). First we crawl, then we scrape, then we parse. This is the work of web-based secondary research. Next, we select an appropriate wrapping for text and associated metadata, JSON or XML perhaps. We build a document store or text corpus for subsequent analysis. Scalable, enterprise-ready databases like PostgreSQL or MongoDB, and data stores built on the Hadoop distributed file system may be employed for this purpose.

Crawling and scraping are fundamental to web and network data science. Modeling and analysis begin with data, and the web is a massive store of data. Learning how to extract relevant data in an efficient manner is an essential skill of data science.

There are a number of excellent reference sources for XML and HTML web scraping with XPath (Tennison 2001; Simpson 2002; Kay 2008; Nolan and Lang 2014). Friedl (2006) covers regular expressions for text parsing. For discussion of focused web crawling and data mining, see Chakrabarti (2003) and Liu (2011).

Exhibit 3.1 shows a Python program for extracting HTML data from one web page, the home page of ToutBay. The program utilizes XPath syntax to parse HTML, extracting the text data within paragraph tags. The program draws on Python packages developed by Behnel (2014), Reitz (2014b), and Richardson (2014). Exhibit 3.2 shows the corresponding R program, which draws on packages developed by Lang (2014a, 2014b).

Exhibit 3.3 shows how to use Python to scrape a single web page using a the Python framework developed by Hoffman et al. (2014) and networking utilities from Lefkowitz et al. (2014).

Building on that Python framework, we show how to construct a focused crawler (affectionately called "the ten-headed spider") that gathers information about sleep, napping, and age from the websites identified earlier. We show the complete crawling, scraping, and parsing code in exhibit 3.4.

What is the tale of the ten-headed spider? Taking naps is not so unusual. Nor is it a sign of poor health. In fact, data acquired from the National Sleep Foundation suggests that I may be in good company. Men such as Winston Churchill, John F. Kennedy, Albert Einstein, and Thomas Edison, among others, were known to be frequent nappers.

The data in favor of napping is strong, and I can collect more data if I like. All I have to do is crawl, scrape, and parse. But first things first. Time for my nap.

Exhibit 3.1. *Extracting and Scraping Web Site Data (Python)*

```
# Extracting and Scraping Web Site Data (Python)

# prepare for Python version 3x features and functions
from __future__ import division, print_function

# import packages for web scraping/parsing
import requests  # functions for interacting with web pages
import lxml   # functions for parsing HTML
from bs4 import BeautifulSoup  # DOM html manipulation

# -------------------------------------------------
# demo using the requests and lxml packages
# -------------------------------------------------
# test requests package on the home page for ToutBay
web_page = requests.get('http://www.toutbay.com/', auth=('user', 'pass'))
# obtain the entire HTML text for the page of interest

# show the status of the page... should be 200 (no error)
web_page.status_code
# show the encoding of the page... should be utf8
web_page.encoding

# show the text including all of the HTML tags... lots of tags
web_page_text = web_page.text
print(web_page_text)

# parse the web text using html functions from lxml package
# store the text with HTML tree structure
web_page_html = lxml.html.fromstring(web_page_text)

# extract the text within paragraph tags using an lxml XPath query
# XPath // selects nodes anywhere in the document  p for paragraph tags
web_page_content = web_page_html.xpath('//p/text()')
# show the resulting text string object
print(web_page_content)  # has a few all-blank strings
print(len(web_page_content))
print(type(web_page_content))  # a list of character strings

# ----------------------------------------------------------
# demo of scraping HTML with beautiful soup instead of lxml
# ----------------------------------------------------------
my_soup = BeautifulSoup(web_page_text)
# note that my_soup is a BeautifulSoup object
print(type(my_soup))

# remove JavaScript code from Beautiful Soup page object
# using a comprehension approach
[x.extract() for x in my_soup.find_all('script')]
# gather all the text from the paragraph tags within the object
# using another list comprehension
soup_content = [x.text for x in my_soup.find_all('p')]
```

```
# show the resulting text string object
print(soup_content)  # note absence of all-blank strings
print(len(soup_content))
print(type(soup_content))  # a list of character strings

# there are carriage return, line feed characters, and spaces
# to delete from the text... but we have extracted the essential
# content of the toutbay.com home page for further analysis
```

Exhibit 3.2. *Extracting and Scraping Web Site Data (R)*

```
# Extracting and Scraping Web Site Data (R)

# install required packages

# bring packages into the workspace
library(RCurl)  # functions for gathering data from the web
library(XML)  # XML and HTML parsing

# gather home page for ToutBay using RCurl package
web_page_text <- getURLContent('http://www.toutbay.com/')

# show the class of the R object and encoding
print(attributes(web_page_text))

# show the text including all of the HTML tags... lots of tags
print(web_page_text)

# scrape the HTML DOM into an internal C data structure for XPath processing
web_page_tree <- htmlTreeParse(web_page_text, useInternalNodes = TRUE,
    asText = TRUE, isHTML = TRUE)
print(attributes(web_page_tree))

# extract the text within paragraph tags using an XPath query
# XPath // selects nodes anywhere in the document  p for paragraph tags
web_page_content <- xpathSApply(web_page_tree, "//p/text()")
print(attributes(web_page_content))
print(head(web_page_content))
print(tail(web_page_content))

# send content to external text file for review
sink("text_file_for_review.txt")
print(web_page_content)
sink()

# there are node numbers, line feed characters, and spaces
# to delete from the text... but we have extracted the essential
# content of the toutbay.com home page for further analysis
```

Exhibit 3.3. *Simple One-Page Web Scraper (Python)*

```python
# Simple One-Page Web Scraper (Python)
#
# prepare for Python version 3x features and functions
from __future__ import division, print_function

# scrapy documentation at http://doc.scrapy.org/

# workspace directory set to outer folder/directory wnds_chapter_3b
# the operating system commands in this example are Mac OS X

import scrapy  # object-oriented framework for crawling and scraping
import os  # operating system commands

# function for walking and printing directory structure
def list_all(current_directory):
    for root, dirs, files in os.walk(current_directory):
        level = root.replace(current_directory, '').count(os.sep)
        indent = ' ' * 4 * (level)
        print('{}{}/'.format(indent, os.path.basename(root)))
        subindent = ' ' * 4 * (level + 1)
        for f in files:
            print('{}{}'.format(subindent, f))


# initial directory should have this form (except for items beginning with .):
#     wnds_chapter_3b
#         run_one_page_scraper.py
#         scrapy.cfg
#         scrapy_application/
#             __init__.py
#             items.py
#             pipelines.py
#             settings.py
#             spiders
#                 __init__.py
#                 one_page_scraper.py

# examine the directory structure
current_directory = os.getcwd()
list_all(current_directory)

# list the avaliable spiders, showing names to be used for crawling
os.system('scrapy list')

# decide upon the desired format for exporting output: csv, JSON, or XML
# run the scraper exporting results as a comma-delimited text file items.csv
os.system('scrapy crawl TOUTBAY -o items.csv')
# run the scraper exporting results as a JSON text file items.json
os.system('scrapy crawl TOUTBAY -o items.json')
# run the scraper exporting results as a dictionary XML text file items.xml
os.system('scrapy crawl TOUTBAY -o items.xml')
```

```
# ---------------------------
# MyItem class defined by
# items.py
# ---------------------------

# location in directory structure:
# wnds_chapter_3b/scrapy_application/items.py

# establishes data fields for scraped items

import scrapy  # object-oriented framework for crawling and scraping

class MyItem(scrapy.item.Item):
    # define the data fields for the item (just one field used here)
    paragraph = scrapy.item.Field()  # paragraph content


# ---------------------------
# MyPipeline class defined by
# pipelines.py
# ---------------------------

# location in directory structure:
# wnds_chapter_3b/scrapy_application/pipelines.py

class MyPipeline(object):
    def process_item(self, item, spider):
        return item


# ---------------------------
# settings for scrapy.cfg
# settings.py
# ---------------------------

# location in directory structure:
# wnds_chapter_3b/scrapy_application/settings.py

BOT_NAME = 'MyBot'
BOT_VERSION = '1.0'

SPIDER_MODULES = ['scrapy_application.spiders']
NEWSPIDER_MODULE = 'scrapy_application.spiders'
USER_AGENT = '%s/%s' % (BOT_NAME, BOT_VERSION)

COOKIES_ENABLED = False
DOWNLOAD_DELAY = 2
RETRY_ENABLED = False
DOWNLOAD_TIMEOUT = 15
REDIRECT_ENABLED = False
DEPTH_LIMIT = 50
```

```
# ----------------------------
# spider class defined by
# script one_page_scraper.py
# ----------------------------
# location in directory structure:
# wnds_chapter_3b/scrapy_application/spiders/one_page_scraper.py

# prepare for Python version 3x features and functions
from __future__ import division, print_function

# each spider class gives code for crawing and scraping

import scrapy  # object-oriented framework for crawling and scraping
from scrapy_application.items import MyItem  # item class

# spider subclass inherits from BaseSpider
# this spider is designed to crawl just one page of one website
class MySpider(scrapy.spider.BaseSpider):
    name = "TOUTBAY"  # unique identifier for the spider
    allowed_domains = ['toutbay.com']  # limits the crawl to this domain list
    start_urls = ['http://www.toutbay.com']  # first url to crawl in domain

    # define the scraping method for the spider
    # note that this function is called "parse," but we are actually scraping
    def parse(self, response):
        html_scraper = scrapy.selector.HtmlXPathSelector(response)
        divs = html_scraper.select('//div')  # identify all <div> nodes
        # XPath syntax to grab all the text in paragraphs in the <div> nodes
        results = []  # initialize list
        this_item = MyItem()  # use this item class
        this_item['paragraph'] = divs.select('.//p').extract()
        results.append(this_item)  # add to the results list
        return results

# Suggestions for the student: Use scrapy to scrape another web page,
# extracting additional DOM content, such as <a> link text and links.
# Utilize new links to move from a one-page scraper to a more complete
# crawler that goes from one page to the next within a web domain.
# Use regular expressions to parse the text from this focused crawl.
```

Exhibit 3.4. *Crawling and Scraping while Napping (Python)*

```
# Crawling and Scraping while Napping (Python)
#
# Focused Crawl with a Ten-Headed Spider Using the Scrapy Framework
#
# prepare for Python version 3x features and functions
from __future__ import division, print_function

# scrapy documentation at http://doc.scrapy.org/

# workspace directory set to outer folder/directory wnds_chapter_3b
# the operating system commands in this example are Mac OS X

import scrapy  # object-oriented framework for crawling and scraping
import os  # operating system commands

# function for walking and printing directory structure
def list_all(current_directory):
    for root, dirs, files in os.walk(current_directory):
        level = root.replace(current_directory, '').count(os.sep)
        indent = ' ' * 4 * (level)
        print('{}{}/'.format(indent, os.path.basename(root)))
        subindent = ' ' * 4 * (level + 1)
        for f in files:
            print('{}{}'.format(subindent, f))

# initial directory should have this form (except for items beginning with .):
#     wnds_chapter_3c
#         run_ten_headed_spider.py
#         scrapy.cfg
#         scrapy_application/
#             __init__.py
#             items.py
#             pipelines.py
#             settings.py
#             spiders
#                 __init__.py
#                 ten_headed_spider.py

# examine the directory structure
current_directory = os.getcwd()
list_all(current_directory)

# list the avaliable spiders, showing names to be used for crawling
os.system('scrapy list')

# decide upon the desired format for exporting output: csv, JSON, or XML

# here we employ JSON for each of the ten sites being crawled
# we run each spider subclass separately so that stored results
# may be identified with the website being crawled
```

```
# Test crawl
os.system('scrapy crawl TEST -o results_TEST.json')

# American Medical Association
os.system('scrapy crawl AMA -o results_AMA.json')

# Harvard Medical School (Press Releases)
os.system('scrapy crawl HARVARD1 -o results_HARVARD1.json')
# Harvard Medical School (Healthy Sleep)
os.system('scrapy crawl HARVARD2 -o results_HARVARD2.json')

# Mayo Foundation for Medical Education and Research
os.system('scrapy crawl MAYO1 -o results_MAYO1.json')
# Mayo Foundation for Medical Education and Research (Napping)
os.system('scrapy crawl MAYO2 -o results_MAYO2.json')

# National Institutes of Health
os.system('scrapy crawl NIH -o results_NIH.json')
# National Sleep Foundation
os.system('scrapy crawl SLEEP -o results_SLEEP.json')
# U.S. Department of Health and Human Services (Sleep and Napping)
os.system('scrapy crawl HHS -o results_HHS.json')
# WebMD (Aging and Sleep)
os.system('scrapy crawl WEBMD -o results_WEBMD.json')
# Wikipedia (Nap)
os.system('scrapy crawl WIKINAP -o results_WIKINAP.json')

# ---------------------------
# MyItem class defined by
# items.py
# ---------------------------
# location in directory structure:
# wnds_chapter_3c/scrapy_application/items.py

# establishes data fields for scraped items

import scrapy  # object-oriented framework for crawling and scraping

class MyItem(scrapy.item.Item):
    # define the data fields for the item (just one field used here)
    paragraph = scrapy.item.Field()  # paragraph content

# ---------------------------
# MyPipeline class defined by
# pipelines.py
# ---------------------------
# location in directory structure:
# wnds_chapter_3c/scrapy_application/pipelines.py

class MyPipeline(object):
    def process_item(self, item, spider):
        return item
```

```
# ----------------------------
# settings for scrapy.cfg
# settings.py
# ----------------------------
# location in directory structure:
# wnds_chapter_3c/scrapy_application/settings.py

BOT_NAME = 'MyBot'
BOT_VERSION = '1.0'

SPIDER_MODULES = ['scrapy_application.spiders']
NEWSPIDER_MODULE = 'scrapy_application.spiders'
USER_AGENT = '%s/%s' % (BOT_NAME, BOT_VERSION)

COOKIES_ENABLED = False
DOWNLOAD_DELAY = 2
RETRY_ENABLED = False
DOWNLOAD_TIMEOUT = 15
REDIRECT_ENABLED = False
DEPTH_LIMIT = 50


# ----------------------------
# spider class defined by
# script ten_headed_spider.py
# ----------------------------
# location in directory structure:
# wnds_chapter_3c/scrapy_application/spiders/ten_headed_spider.py

# prepare for Python version 3x features and functions
from __future__ import division, print_function

# each spider class gives code for crawing and scraping
import scrapy  # object-oriented framework for crawling and scraping
from scrapy_application.items import MyItem  # item class

# each spider subclass inherits from BaseSpider
# each spider subclass is designed to crawl one website
# each spider can have its own parsing logic based on the
# DOM of the website being crawled snd scraped...
class MySpiderTEST(scrapy.spider.BaseSpider):
    name = "TEST"  # unique identifier for the spider
    allowed_domains = ['toutbay.com']  # limits the crawl to this domain list
    start_urls = ['http://www.toutbay.com']  # first url to crawl in domain
    # define the parsing method for the spider
    def parse(self, response):
        html_scraper = scrapy.selector.HtmlXPathSelector(response)
        divs = html_scraper.select('//div')  # identify all <div> nodes
        # XPath syntax to grab all the text in paragraphs in the <div> nodes
        results = []  # initialize list
        this_item = MyItem()  # use this item class
        this_item['paragraph'] = divs.select('.//p').extract()
        results.append(this_item)  # add to the results list
        return results
```

```
# American Medical Association
class MySpiderAMA(scrapy.spider.BaseSpider):
    name = "AMA"  # unique identifier for the spider
    # limit the crawl to this domain list
    allowed_domains = ['ama-assn.org']
    # first url to crawl in domain
    start_urls = ['http://jama.jamanetwork.com/solr/searchresults.aspx?\
        q=sleep&fd_JournalID=67&f_JournalDisplayName=JAMA&SearchSourceType=3']
    # define the parsing method for the spider
    def parse(self, response):
        html_scraper = scrapy.selector.HtmlXPathSelector(response)
        divs = html_scraper.select('//div')  # identify all <div> nodes
        # XPath syntax to grab all the text in paragraphs in the <div> nodes
        results = []  # initialize list
        this_item = MyItem()  # use this item class
        this_item['paragraph'] = divs.select('.//p').extract()
        results.append(this_item)  # add to the results list
        return results


# Harvard Medical School (Press Releases)
class MySpiderHARVARD1(scrapy.spider.BaseSpider):
    name = "HARVARD1"  # unique identifier for the spider
    # limits the crawl to this domain list
    allowed_domains = ['health.harvard.edu']
    # first url to crawl in domain
    start_urls = ['http://www.health.harvard.edu/press_releases/\
        snoozing-without-guilt--a-daytime-nap-can-be-good-for-health']
    # define the parsing method for the spider
    def parse(self, response):
        html_scraper = scrapy.selector.HtmlXPathSelector(response)
        divs = html_scraper.select('//div')  # identify all <div> nodes
        # XPath syntax to grab all the text in paragraphs in the <div> nodes
        results = []  # initialize list
        this_item = MyItem()  # use this item class
        this_item['paragraph'] = divs.select('.//p').extract()
        results.append(this_item)  # add to the results list
        return results


# Harvard Medical School (Healthy Sleep)
class MySpiderHARVARD2(scrapy.spider.BaseSpider):
    name = "HARVARD2"  # unique identifier for the spider
    allowed_domains = ['med.harvard.edu']  # limits the crawl to this domain list
    # first url to crawl in domain
    start_urls = ['http://healthysleep.med.harvard.edu/healthy/science/\
        variations/changes-in-sleep-with-age']
    # define the parsing method for the spider
    def parse(self, response):
        html_scraper = scrapy.selector.HtmlXPathSelector(response)
        divs = html_scraper.select('//div')  # identify all <div> nodes
        # XPath syntax to grab all the text in paragraphs in the <div> nodes
```

```
        results = []  # initialize list
        this_item = MyItem()  # use this item class
        this_item['paragraph'] = divs.select('.//p').extract()
        results.append(this_item)  # add to the results list
        return results

# Mayo Foundation for Medical Education and Research
class MySpiderMAYO1(scrapy.spider.BaseSpider):
    name = "MAYO1"  # unique identifier for the spider
    # limit the crawl to this domain list
    allowed_domains = ['mayoclinic.org']
    # first url to crawl in domain
    start_urls = ['http://www.mayoclinic.org']
    # define the parsing method for the spider
    def parse(self, response):
        html_scraper = scrapy.selector.HtmlXPathSelector(response)
        divs = html_scraper.select('//div')  # identify all <div> nodes
        # XPath syntax to grab all the text in paragraphs in the <div> nodes
        results = []  # initialize list
        this_item = MyItem()  # use this item class
        this_item['paragraph'] = divs.select('.//p').extract()
        results.append(this_item)  # add to the results list
        return results

# Mayo Foundation for Medical Education and Research (Napping)
class MySpiderMAYO2(scrapy.spider.BaseSpider):
    name = "MAYO2"  # unique identifier for the spider
    # limit the crawl to this domain list
    allowed_domains = ['mayoclinic.org']
    # first url to crawl in domain
    start_urls = ['http://www.mayoclinic.org/napping/ART-20048319?p=1']
    # define the parsing method for the spider
    def parse(self, response):
        html_scraper = scrapy.selector.HtmlXPathSelector(response)
        divs = html_scraper.select('//div')  # identify all <div> nodes
        # XPath syntax to grab all the text in paragraphs in the <div> nodes
        results = []  # initialize list
        this_item = MyItem()  # use this item class
        this_item['paragraph'] = divs.select('.//p').extract()
        results.append(this_item)  # add to the results list
        return results

# National Institutes of Health
class MySpiderNIH(scrapy.spider.BaseSpider):
    name = "NIH"  # unique identifier for the spider
    allowed_domains = ['nih.gov']  # limits the crawl to this domain list
    start_urls = ['http://nih.gov']  # first url to crawl in domain
    # define the parsing method for the spider
    def parse(self, response):
        html_scraper = scrapy.selector.HtmlXPathSelector(response)
        divs = html_scraper.select('//div')  # identify all <div> nodes
        # XPath syntax to grab all the text in paragraphs in the <div> nodes
```

```
        results = []  # initialize list
        this_item = MyItem()  # use this item class
        this_item['paragraph'] = divs.select('.//p').extract()
        results.append(this_item)  # add to the results list
        return results


# National Sleep Foundation
class MySpiderSLEEP(scrapy.spider.BaseSpider):
    name = "SLEEP"  # unique identifier for the spider
    # limit the crawl to this domain list
    allowed_domains = ['sleepfoundation.org']
    # first url to crawl in domain
    start_urls = ['http://sleepfoundation.org/sleep-topics/napping']
    # define the parsing method for the spider
    def parse(self, response):
        html_scraper = scrapy.selector.HtmlXPathSelector(response)
        divs = html_scraper.select('//div')  # identify all <div> nodes
        # XPath syntax to grab all the text in paragraphs in the <div> nodes
        results = []  # initialize list
        this_item = MyItem()  # use this item class
        this_item['paragraph'] = divs.select('.//p').extract()
        results.append(this_item)  # add to the results list
        return results


# U.S. Department of Health and Human Services (Sleep and Napping)
class MySpiderHHS(scrapy.spider.BaseSpider):
    name = "HHS"  # unique identifier for the spider
    # limit the crawl to this domain list
    allowed_domains = ['healthfinder.gov']
    # first url to crawl in domain
    start_urls = ['http://healthfinder.gov/search/?q=sleep+and+napping']
    # define the parsing method for the spider
    def parse(self, response):
        html_scraper = scrapy.selector.HtmlXPathSelector(response)
        divs = html_scraper.select('//div')  # identify all <div> nodes
        # XPath syntax to grab all the text in paragraphs in the <div> nodes
        results = []  # initialize list
        this_item = MyItem()  # use this item class
        this_item['paragraph'] = divs.select('.//p').extract()
        results.append(this_item)  # add to the results list
        return results


# WebMD (Aging and Sleep)
class MySpiderWEBMD(scrapy.spider.BaseSpider):
    name = "WEBMD"  # unique identifier for the spider
    # limit the crawl to this domain list
    allowed_domains = ['webmd.com']
    # first url to crawl in domain
    start_urls =\
        ['http://www.webmd.com/sleep-disorders/guide/aging-affects-sleep']
    # define the parsing method for the spider
```

```
    def parse(self, response):
        html_scraper = scrapy.selector.HtmlXPathSelector(response)
        divs = html_scraper.select('//div')  # identify all <div> nodes
        # XPath syntax to grab all the text in paragraphs in the <div> nodes
        results = []  # initialize list
        this_item = MyItem()  # use this item class
        this_item['paragraph'] = divs.select('.//p').extract()
        results.append(this_item)  # add to the results list
        return results


# Wikipedia (Nap)
class MySpiderWIKINAP(scrapy.spider.BaseSpider):
    name = "WIKINAP"  # unique identifier for the spider
    allowed_domains = ['wikipedia.org']  # limits the crawl to this domain list
    # first url to crawl in domain
    start_urls = ['http://en.wikipedia.org/wiki/Nap']
    # define the parsing method for the spider
    def parse(self, response):
        html_scraper = scrapy.selector.HtmlXPathSelector(response)
        divs = html_scraper.select('//div')  # identify all <div> nodes
        # XPath syntax to grab all the text in paragraphs in the <div> nodes
        results = []  # initialize list
        this_item = MyItem()  # use this item class
        this_item['paragraph'] = divs.select('.//p').extract()
        results.append(this_item)  # add to the results list
        return results


# Suggestions for the student: Use source code and element inspection
# utilities provided in modern browsers such as Firefox and Chrome
# to examine the DOM of each website being crawled/scraped. Modify the
# scraping logic of the spider for that website as defined in the
# def parse functions for the spider class for that website.
# Add URLs to the spider classes by adding to the start.urls attribute
# of each spider class. Ensure that each website is crawled thoroughly
# using <a> links to additional pages within the website.
# Note that some of the start.urls used in the current spiders merely
# provide links to other sources. We need to drill down into those
# sources to find journal or web articles with titles relating
# to the problem at hand: sleep, napping, and age.
# Run the focused crawl with these def parse enhancements and
# examine the results. Repeat this process as needed.
# Use regular expressions to parse text from the final focused crawl.
# Build a text corpus for subsequent text analysis.
```

4

Testing Links, Look, and Feel

Debbie: "Wow...that's bitchin' tuck and roll.
You know, I really love the feel of tuck and roll upholstery."

Terry: "You do?"

Debbie: "Yeah."

Terry: "Well, come in...I'll let you feel it.
I mean, you can touch it if you want...
Um...I'll let you feel the upholstery."

—CANDY CLARK AS DEBBIE DUNHAM
AND CHARLES MARTIN SMITH AS TERRY "THE TOAD" FIELDS
IN *American Graffiti* (1973)

My father would tell about how he learned to swim. Older boys in the neighborhood of Schwenksville, Pennsylvania, threw him in the Perkiomen Creek and shouted, "Sink or swim." This was quite literally an immersion approach to learning.

My son Daniel, named after my father, had a distinctly different experience at Swim West in Madison, Wisconsin. He learned how to swim across ten class levels from jellyfish to shark. I see value in both extremes of pedagogy. Small steps work well when learning a new language or new way of doing things. But not too far into the process, I think immersion is best.

Extremes play out in many areas of endeavor, including business on the web. Some companies jump into online marketing with little forethought. Others consult with search engine experts and execute a detailed plan of action with the goal of being recognized online. Whatever the approach, data tell the story of success or failure, and data guide the course of website revision.

A story about Marissa Mayer, currently president and CEO of Yahoo!, concerns her reliance on data to make decisions. While at Google, she is known to have ordered the testing of forty shades of blue to see which would yield the highest click-through rate. Disputes among information technologists and graphics artists were resolved by experiments. Tests and data ruled the day, not the judgement or opinion of website designers.

Testing is critical to all aspects of website design and maintenance. Think in terms of testing the links, look, and feel of a website. To test links—the number and quality of links a website gets from others—we evaluate performance in search. To test the look of a website, we evaluate web pages varying in content and design to see which performs best in terms of click-through rate or session duration. And by testing feel, we refer to usability testing and the responsiveness of a website—its performance.

Consultants in the web arena may claim to know the answers. But in the fast-paced world of the web, consultant answers are often yesterday's answers. To know facts rather than opinions, and to keep up with new facts that emerge each day, we must measure and measure often. We must experiment. We must test. The truth is in the data.

We have discussed website performance from the server and client sides, focusing on one website at a time. When considering search, however, we are judging a website based on its performance against other websites.

Organizations want to know about their web presence, the degree to which they are getting recognized on the web. Well-designed search queries provide useful competitive information. Suppose a researcher wanted to assess the web presence of Yahoo! in fantasy sports. The researcher is not interested in Yahoo! as a brand, the Yahoo! website, or Yahoo! Finance. General website statistics are not relevant to Yahoo! fantasy sports.

One approach to understanding how Yahoo! is doing relative to other fantasy sports providers would be to construct prototypical search queries for fantasy sports. We could enter those queries into search engines and observe search results. There are two classes of search results: *organic search results* and advertisements (paid search). When testing links, the focus is on organic search results, which are natural search results with the search engine finding the most relevant matches to a user's query. By far the most important component of relevance is the links a web page receives from other web pages.

As of December 2013, there were an estimated 2.8 billion users of the Internet, or about 39 percent of the world's population. In North America there were an estimated 300 million users or 84.9 percent of the population of the region, and in Europe 566 million users or 68.6 percent of the population (Miniwatts Marketing Group 2014). These numbers translate into billions of web searches every day, not mentioning social media access and game play.

Getting recognized through web-based applications is like putting up a sign in front of a shop. The company wants to be discovered. The company wants to be recognized for being what it is and doing what it does. When working on the web, getting recognized is a first step toward making a sale.

The rank of a web page in organic search reflects its success in attracting visitors to the site. Higher ranks align with greater visibility. More people finding a site translates into more people seeing its message. And for online retailers, we expect that people finding a site translates into sales.

Typing a web address into a browser address field provides a direct path to a website. But most users find their way to a site through search engines. Rather than typing "www.toutbay.com" in the browser's address field, they type "toutbay" in the search box, and the default search engine does the rest. What the user types is the search query, and what he or she gets back is the search response.

Search results are short descriptions of listed links. There are paid-for links and organic links. How users respond to search results depends on a number of factors, including the relative rank of each search item in the list and, most importantly, its perceived relevance to the query. Highlighted keywords and titles influence user choices.

A general-purpose search engine will continuously crawl the addressable web. For each page that permits indexing, the search engine identifies keywords and places them in its index. Keywords constitute text within hypertext tags. Keywords appear in plain text fields or in meta-text describing videos, images, databases, and JavaScript code. For the website provider, being indexed is a prerequisite to being found.

Search engine providers match user queries with lists of keywords in their search indexes. These lists provide links to web pages, which in turn link to other web pages. To be included in organic search results, the keywords from a website page must match keywords in the user's query. This ensures that search results are relevant to the user's query.

What determines the position of a web page in organic search results? All other things being equal, web pages with higher *keyword density* should be ranked higher in organic search results than web pages with lower keyword density, where keyword density is the ratio of query terms to all terms on a web page. But all other things are not equal in the world of search. To a large extent, the position of a web page in a search list depends on links from other web pages.

Algorithms for search represent closely guarded trade secrets that vary from one search provider to the next and from one day to the next. We know a good deal about these algorithms from the search literature. We know that they capitalize on the structure of the web. And we know that PageRank (Brin and Page 1998) is at the heart of many algorithms. Links from some pages contribute more than links from other pages.

Just as we judge a person's importance or prestige in a social network by the people he or she knows or has connections with, we can judge a web page's importance by its links from other web pages. Think of every web page as having some level of importance or relevance to the user's query. And every web page has links to other web pages, which in turn have some level of importance or relevance to the user's query. The PageRank algorithm for search takes all of these measures of importance or relevance into consideration. Web pages highest in the organic search list are deemed highest in importance or relevance to the user's query.

When consultants use the term "search engine optimization (SEO)," they are not talking about improving the performance of search engines. (That is the job of search engine providers.) Nor are they talking about "optimization" as we normally understand the term in data science or operations research. Search engine optimization is one aspect of an organization's web presence. Rather than talking about "search engine optimization," we should be talking about testing links—testing web presence within the context of search.

Having a strong web presence means doing a lot of little things right with websites and social media. And it means doing the right things across various media and devices. In order to grab the brass ring, website managers must endure the inanity and repetition of the merry-go-round.

SEOMoz, Inc. (2013), a leading search performance training and consulting firm, surveyed more than one hundred search consultants to obtain their opinions about factors affecting search performance. Among ten categories of factors, link authority factors, such as the number of links to a page, were ranked most highly. Referrals from social media were second in importance, followed by page anchor text and keyword factors. We summarize the survey results in table 4.1.

Suppose we were working for the Los Angeles Dodgers. The owners want to increase game attendance and associated revenues. In the Los Angeles area, the Dodgers are one of two Major League Baseball teams, baseball is one of many sports activities, and one of many forms of entertainment. How shall we assess the Dodgers' web presence in Los Angeles?

Consider search queries at these logical extremes: a generic query "find fun things to do in Los Angeles" versus the specific query "buy Dodgers tickets." For search hits on the Dodgers home page, we would expect the generic query to have considerably lower organic search performance than the specific query.

Between the generic and specific, there are many possible queries. Guidance in constructing search queries may be gleaned from an analysis of Google keyword data, as described in *Keyword Games: Dodgers and Angels* (page 288).

Suppose we consider a search competition between the Dodgers and the Angels. We construct search query strings that refer to baseball tickets, but not to the Dodgers or Angels. That is, the words used in the queries do

Table 4.1. *Search Engine Ranking Factors*

(Rank) Factor	Description
(1) Page Link Authority Features	Link metrics to the individual page, such as number of links to the page. (PageRank)
(2) Page Level Social	Metrics associated with social media referrals, such as Facebook, Twitter, and Google+.
(3) Page Level Anchor Text	Text metrics for the individual page, including the anchor text and URL. (keyword position)
(4) Page Level Keyword Usage	Keywords within parts of the page, including the title. (keyword density)
(5) Page Level Keyword Agnostic	Non-keyword usage and non-link metrics of pages, such as length of the page and load speed.
(6) Domain Link Authority Features	Link metrics about the root domain hosting the page.
(7) Domain Level Anchor Text	Anchor text metrics for the root domain hosting the page.
(8) Domain Level Keyword Usage	How keywords are used in the root domain or subdomain.
(9) Domain Keyword Agnostic	Non-keyword features relating to the root domain, such as the character length of the domain name.
(10) Domain Level Brand Metrics	Features of the root domain that reflect branding. (reputation)

Source. SEOMoz, Inc. (2013) at `http://moz.com/search-ranking-factors`

not in any way favor the Dodgers or Angels. We can automate the testing process, using a series of search requests and receiving associated search responses. The responses are character strings we can scrape and parse to compute various performance measures.

We can use data generated from the Google AdWords Keyword Planner to identify keywords for baseball ticket sales, as required for evaluating the Dodgers and Angels sites. The Google AdWords Keyword Planner is Google's way of distributing its pricing model for advertisers. It shows keywords corresponding to coded versions of the words that are observed in user queries—words from Google keyword indexes. For each keyword, Google assigns values for competition and bid. Competition is a scaled value between zero and one that is said to relate to the expected number of advertisers who will bid on the keyword. And bid is the price per click that Google is suggesting for the keyword (a price that has a good chance to win in a competitive bidding process). The Keyword Planner also shows what are described as traffic estimates. These values are not observed data. Instead, they are hypothetical values that Google provides to advertisers to guide the planning process.

While the Keyword Planner tells us little about user search behavior, we can use keywords that Google provides to set up tests of website performance in search. If we rank keywords by estimated traffic, we can identify what could be the top keywords employed by users, as identified, coded, and indexed by Google. Suppose we select the top thirty keywords relevant to searches for Dodgers or Angels tickets and use these keywords in a search engine testing script.

Organic search rank is one measure of web presence, and it can be an indicator of online market position. To construct a modest assessment example, we consider online users of desktop or laptop computers and search as the online activity. With this device type and activity in mind, we can construct a series of test items. Each item consists of a request stimulus and a search engine response.

Performance in search is part of the larger task of getting recognized on the web. Organizations want to know about their web presence, and assessing web presence is a nontrivial measurement problem. People use many applications in addition to search engines.

In the old days of mass media, devices monitored what viewers saw on their televisions. Doing something comparable today would require monitoring user behavior across various devices—televisions, desktop computers, laptops, tablets, mobile phones, and wearables. We expect online behavior to vary across devices.

We should also expect online behavior to vary with time and place of usage, and with the application. People communicate with one another using e-mail, text, and social media. They listen to music, download books, watch videos, and play games. Because much of what people do today is done online, a complete assessment of web presence would entail behavior sampling across people's day-to-day activities. This would be online behavioral, primary research on a grand scale. But perhaps that is what is needed to assess fully an organization's web presence.

For additional information about website analytics, see Kaushik (2010) and Clifton (2012). For advice on the design of websites for search performance, see Moran and Hunt (2009) and Dover (2011). Croll and Power (2009) provide an overview of website monitoring and performance measurement.

Getting recognized on the web is the first step toward getting a message across and (if you are an online retailer) making a sale. The next step or steps involve website design. There is the look of a website (aesthetics) and the feel of a website (usability). For aesthetics and graphics design, useful references include Samara (2007) and Golombisky and Hagen (2013). And for usability, there is the classic work by Steve Krug (2014), *Don't Make Me Think!: A Common Sense Approach to Web Usability*.

Exhibit 4.1. *Identifying Keywords for Testing Performance in Search (R)*

```
# Identifying Keywords for Testing Performance in Search (R)

# begin by installing necessary package RJSONIO

# load package into the workspace for this program
library(RJSONIO)  # JSON to/from R objects

# read Angels keyword data from Google AdWords Keyword Planner
angels_1 <- read.csv("tickets_angels_arts_entertainment.csv",
    stringsAsFactors = FALSE)
angels_2 <- read.csv("tickets_angels_baseball.csv",
    stringsAsFactors = FALSE)
angels_3 <- read.csv("tickets_angels_sports_entertainment.csv",
    stringsAsFactors = FALSE)
angels_4 <- read.csv("tickets_angels_sports_events_ticketing.csv",
    stringsAsFactors = FALSE)
angels_5 <- read.csv("tickets_angels_sports_fitness.csv",
    stringsAsFactors = FALSE)
angels_6 <- read.csv("tickets_angels_sports.csv",
    stringsAsFactors = FALSE)

# read Dodgers keyword data from Google AdWords Keyword Planner
dodgers_1 <- read.csv("tickets_dodgers_arts_entertainment.csv",
    stringsAsFactors = FALSE)
dodgers_2 <- read.csv("tickets_dodgers_baseball.csv",
    stringsAsFactors = FALSE)
dodgers_3 <- read.csv("tickets_dodgers_sports_entertainment.csv",
    stringsAsFactors = FALSE)
dodgers_4 <- read.csv("tickets_dodgers_sports_events_ticketing.csv",
    stringsAsFactors = FALSE)
dodgers_5 <- read.csv("tickets_dodgers_sports_fitness.csv",
    stringsAsFactors = FALSE)
dodgers_6 <- read.csv("tickets_dodgers_sports.csv",
    stringsAsFactors = FALSE)

# check column names to ensure matches
names(angels_1) == names(angels_2)
names(angels_1) == names(angels_3)
names(angels_1) == names(angels_4)
names(angels_1) == names(angels_5)
names(angels_1) == names(angels_6)
names(angels_1) == names(dodgers_1)
names(angels_1) == names(dodgers_2)
names(angels_1) == names(dodgers_3)
names(angels_1) == names(dodgers_4)
names(angels_1) == names(dodgers_5)
names(angels_1) == names(dodgers_6)

# define simple column names prior to merging data frames
names(angels_1) <- names(angels_2) <- names(angels_3) <-
    names(angels_4) <- names(angels_5) <- names(angels_6) <-
```

```
        names(dodgers_1) <- names(dodgers_2) <- names(dodgers_3) <-
        names(dodgers_4) <- names(dodgers_5) <- names(dodgers_6) <-
        c("group", "keyword", "currency",
        "traffic", "october", "november", "december", "january",
        "february", "march", "april", "may", "june", "july",
        "august", "september", "competition", "cpcbid")

# add study category to each record of each data frame
angels_1$study <- rep("Arts and Entertainment", length = nrow(angels_1))
angels_2$study <- rep("Baseball", length = nrow(angels_2))
angels_3$study <- rep("Sports Entertainment", length = nrow(angels_3))
angels_4$study <- rep("Sports Events Ticketing", length = nrow(angels_4))
angels_5$study <- rep("Sports and Fitness", length = nrow(angels_5))
angels_6$study <- rep("Sports", length = nrow(angels_6))
dodgers_1$study <- rep("Arts and Entertainment", length = nrow(dodgers_1))
dodgers_2$study <- rep("Baseball", length = nrow(dodgers_2))
dodgers_3$study <- rep("Sports Entertainment", length = nrow(dodgers_3))
dodgers_4$study <- rep("Sports Events Ticketing", length = nrow(dodgers_4))
dodgers_5$study <- rep("Sports and Fitness", length = nrow(dodgers_5))
dodgers_6$study <- rep("Sports", length = nrow(dodgers_6))

# add team name to each record of each data frame
angels_1$team <- rep("Angels", length = nrow(angels_1))
angels_2$team <- rep("Angels", length = nrow(angels_2))
angels_3$team <- rep("Angels", length = nrow(angels_3))
angels_4$team <- rep("Angels", length = nrow(angels_4))
angels_5$team <- rep("Angels", length = nrow(angels_5))
angels_6$team <- rep("Angels", length = nrow(angels_6))
dodgers_1$team <- rep("Dodgers", length = nrow(dodgers_1))
dodgers_2$team <- rep("Dodgers", length = nrow(dodgers_2))
dodgers_3$team <- rep("Dodgers", length = nrow(dodgers_3))
dodgers_4$team <- rep("Dodgers", length = nrow(dodgers_4))
dodgers_5$team <- rep("Dodgers", length = nrow(dodgers_5))
dodgers_6$team <- rep("Dodgers", length = nrow(dodgers_6))

# combine the data frames
keyword_data_frame <- rbind(angels_1, angels_2, angels_3,
    angels_4, angels_5, angels_6, dodgers_1, dodgers_2,
    dodgers_3, dodgers_4, dodgers_5, dodgers_6)

# drop currency variable because everything is in US dollars
keyword_data_frame <- keyword_data_frame[, -3]   # currency is thrid column

# drop cases with missing values
keyword_data_frame <- na.omit(keyword_data_frame)

# examine the structure of the data frame
print(str(keyword_data_frame))

# select Sports category for both Dodgers and Angels
sports_data_frame <- subset(keyword_data_frame,
    subset = (study == "Sports"))
print(str(sports_data_frame))
```

```
# check on the keywords used for Sports (sports_data_frame)
with(sports_data_frame, print(table(keyword)))  # many not relevant


# distribution of cost-per-click bids
with(sports_data_frame, plot(density(cpcbid)))  # a few very high values


# relationship between traffic and cost-per-click bids
# weak positive relationship
with(sports_data_frame,
    cat("\n\nCorrelation between traffic and suggested CPC bid:",
        cor(traffic, cpcbid)))
with(sports_data_frame, plot(traffic, cpcbid))


# relationship between competition and cost-per-click bids
# moderate positive relationship
with(sports_data_frame,
    cat("\n\nCorrelation between competitors and CPC and suggested CPCbid:",
        cor(competition, cpcbid)))
with(sports_data_frame, plot(competition, cpcbid))


# select Baseball category for both Dodgers and Angels
baseball_data_frame <- subset(keyword_data_frame,
    subset = (study == "Baseball"))
print(str(baseball_data_frame))
# check on the keywords used for Baseball (baseball_data_frame)
with(baseball_data_frame, print(table(keyword)))  # many not relevant


# traffic estimates for keyword: "baseball tickets for sale"
# note identical values for Dodgers and Angels
baseball_tickets_for_sale_data_frame <- subset(baseball_data_frame,
        subset = (keyword == "baseball tickets for sale"))
print(baseball_tickets_for_sale_data_frame)


# traffic estimates for keyword: "dodgers tickets"
# note identical values for Dodgers and Angels
dodgers_tickets_data_frame <- subset(baseball_data_frame,
        subset = (keyword == "dodgers tickets"))
print(dodgers_tickets_data_frame)


# traffic estimates for keyword: "angels tickets"
# note identical values for Dodgers and Angels
# interesting that "angels tickets" has lower traffic
# estimates than "dodgers tickets" but a higher CPC
angels_tickets_data_frame <- subset(baseball_data_frame,
        subset = (keyword == "angels tickets"))
print(angels_tickets_data_frame)


# what about "baseball tickets" across all the studies
# this occurs in three of the categories for both Dodgers and Angels
# note the expected seasonal pattern in traffic
# also note identical traffic estimates for March, April, and May
# and identical values for June and July... a clear indication
# that these are not actual data... nor are they likely to have
```

```
# come from a data-based predictive model... there is too much
# regularity across the time series of monthly traffic estimates
baseball_tickets_data_frame <- subset(keyword_data_frame,
        subset = (keyword == "baseball tickets"))
print(baseball_tickets_data_frame)

# select Sports Entertainment for keyword search
working_data_frame <- subset(keyword_data_frame,
    subset = (study == "Sports Entertainment"))
print(str(working_data_frame))

# rank keyword records by traffic estimate (highest first)
sorted_data_frame <-
    working_data_frame[sort.list(working_data_frame$traffic,
        decreasing = TRUE),]

# consider only unique keywords in the sorted list
preliminary_keyword_list <- unique(sorted_data_frame$keyword)
cat("\n\n", length(preliminary_keyword_list),
  "keywords in preliminary list\n")

# output the list for review with the intention of selecting
# a subset of keywords relevant to both the Dodgers and Angels
# we use a JSON file for this purpose
json_string <- toJSON(preliminary_keyword_list)
# remove backslashes from string
sink("preliminary_json.txt")
json_string
sink()
```

5

Watching Competitors

"Keep your friends close, but your enemies closer."

—AL PACINO AS MICHAEL CORLEONE IN *The Godfather Part II* (1974)

Recently I flew Spirit Airlines from Los Angeles to Chicago. Known for its low fares, Spirit charges a base fare and then adds amenities. My first and only amenity was a seat. I bought seat 10F in each direction: $2 \times \$14 = \$28$. I could have purchased seats further forward at $25 each, but that seemed a bit excessive. Ordering online was a snap except for a barrage of pop-up ads. I was asked if I wanted to pay extra to board the plane early. I did not. And I accepted no luggage charges.

At the airport I spent twenty minutes rearranging my bags, converting two canvas bags of books and clothing into one. This benefitted me because, according to the gate agent, I would be charged one hundred dollars at the gate if I had two carry-on bags. I put my jacket on and stuffed its pockets with a computer mouse, extra glasses, Python pocket guide, and such.

Shortly after takeoff, an infant began his crying. This lasted for much of the duration of the flight, three hours and twenty minutes. This was not Spirit's fault. But I half expected the flight attendant to announce, "And for your flying enjoyment today, we offer ear plugs for twenty dollars and noise-canceling headphones for one hundred."

I got to Chicago safely and on time, and two days later the return trip was equally efficient, sans the screaming infant. But having to buy my seat and the inconvenience with the bags made me wonder about flying Spirit again. There are many carriers serving the LA–Chicago route. I wonder about Spirit's competitive position.

Inveterate crawlers and scrapers that we are, we could start the research with the web. But, suppose we step back from that obvious activity for a moment to consider the work of competitive intelligence in general.

Competition is evident in most areas of business, whether we characterize it in economic terms as perfect competition, monopolistic competition, or rivalry among a few leading firms (oligopoly). Whatever form competition takes, most managers would agree that there are decided advantages to knowing the competition.

Compared to other types of research, competitive intelligence is little understood, misunderstood, often maligned, and sometimes feared. Stories of unethical business practices, corporate espionage, and spying make for good reading in the popular press (Penenberg and Barry 2000), but the real work of competitive intelligence is mundane.

Recall the whimsical quotation attributed to Yogi Berra: *You can observe a lot just by watching*. Such is the essence of competitive intelligence. We study, we observe, we learn.

It is fine to begin a competitive intelligence project with little understanding of the competitors. In fact, it may be an advantage to have few preconceptions, to enter the marketplace with open eyes.

Competitive intelligence is forward looking. It is not enough to know what competitors have done in the past. We must anticipate the future. We make predictions about the future actions of firms, their products, services, and prices. Done right, competitive intelligence guides business strategy.

The pace of business is fast, and firms must respond quickly to changes in the competitive landscape, the political and regulatory environment, and developments in technology. Increasingly, firms face global competition, so we must study, look, and learn worldwide.

Many game theory models assume that players have common knowledge and equal access to information. In practice, asymmetric information is common. A player in a competitive game with less information than others has a competitive disadvantage.

Sources of competitive information are many, and much of this information is online. We attend to trade publications and look for lists of association members. We interview business insiders. It may be surprising to see how much information is in the public domain, available to anyone with access to a web browser and a good search tool. Many firms reveal more than they should about themselves on their corporate web sites. There will be information about current and future products, organizational structure, biographies of key employees, and mission statements.

Woodward and Bernstein could "follow the money," and so can we. Business transactions reveal information. When a corporate competitor decides to build a new manufacturing facility, it engages in numerous business transactions. Many people are involved, including corporate managers, sellers of land, contractors, realtors, lawyers, and bankers. There are public records, including permits and filings with local and state governments, bank filings, and environmental impact statements. There could be news reports and community gossip. Information revealed in public sources will point to the exact location of the facility and its size. Through environmental impact statements, it may be possible to infer the type of materials being produced and the quantity of production.

Registered business process patents are public records; these provide information about current and future products. The federal government provides access to patent information, as do many corporate websites.

Bids on governmental projects are open for public inspection. These provide information about products and pricing. Additional information about competitors may be gleaned from published price guides, press releases, newspaper and magazine articles, conference proceedings, white papers, technical reports, annual reports, and financial analyst reports.

News media provide extensive information about competitors. There are recorded radio and television interviews with key employees and industry consultants. Competitive intelligence professionals know to look locally, searching small news sources, as well as the national media.

Additional sources of information include syndicated data sources with restricted access to subscribers. Extensive legal and business databases are searchable within the syndicated offerings.These fee-based information services can be quite useful in competitive intelligence.

In competitive intelligence, professionals distinguish between primary and secondary sources (not to be confused with primary and secondary research). A primary source is a direct source of information, a representative of the competitive firm being studied. A secondary source is someone who gets information from a primary source or another secondary source, and may have altered that information, either intentionally or unintentionally. In journalism and law it is normal procedure to view primary sources as being more trustworthy and accurate than secondary sources.

Moving beyond published sources of information, competitive intelligence professionals turn to one of their favorite activities—talking with people. Employees and former employees, salespeople, and customers—these represent potential sources of competitive intelligence.

We assume, as we do for all types of research, that the work of competitive intelligence will be done in a legal and ethical manner. Ethical questions often arise when working in the area of competitive intelligence. It is entirely legal and ethical to ask questions as long as the researcher identifies herself properly. She uses the skills of a journalist while explaining that she works for a particular firm. Competitive intelligence builds on the free and open exchange of information among business professionals. Trade shows and professional conferences are good places to find people "in the know." But we can find these people online as well.

How far can one go in getting information about the competition? While eavesdropping, bribery, and acts of deception are clear violations of law and ethics, there are many gray areas in which actions are questionable but not obviously illegal or unethical. Competitive intelligence professionals must be careful, avoiding any implication of wrongdoing.

When a company puts information on the web, it posts that information to the world. To protect information of competitive value—proprietary information and trade secrets—firms should be wary of the web. Of course, this is more easily said than done in today's world.

Much of what we know about competitive intelligence in general applies to competitive intelligence online. Some books about online search focus on competitive intelligence sources (Miller 2000a; Vine 2000; Campbell and Swigart 2014). A few researchers discuss competitive intelligence spiders (Chau and Chen 2003; Hemenway and Calishain 2004), and most of what we have learned about focused crawlers or spiders applies to their use in competitive intelligence. Thompson (2000) and Houston, Bruzzese, and Weinberg (2002) provide advice to journalists involved in business reporting. Their advice is equally relevant to competitive intelligence professionals. Training in methods of inquiry for competitive intelligence is offered by the Strategic and Competitive Intelligence Professionals (SCIP) and by a number of universities under library and information systems programs.

We demonstrate the competitive intelligence process, focusing on Spirit Airlines. Here is the scenario. We are one of the airlines competing with Spirit on the route between Los Angeles and Chicago. We are concerned about Spirit's growth and suspect that its low fares may be attracting many airline passengers. We want to learn as much as we can about Spirit, and we want to learn quickly. We are working online, as usual.[1] A summary of our interactive work is provided in table 5.1.

We want to make the process of learning about Spirit Airlines as efficient as possible. We start with an overview from Wikipedia: date of founding, top executives, headquarters, and the like. We also see the top ten airports in its network and the number of daily flights from each.

Next, we move to the organization. We have the name of the president and CEO, Ben Baldanza, so we could start with him and trace his links, mapping an ego-centric network. But suppose we go to a source of such professional links, LinkedIn, and find out how many of Spirit's employees are listed with the site. We can click on "See all" to see them all along with a listing of where they are located: Miami/Fort Lauderdale (656), greater Detroit area (123), Dallas/Fort Worth area (81), and so on. These data may be used to estimate the proportion of employees at each site, providing an idea of the airline's presence in the areas it serves.

Job postings may be a good place to learn about growth plans of the airline. We start with an advanced job search for all jobs at Spirit Airlines. Then

---

[1] Data we review here were collected November 4, 2014.

Table 5.1. *Competitive Intelligence Sources for Spirit Airlines*

Source	Objective (Method)	Web Address (Query String)
Wikipedia	Overview	`http://en.wikipedia.org/wiki/Spirit_Airlines`
LinkedIn	Organization	`https://www.linkedin.com/company/spirit-airlines`

indeed.com	Job Postings (Advanced Job Search)	`http://www.indeed.com/`
		`FindJobs:what=company:`
		`    (SpiritAirlineswhere=blank)`

	(Advanced Job Search)	`FindJobs:what=company:` `    (SpiritAirlineswhere=Miami,FL)`
	(Advanced Job Search)	`FindJobs:what=company:` `    (SpiritAirlineswhere=Detriot,MI)`
	(Advanced Job Search)	`FindJobs:what=company:` `    (SpiritAirlineswhere=Dallas,TX)`

Glassdoor	Job Postings (Search) (Search)	`http://www.glassdoor.com/index.htm`
		`Jobs.Company:SpiritAirlines`
		`Salaries`

Compete.com	Website Activity (Search)	`https://www.compete.com/` `Website:http://www.spirit.com/`
Yahoo! Finance	Financial Data (Search)	`http://finance.yahoo.com/` `SearchFinance:SAVE`
Google Finance	Financial Data (Search)	`https://www.google.com/finance` `Search:NASDAQ:SAVE`
Bloomberg	Financial News (Search)	`http://www.bloomberg.com/` `http://www.bloomberg.com/quote/SAVE:US`
Quandl	Financial Data (Financial Statements)	`https://www.quandl.com/` `https://www.quandl.com/c/stocks/save`
Spirit Airlines	Company Website (Investor Relations)	`http://www.spirit.com/` `http://ir.spirit.com/`

we add the locations of the major cities served by the airline. So we go to indeed.com and conduct searches across major cities served by the airline. We see 50 jobs listed, including 27 in Miami, 3 in Detroit, and 3 in Dallas.

Glassdoor is another jobs site. It lists 272 jobs at Spirit Airlines on the date of our search. Clicking on the "Salaries" link, we see the salary breakdown for both salaried and hourly workers. A web developer at Spirit might make $86 thousand a year, but a passenger service agent is making just under $10 an hour. We could continue on to compare these with compensation at other airlines.

Website statistics reflect customer interest in the services provided by a firm. This may be less true for Spirit Airlines because many of its sales will be through travel and ticketing service providers. We go to Compete.com for website data. We note that there were more than 1.8 million unique visitors to the Spirit Airlines site in September 2014. More importantly, we see a level pattern of activity (no growth) over the last year. It is not the website itself we are interested in here. It is the level of activity, which is an indicator of customer interest in the services that the airline provides.

We could also go to social media sites such as Twitter, Facebook, and Google+ to see what people are saying about the airline. This would be best accomplished by using an Application Programming Interface (API) for each of the social media sources (Russell 2014).

Financial data for Spirit Airlines (NASDAQ stock symbol: SAVE) are available through a variety of sources, including Yahoo! Finance Google Finance, and Bloomberg. Even better, we can gather extensive data about the firm from Quandl. Using R, we can download these data in an efficient manner, display time series, and create financial forecasts.

Of course, our competitive work would not be finished without looking at the Spirit Airlines website itself. There we will find extensive information about products and services. We can review the firm's annual report to shareholders to get general information about total revenues and numbers of employees. We can read press releases, such as this one from October 28, 2014: "Spirit Airlines Announces Record Third Quarter 2014 Results: Third Quarter 2014 Adjusted Net Income Increases 27.6 Percent to $73.9 Million."

Figure 5.1. Competitive Intelligence: Spirit Airlines Flying High

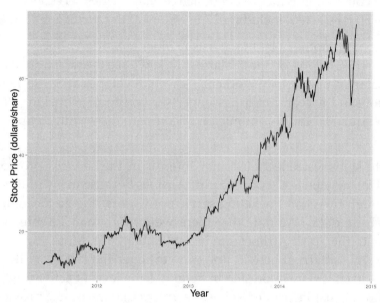

Perhaps other airlines should have reason to be concerned about Spirit and its low-fare/amenity pricing approach to air travel. Note the trend in stock price over the last four years in figure 5.1.

Sean Campbell and Scott Swigart (2014) of Cascade Insights provide an excellent review of online methods for competitive intelligence, such as those we have demonstrated with the Spirit Airlines example. Research about Spirit Airlines is more fun than flying. The pointing and clicking is easy, and anything we do interactively can be programmed for automated data acquisition from the web. Competitive intelligence is a low-fare trip with no screaming infants.

Exhibit 5.1 shows the R program for gathering competitive intelligence for Spirit Airlines from the web and storing those data for future analysis. Airlines are particularly responsive to economic cycles. The program draws on numerous R packages for financial data acquisition, manipulation, analysis, and display (Grolemund and Wickham 2014; Lang 2014a; Lang 2014b; McTaggart and Daroczi 2014; Ryan 2014; Ryan and Ulrich 2014; Wickham and Chang 2014; Zeileis, Grothendieck, and Ryan 2014).

Exhibit 5.1. *Competitive Intelligence: Spirit Airlines Financial Dossier (R)*

```
# Competitive Intelligence: Spirit Airlines Financial Dossier (R)

# install required packages

# bring packages into the workspace
library(RCurl)  # functions for gathering data from the web
library(XML)  # XML and HTML parsing
library(quantmod) # use for gathering and charting economic data
# online documentation for quantmod at <http://www.quantmod.com/>
library(Quandl)  # extensive financial data online
# online documentation for Quandl at <https://www.quandl.com/>
library(lubridate) # date functions
library(zoo)  # utilities for working with time series
library(xts)   # utilities for working with time series
library(ggplot2)  # data visualization

# ---------------------------------
# Text data acquisition and storage
# ---------------------------------
# get current working directory (commands for Mac OS or Linux)
cwd <- getwd()
# create directory for storing competitive intelligence data
ciwd <- paste(cwd, "/ci_data/", sep ="")
dir.create(ciwd)

# gather Wikipedia data using RCurl package
wikipedia_web_page <-
    getURLContent('http://en.wikipedia.org/wiki/Spirit_Airlines')
# store data in directory ciwd for future processing
sink(paste(ciwd, "/", "wikipedia_web_page", sep = ""))
wikipedia_web_page
sink()

# similar procedures may be used for all acquired data
# for the Spirit Airlines competitive intelligence study
# use distinct file names to identify the data sources

# ----------------------------------------------------------------
# Yahoo! Finance for Spirit Airlines (NASDAQ stock symbol: SAVE)
# ----------------------------------------------------------------
# stock symbols for companies can be obtained from Yahoo! Finance
# <http://finance.yahoo.com/lookup>

# get Spirit Airlines stock price data
getSymbols("SAVE", return.class = "xts", src = "yahoo")
print(str(SAVE)) # show the structure of this xts time series object
# plot the series stock price
chartSeries(SAVE,theme="white")
# examine the structure of the R data object
print(str(SAVE))
print(SAVE)
```

```
# convert character string row names to decimal date for the year
Year <- decimal_date(ymd(row.names(as.data.frame(SAVE))))
# Obtain the closing price of Spirit Airlines stock
Price <- as.numeric(SAVE$SAVE.Close)
# create data frame for Spirit Airlines Year and Price for future plots
SPIRIT_data_frame <- data.frame(Year, Price)

# similar procedures may be used for all airline competitors

# ----------------------------------------------------------------
# Google! Finance for Spirit Airlines (NASDAQ stock symbol: SAVE)
#    (basically the same data as from Yahoo! Finance)
# ----------------------------------------------------------------

# get Spirit Airlines stock price data
getSymbols("SAVE", return.class = "xts", src = "google")
print(str(SAVE)) # show the structure of this xts time series object
# plot the series stock price
chartSeries(SAVE,theme="white")
# examine the structure of the R data object
print(str(SAVE))
print(SAVE)


# -------------------------------------------
# ggplot2 time series plotting, closing
# price of Spirit Airlines common stock
# -------------------------------------------
# with a data frame object in hand... we can go on to use ggplot2
# and methods described in Chang (2013) R Graphics Cookbook

# use data frame defined from Yahoo! Finance
# the Spirit Airlines closing price per share SPIRIT_data_frame
# now we use that data structure to prepare a time series plot
# again, let's highlight the Great Recession on our plot for this time series
plotting_object <- ggplot(SPIRIT_data_frame, aes(x = Year, y = Price)) +
    geom_line() +
    ylab("Stock Price (dollars/share)") +
    ggtitle("Spirit Airlines Stock Price")
print(plotting_object)
# send the plot to an external file (sans title, larger axis labels)
pdf("fig_competitive_intelligence_spirit.pdf", width = 11, height = 8.5)
plotting_object <- ggplot(SPIRIT_data_frame, aes(x = Year, y = Price)) +
    geom_line() + ylab("Stock Price (dollars/share)") +
    theme(axis.title.y = element_text(size = 20, colour = "black")) +
    theme(axis.title.x = element_text(size = 20, colour = "black"))
print(plotting_object)
dev.off()


# -------------------------------------------
# FRED for acquiring general financial data
# -------------------------------------------
# general financial data may be useful in understanding what is
# happening with a company over time... here is how to get those data
```

```
# demonstration of R access to and display of financial data from FRED
# requires a connection to the Internet
# ecomonic research data from the Federal Reserve Bank of St. Louis
# see documentation of tags at http://research.stlouisfed.org/fred2/
# choose a particular series and click on it, a graph will be displayed
# in parentheses in the title of the graph will be the symbol
# for the financial series some time series are quarterly, some monthly
# ... others weekly... so make sure the time series match up in time
# see the documentation for quantmod at
# <http://cran.r-project.org/web/packages/quantmod/quantmod.pdf>


# here we show how to download the Consumer Price Index
# for All Urban Consumers: All Items, Not Seasonally Adjusted, Monthly
getSymbols("CPIAUCNS",src="FRED",return.class = "xts")
print(str(CPIAUCNS)) # show the structure of this xts time series object
# plot the series
chartSeries(CPIAUCNS,theme="white")


# Real Gross National Product in 2005 dollars
getSymbols("GNPC96",src="FRED",return.class = "xts")
print(str(GNPC96)) # show the structure of this xts time series object
# plot the series
chartSeries(GNPC96,theme="white")


# National Civilian Unemployment Rate,
#    not seasonally adjusted (monthly, percentage)
getSymbols("UNRATENSA",src="FRED",return.class = "xts")
print(str(UNRATENSA)) # show the structure of this xts time series object
# plot the series
chartSeries(UNRATENSA,theme="white")


# University of Michigan: Consumer Sentiment,
#    not seasonally adjusted (monthly, 1966 = 100)
getSymbols("UMCSENT",src="FRED",return.class = "xts")
print(str(UMCSENT)) # show the structure of this xts time series object
# plot the series
chartSeries(UMCSENT,theme="white")


# New Homes Sold in the US, not seasonally adjusted (monthly, thousands)
getSymbols("HSN1FNSA",src="FRED",return.class = "xts")
print(str(HSN1FNSA)) # show the structure of this xts time series object
# plot the series
chartSeries(HSN1FNSA,theme="white")


# ----------------------------------------
# Multiple time series plots
# ----------------------------------------
# let's try putting consumer sentiment and new home sales on the same plot

# University of Michigan Index of Consumer Sentiment (1Q 1966 = 100)
getSymbols("UMCSENT", src="FRED", return.class = "xts")
ICS <- UMCSENT # use simple name for xts object
dimnames(ICS)[2] <- "ICS" # use simple name for index
```

```
chartSeries(ICS, theme="white")
ICS_data_frame <- as.data.frame(ICS)
ICS_data_frame$date <- ymd(rownames(ICS_data_frame))
ICS_time_series <- ts(ICS_data_frame$ICS,
  start = c(year(min(ICS_data_frame$date)), month(min(ICS_data_frame$date))),
  end = c(year(max(ICS_data_frame$date)),month(max(ICS_data_frame$date))),
  frequency=12)

# New Homes Sold in the US, not seasonally adjusted (monthly, millions)
getSymbols("HSN1FNSA",src="FRED",return.class = "xts")
NHS <- HSN1FNSA
dimnames(NHS)[2] <- "NHS" # use simple name for index
chartSeries(NHS, theme="white")
NHS_data_frame <- as.data.frame(NHS)
NHS_data_frame$date <- ymd(rownames(NHS_data_frame))
NHS_time_series <- ts(NHS_data_frame$NHS,
  start = c(year(min(NHS_data_frame$date)),month(min(NHS_data_frame$date))),
  end = c(year(max(NHS_data_frame$date)),month(max(NHS_data_frame$date))),
  frequency=12)

# define multiple time series object
economic_mts <- cbind(ICS_time_series,
  NHS_time_series)
  dimnames(economic_mts)[[2]] <- c("ICS","NHS") # keep simple names
modeling_mts <- na.omit(economic_mts) # keep overlapping time intervals only

# examine the structure of the multiple time series object
# note that this is not a data frame object
print(str(modeling_mts))

# -----------------------------------------
# Prepare data frame for ggplot2 work
# -----------------------------------------
# for zoo examples see vignette at
# <http://cran.r-project.org/web/packages/zoo/vignettes/zoo-quickref.pdf>
modeling_data_frame <- as.data.frame(modeling_mts)
modeling_data_frame$Year <- as.numeric(time(modeling_mts))

# examine the structure of the data frame object
# notice an intentional shift to underline in the data frame name
# this is just to make sure we keep our object names distinct
# also you will note that programming practice for database work
# and for work with Python is to utilize underlines in variable names
# so it is a good idea to use underlines generally
print(str(modeling_data_frame))
print(head(modeling_data_frame))

# -----------------------------------------------
# ggplot2 time series plotting of economic data
# -----------------------------------------------
# according to the National Bureau of Economic Research the
# Great Recession extended from December 2007 to June 2009
# using our Year variable this would be from 2007.917 to 2009.417
```

```
# let's highlight the Great Recession on our plot
plotting_object <- ggplot(modeling_data_frame, aes(x = Year, y = ICS)) +
    geom_line() +
    annotate("rect", xmin = 2007.917, xmax = 2009.417,
        ymin = min(modeling_data_frame$ICS),
        ymax = max(modeling_data_frame$ICS),
        alpha = 0.3, fill = "red") +
    ylab("Index of Consumer Sentiment") +
    ggtitle("Great Recession and Consumer Sentiment")
print(plotting_object)


# ----------------------------------------------------------------
# Quandl for Spirit Airlines (NASDAQ stock symbol: SAVE)
# obtain more extensive financial data for Spirit Airlines
# more documentation at http://blog.quandl.com/blog/using-quandl-in-r/
# ----------------------------------------------------------------
Spirit_Price <- Quandl("GOOG/NASDAQ_SAVE", collapse="monthly", type="ts")
plot(stl(Spirit_Price[,4],s.window="periodic"),
    main = "Time Series Decomposition for Spirit Airlines Stock Price")

# Suggestions for the student: Employ search and crawling code to access
# all of the competitive intelligence reports cited in the chapter.
# Save these data in the directory, building a text corpus for further study.
# Scrape and parse the text documents using XPath and regular expressions.
# Obtain stock price series for all the major airlines and compare those
# series with Spirit's. See if there are identifiable patterns in these
# data and if those patterns in any way correspond to economic conditions.
# (Note that putting the economic/monthly and stock price/daily data
#   into the same analysis will require a periodicity change using the
#   xts/zoo data for Spirit Airlines. Refer to documentation at
#   <http://www.quantmod.com/Rmetrics2008/quantmod2008.pdf>
#   or from Quandl at <http://blog.quandl.com/blog/using-quandl-in-r/>.)
# Conduct a study of ticket prices on a round trips between two cities
# (non-stop flights between Los Angeles and Chicago, perhaps).
# Crawl and scrape the pricing data to create a pricing database across
# alternative dates into the future, taking note of the day and time
# of each flight and the airline supplying the service. Develop a
# competitive pricing model for airline travel between these cities.
# Select one of the competitive airlines as your client, and report
# the results of your competitive intelligence and competitive pricing
# research. Make strategic recommendations about what to do about
# the low-fare/amenities pricing approach of Spirit Airlines.
```

6

Visualizing Networks

"All right, Mr. DeMille, I'm ready for my close-up."

—Gloria Swanson as Norma Desmond in *Sunset Boulevard*
(1950)

The Pacific Surfliner goes from San Luis Obispo to San Diego. I love this train. I board at Glendale and start my writing. The first ten minutes to Los Angeles are dismal, as we pass electrical transformer distributors and recycling firms. But the path from San Juan Capistrano to Oceanside is as fine as a trip can be.

I like to know where I am going, where I have been, and the map or path between. But with an ocean and surfers in sight, I like what is happening now the most.

The more we study networks, the more we see them in the experience of life—the places we go, things we experience, what we read, and people we know. Things are linked together.

Networks come in all shapes and sizes. Each network is composed of nodes and links. In mathematical jargon, a network is a graph with vertices and edges. A plot of a network can help us to discover structure across its nodes and links. For any given network of sufficient size, there are many possible plots. We try alternative visualizations to see which does the best job of revealing structure.

The web is a network, one page linked to the next. A software system emerges from programmers' code, objects and methods linking in a directed network. We see one package depending on another, functions calling other functions.

A transportation network consists of links to links with virtual nodes defined at intersections. The length and geographic direction of each link has meaning. The visualization is a map, and a common objective is to find the shortest path from one intersection to another.

A scientific domain documented in publications has articles as nodes and citations as links—a network of knowledge. We might judge an article by the number or quality of its references or, better still, by the number of times other articles reference it.

The number and quality of connections is key in academia, politics, and business. Power and influence in these endeavors often flow from working social networks, finding influential members, and identifying important subgroups or cliques.

A convenient way to describe a network is in terms of its *degree distribution*. Each node in an undirected network has a degree or number of links to other nodes. The degree distribution shows the degree to which nodes are connected to one another. The average degree or *degree centrality* is a summary statistic for networks.

To compute summary statistics for a network such as degree centrality, we represent the network as a matrix. We let nodes define the rows and columns of a square matrix and obtain cell entries by considering characteristics for each pair of nodes. For example, we obtain the *adjacency matrix* of a network by considering links between nodes. We let $x_{ij} = 1$ if node i is linked to node j and let $x_{ij} = 0$ otherwise. Undirected networks yield symmetric matrices. Directed networks yield asymmetric matrices. These matrices are sometimes called sociomatrices.

Simple network structures include star, circle, line, and tree structures. Figure 6.1 shows a star network. The degree distribution shows that one individual (Amy) is central to this undirected network, decidedly better connected than other individuals in this network of seven people.

Figure 6.1. *A Simple Star Network*

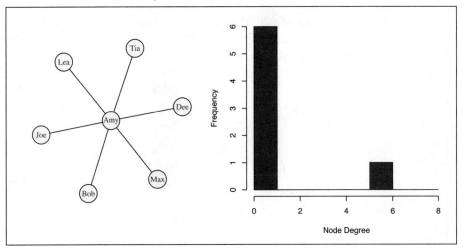

Figure 6.2 shows a circle network. In the circle, each of the individual nodes is equally important, with links to exactly two other nodes. This is reflected in its simple degree distribution.

In the line or chain network shown in figure 6.3, all but the end nodes have links to two other nodes. Note that the topology of the line is similar to the circle. In this example, adding one link between the end nodes (Roy and Zoe) would create a circle. One could argue that the node at the center of the line network is more important that other nodes because it is on the path between more pairs of nodes in the network. This is the notion of *betweenness centrality*.

If there are n nodes in a network, the minimum number of links needed for all nodes to be connected to at least one other node is $(n - 1)$. We can see this from the plots of the simple star and line networks.

A *clique* or complete graph is a fully connected network—every node is connected to every other node. An n-node clique will have $\frac{n(n-1)}{2}$ links. Figure 6.4 shows a seven-node clique. Each of the seven nodes has six links (one link to each of the other nodes). The network has $\frac{7(7-1)}{2} = 21$ links. Cliques are important in social network analysis because they help us to identify the core members of a community.

Figure 6.2. *A Simple Circle Network*

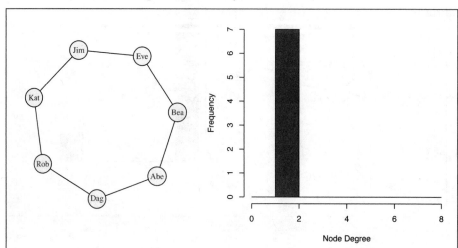

Figure 6.3. *A Simple Line Network*

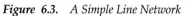
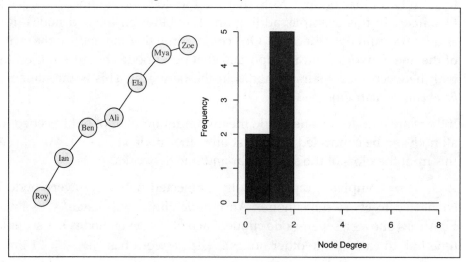

Figure 6.4. *A Clique or Fully Connected Network*

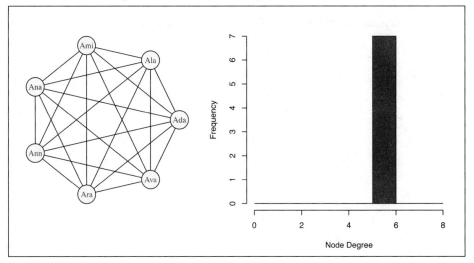

Finding the right layout for a network can be a hit-and-miss proposition. It is often best to try various algorithms to see which works best in revealing network structure. Figure 6.5 shows how hard it can be to discover a tree or hierarchical structure from standard forms of network visualization, even when working with a network of only eight nodes. So we have a new twist on the saying, "Can't see the forest for the trees." Here we cannot see the tree from the links.

When the number of nodes grows large, it can be especially difficult to detect patterns by sight. Visualization is most useful when we focus on smaller networks or subset networks.

With large networks, we can look for the links that are most heavily traveled or the strongest links between nodes. We could look at a subset of nodes that have the highest degree centrality or are viewed as the most important nodes in the network. Visualization can be useful in the analysis of *ego-centric* networks or subset networks that begin with a particular node of interest. We can look for clusters of nodes on the web. We can look for communities or cliques within social networks. Modeling techniques guide us to a subset network of interest, and then we use data visualization to explore that subset further.

Figure 6.5. *Cannot See the Tree from the Links*

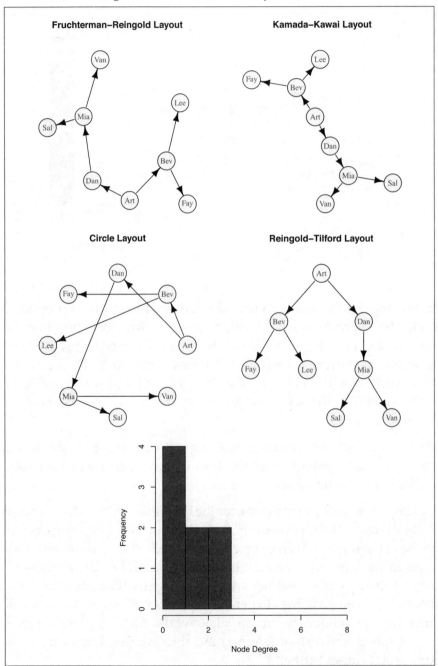

To illustrate the use of visualization in the analysis of a social network, we turn to the Enron e-mail archive described in appendix C (page 291). This is a network with more than 36 thousand nodes and more than 186 thousand links, so we look for subsets of the network (subgraphs) to gain understanding of communications among the Enron executives.

One approach is to examine an ego-centric network or subgraph starting with a key node, its immediate neighbor nodes (nodes defined by the from and to e-mail links), and the links among those nodes. Visualizations for the ego-centric network for Enron executive *1* is shown in figure 6.6. Groups of nodes are easily detected with the Kamada-Kawai algorithm, and nodes are sufficiently separated in space for labeling. So we use that algorithm for making a larger plot with node names, as shown in figure 6.7.

The Enron e-mail archive contains e-mail from and to Enron executives. This means that non-Enron executives are sinks or sources. If we were to use uppercase letters A and B to represent two Enron executives and lower-case letters a and b to represent two non-Enron individuals, then we might see any of the following e-mail from-and-to e-mail messages: $(A \Rightarrow B)$, $(B \Rightarrow A)$, $(A \Rightarrow a)$, $(A \Rightarrow b)$, $(a \Rightarrow A)$, $(b \Rightarrow A)$, $(a \Rightarrow B)$, or $(b \Rightarrow B)$. But we would not see messages of the form $(a \Rightarrow b)$ or $(b \Rightarrow a)$. Which is to say that the Enron e-mail archive is Enron-centric with an uncommon degree distribution. Enron executives are active members of the network, but there are thousands of actor nodes that are of limited importance to the network.

Suppose we identify the top fifty e-mail users in terms of total links (degree centrality). Figure 6.8 shows alternative plots for this subgraph of fifty executives, and figure 6.9 shows the Kamada-Kawai layout with node names.

Next we use graph theory algorithms to drill down to core groups of executives. One way to find the core is to identify cliques or groups of executives that are fully interconnected. A clique in the case of Enron is a subset of nodes having to- or from-links (out-box or in-box e-mail) for all pairs of nodes within the subset. Figure 6.10 shows the top fifty executives with a core group of fourteen executives at the center.

One objective of social network analysis is to find the sources of power, influence, or importance. In social network analysis these concepts are reflected in measures of centrality.

Figure 6.6. *Four Views of an Ego-Centric Network*

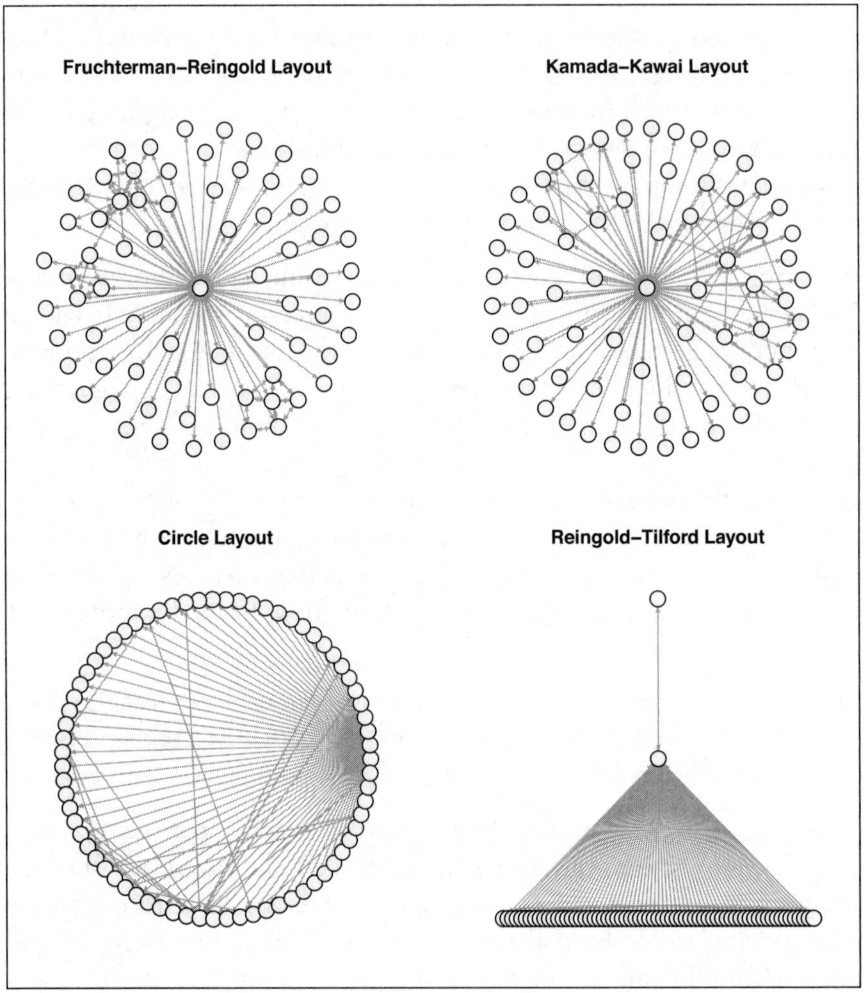

Fruchterman–Reingold Layout

Kamada–Kawai Layout

Circle Layout

Reingold–Tilford Layout

Figure 6.7. *An Ego-Centric Network*

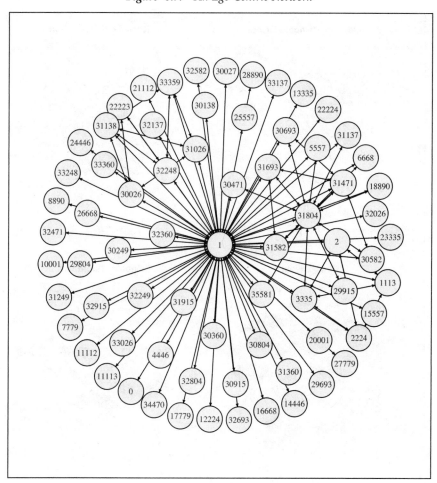

Figure 6.8. *Four Views of the Most Active Members of a Community*

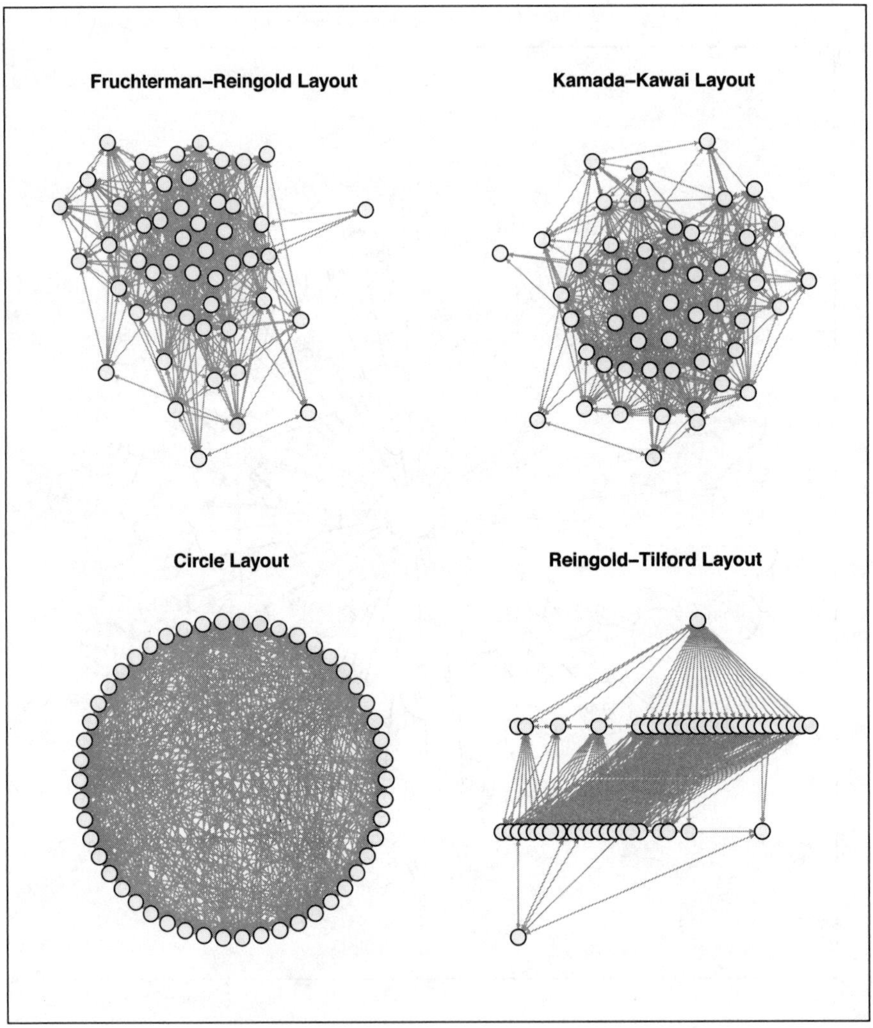

Figure 6.9. *Identifying the Most Active Members of a Community*

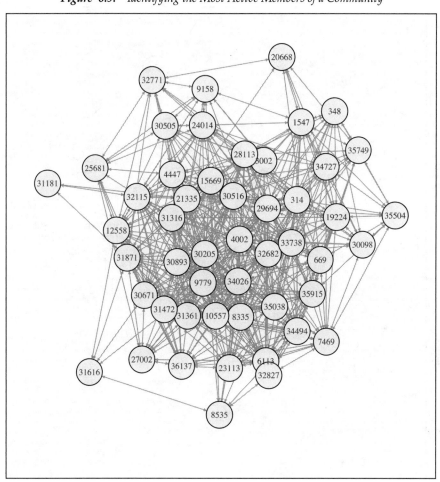

Figure 6.10. *Cliques and Core Members of a Community*

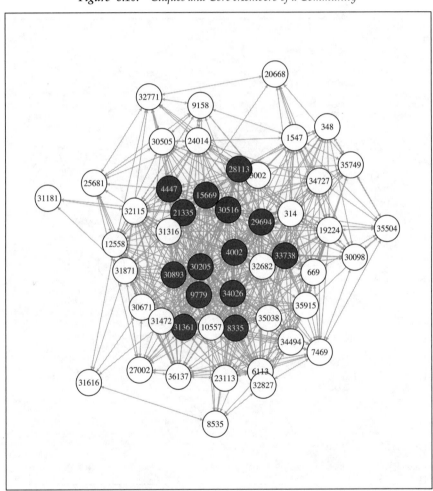

In the Enron example, we can obtain measures of centrality for each node among the fifty nodes with the highest degree centrality. Will network plots and measures of centrality lead us to the people with the power? Or are the top executives outside the group we have identified as core? Additional research and work with the Enron e-mail corpus itself may help to answer these questions.[1]

Our overview of visualization shows its utility as a technology for understanding social networks and the structure of organizations. We also see how alternative visualization algorithms can lead to decidedly different visualizations.

We want plots that look good—aesthetic graphs. We want plots that are readable and understandable. The goal may be to detect topological structure, to discover communities, subgroups, or cliques, or to identify outliers or anomalies. There are often restrictions on the size of the visualization. We like graphs that fill the available plotting area, while displaying all nodes and links with sufficient size to be visible or identifiable. Links that cross over one another detract from these objectives, as do overlapping nodes and labels. It is an added benefit, of course, when plotting objectives can be met within the software environment we use for modeling and data science.

There are those who are critical of large network/graph visualizations. In the words of Ben Fry (2008), "Everything looks like a graph, but almost nothing should ever be drawn as one" (240). Used in conjunction with modeling techniques, however, network visualizations can be most informative. The secret is to simplify, identify groups and key players, and focus on subnetworks within the larger network.

Di Battista et al. (1999) and contributors to Kratochvíl (1999) review the mathematics and theory behind network visualization. Common algorithms for network visualization have been presented by Kamada and Kawai (1989) and Fruchterman and Reingold (1991). Freeman (2005) reviews network vi-

---

[1] Additional analyses of the Enron e-mail archive have been provided by numerous researchers (Berry and Browne 2005; Chapanond, Krishnamoorthy, and Yener 2005; Diesner, Frantz, and Carley 2005; Keila and Skillicorn 2005). Moving beyond standard social network analyses, we see time-evolving network models, outlier/anomaly detection methods, and wave propagation algorithms being explored with data from the Enron corpus. Exemplary studies have been provided by Priebe et al. (2005), Uddin, Hamra, and Hossain (2013), Wang et al. (2013), Savage et al. (2014), Uddin et al. (2014), and Ho and Xing (2014).

sualization methods, paying special attention to multivariate methods, including multidimensional scaling, correspondence analysis, and principal components analysis.

Trees or hierarchies present interesting and challenging problems in data visualization. Specialized algorithms for trees have been discussed by numerous researchers (Lengauer and Tarjan 1979; Reingold and Tilford 1981; Supowit and Reingold 1983; Rusu 2014). Link/edge bundling, for example, offers promise as a method for visualizing trees and hierarchical networks and for subgroup discovery within networks (Holden 2006; Crnovrsanin et al. 2014).

In addition to Python and R packages for data and network visualization, there are many programs for drawing graphs (Jünger and Mutzel 2004). A comprehensive survey of two-dimensional network visualization techniques is provided by Gibson, Faith, and Vickers (2013).

Exhibit 6.1 lists a Python program for defining and visualizing simple networks. It shows how to plot networks, determine their degree distributions, and convert network or graph structures into matrices for further analysis. The program draws on the work of Hagbert and Schult (2014).

Exhibit 6.2 shows the R program for defining and visualizing simple networks. And exhibit 6.3 shows the R program for analyzing the Enron e-mail network. These programs draw on the work of Bojanowski (2014), Butts (2014a), and Csardi (2014b).

Exhibit 6.1. *Defining and Visualizing Simple Networks (Python)*

```
# Defining and Visualizing Simple Networks (Python)

# prepare for Python version 3x features and functions
from __future__ import division, print_function

# load packages into the workspace for this program
import networkx as nx
import matplotlib.pyplot as plt  # 2D plotting
import numpy as np


# -------------------------
# star (undirected network)
# -------------------------
# define graph object for undirected star network
# adding one link at a time for pairs of nodes
star = nx.Graph()
star.add_edge('Amy', 'Bob')
star.add_edge('Amy', 'Dee')
star.add_edge('Amy', 'Joe')
star.add_edge('Amy', 'Lea')
star.add_edge('Amy', 'Max')
star.add_edge('Amy', 'Tia')

# examine the degree of each node
print(nx.degree(star))

# plot the star network and degree distribution
fig = plt.figure()
nx.draw(star, node_size = 2000, node_color = 'yellow')
plt.show()

fig = plt.figure()
plt.hist(nx.degree(star).values())
plt.axis([0, 8, 0, 8])
plt.xlabel('Node Degree')
plt.ylabel('Frequency')
plt.show()

# create an adjacency matrix object for the star network
# use nodelist argument to order the rows and columns
star_mat = nx.adjacency_matrix(star,\
    nodelist = ['Amy', 'Bob', 'Dee', 'Joe', 'Lea', 'Max', 'Tia'])

print(star_mat)  # undirected networks are symmetric

# determine the total number of links for the star network (n-1)
print(np.sum(star_mat)/2)


# ---------------------------
# circle (undirected network)
# ---------------------------
```

```
# define graph object for undirected circle network
# using a list of links for pairs of nodes
circle = nx.Graph()
circle.add_edges_from([('Abe', 'Bea'), ('Abe', 'Rob'), ('Bea', 'Dag'),\
    ('Dag', 'Eve'), ('Eve', 'Jim'), ('Jim', 'Kat'), ('Kat', 'Rob')])

# examine the degree of each node
print(nx.degree(circle))

# plot the circle network and degree distribution
fig = plt.figure()
nx.draw(circle, node_size = 2000, node_color = 'yellow')
plt.show()

fig = plt.figure()
plt.hist(nx.degree(circle).values())
plt.axis([0, 8, 0, 8])
plt.xlabel('Node Degree')
plt.ylabel('Frequency')
plt.show()

# create an adjacency matrix object for the circle network
# use nodelist argument to order the rows and columns
circle_mat = nx.adjacency_matrix(circle,\
    nodelist = ['Abe', 'Bea', 'Dag', 'Eve', 'Jim', 'Kat', 'Rob'])
print(circle_mat)  # undirected networks are symmetric

# determine the total number of links for the circle network
print(np.sum(circle_mat)/2)

# ------------------------
# line (undirected network)
# ------------------------
# define graph object for undirected line network
# using a list of links for pairs of nodes
line = nx.Graph()
line.add_edges_from([('Ali', 'Ben'), ('Ali', 'Ela'), ('Ben', 'Ian'),\
    ('Ela', 'Mya'), ('Ian', 'Roy'), ('Mya', 'Zoe')])

# examine the degree of each node
print(nx.degree(line))

# plot the line network and degree distribution
fig = plt.figure()
nx.draw(line, node_size = 2000, node_color = 'yellow')
plt.show()

fig = plt.figure()
plt.hist(nx.degree(line).values())
plt.axis([0, 8, 0, 8])
plt.xlabel('Node Degree')
plt.ylabel('Frequency')
plt.show()
```

```
# create an adjacency matrix object for the line network
# use nodelist argument to order the rows and columns
line_mat = nx.adjacency_matrix(line,\
    nodelist = ['Ali', 'Ben', 'Ela', 'Ian', 'Mya', 'Roy', 'Zoe'])
print(line_mat)  # undirected networks are symmetric

# determine the total number of links for the line network (n-1)
print(np.sum(line_mat)/2)


# --------------------------
# clique (undirected network)
# --------------------------
# define graph object for undirected clique
a_clique = nx.Graph()
a_clique.add_edges_from([('Ada', 'Ala'), ('Ada', 'Ami'), ('Ada', 'Ana'),\
('Ada', 'Ann'), ('Ada', 'Ara'), ('Ada', 'Ava'), ('Ala', 'Ami'),\
('Ala', 'Ana'), ('Ala', 'Ann'), ('Ala', 'Ara'), ('Ala', 'Ava'),\
('Ami', 'Ana'),('Ami', 'Ann'), ('Ami', 'Ara'), ('Ami', 'Ava'),\
('Ana', 'Ann'), ('Ana', 'Ara'), ('Ana', 'Ava'),\
('Ann', 'Ara'), ('Ann', 'Ava'), ('Ara', 'Ava')])

# examine the degree of each node
print(nx.degree(a_clique))

# plot the clique and degree distribution
fig = plt.figure()
nx.draw_circular(a_clique, node_size = 2000, node_color = 'yellow')
plt.show()

fig = plt.figure()
plt.hist(nx.degree(a_clique).values())
plt.axis([0, 8, 0, 8])
plt.xlabel('Node Degree')
plt.ylabel('Frequency')
plt.show()

# create an adjacency matrix object for the line network
# use nodelist argument to order the rows and columns
a_clique_mat = nx.adjacency_matrix(a_clique,\
    nodelist = ['Ada', 'Ala', 'Ami', 'Ana', 'Ann', 'Ara', 'Ava'])
print(a_clique_mat)  # undirected networks are symmetric

# determine the total number of links for the clique n(n-1)/2
print(np.sum(a_clique_mat)/2)

# -----------------------------
# tree (directed network/digraph)
# -----------------------------
# define graph object for undirected tree network
# using a list of links for pairs of from-to nodes
tree = nx.DiGraph()
tree.add_edges_from([('Art', 'Bev'), ('Art', 'Dan'), ('Bev', 'Fay'),\
    ('Bev', 'Lee'), ('Dan', 'Mia'), ('Mia', 'Sal'), ('Mia', 'Van')])
```

```
# examine the degree of each node
print(nx.degree(tree))

# create an adjacency matrix object for the line network
# use nodelist argument to order the rows and columns
tree_mat = nx.adjacency_matrix(tree,\
    nodelist = ['Art', 'Bev', 'Dan', 'Fay', 'Lee', 'Mia', 'Sal', 'Van'])
print(tree_mat)  # directed networks are not symmetric

# determine the total number of links for the tree
# upper triangle only has values
print(np.sum(tree_mat))

# plot the degree distribution
fig = plt.figure()
plt.hist(nx.degree(tree).values())
plt.axis([0, 8, 0, 8])
plt.xlabel('Node Degree')
plt.ylabel('Frequency')
plt.show()

# examine alternative layouts for plotting the tree
# plot the network/graph with default layout
fig = plt.figure()
nx.draw(tree, node_size = 2000, node_color = 'yellow')
plt.show()

# spring layout
fig = plt.figure()
nx.draw_spring(tree, node_size = 2000, node_color = 'yellow')
plt.show()

# circlular layout
fig = plt.figure()
nx.draw_circular(tree, node_size = 2000, node_color = 'yellow')
plt.show()

# concentric circles layout
fig = plt.figure()
nx.draw_shell(tree, node_size = 2000, node_color = 'yellow')
plt.show()

# note. plotting as tree may require pygraphviz

# Suggestions for the student: Define alternative network structures.
# Use matplotlib to create their plots. Create the corresponding
# adjacency matrices and compute network descriptive statistics,
# beginning with degree centrality. Plot the degree distribution
# for each network.  Read about pygraphviz and try to plot a tree.
```

Exhibit 6.2. *Defining and Visualizing Simple Networks (R)*

```
# Defining and Visualizing Simple Networks (R)

# begin by installing necessary package igraph

# load package into the workspace for this program
library(igraph)  # network/graph functions

# -------------------------
# star (undirected network)
# -------------------------
# define graph object for undirected star network
# using links for pairs of nodes by number
star <- graph.formula(1-2, 1-3, 1-4, 1-5, 1-6, 1-7)
# examine the graph object
print(str(star))

# name the nodes (vertices)
V(star)$name <- c("Amy", "Bob", "Dee", "Joe", "Lea", "Max", "Tia")

# examine the degree of each node
print(degree(star))

# determine the total number of links for the star (n-1)
print(sum(degree(star))/2)

# create an adjacency matrix object for the star network
star_mat <- get.adjacency(star)
print(star_mat)  # undirected networks are symmetric

# plot the network/graph to the console
plot(star, vertex.size = 20, vertex.color = "yellow")
# plot the network/graph and degree distribution to an external file
pdf(file = "fig_star_network.pdf", width = 5.5, height = 5.5)
plot(star, vertex.size = 25, vertex.color = "yellow", edge.color = "black")
hist(degree(star), col = "darkblue",
    xlab = "Node Degree", xlim = c(0,8), main = "",
    breaks = c(0,1,2,3,4,5,6,7,8))
dev.off()

# ---------------------------
# circle (undirected network)
# ---------------------------
# define graph object for undirected circle network
# using links for pairs of nodes by number
circle <- graph.formula(1-2, 1-7, 2-3, 3-4, 4-5, 5-6, 6-7)
# examine the graph object
print(str(circle))
# name the nodes (vertices)
V(circle)$name <- c("Abe", "Bea", "Dag", "Eve", "Jim", "Kat", "Rob")
# examine the degree of each node
print(degree(circle))
```

```
# determine the total number of links
print(sum(degree(circle))/2)

# create an adjacency matrix object for the circle network
circle_mat <- get.adjacency(circle)
print(circle_mat)  # undirected networks are symmetric

# plot the network/graph to the console
plot(circle, vertex.size = 20, vertex.color = "yellow")
# plot the network/graph and degree distribution to an external file
pdf(file = "fig_circle_network.pdf", width = 5.5, height = 5.5)
plot(circle, vertex.size = 25, vertex.color = "yellow", edge.color = "black")
hist(degree(circle), col = "darkblue",
    xlab = "Node Degree", xlim = c(0,8), main = "",
    breaks = c(0,1,2,3,4,5,6,7,8))
dev.off()


# -------------------------
# line (undirected network)
# -------------------------
# define graph object for undirected line network
# using links for pairs of nodes by number
line <- graph.formula(1-2, 1-3, 2-4, 3-5, 4-6, 5-7)
# examine the graph object
print(str(line))

# name the nodes (vertices)
V(line)$name <- c("Ali", "Ben", "Ela", "Ian", "Mya", "Roy", "Zoe")

# examine the degree of each node
print(degree(line))

# determine the total number of links (n-1)
print(sum(degree(line))/2)
# create an adjacency matrix object for the line network
line_mat <- get.adjacency(line)
print(line_mat)  # undirected networks are symmetric

# plot the network/graph to the console
plot(line, vertex.size = 20, vertex.color = "yellow")
# plot the network/graph and degree distribution to an external file
pdf(file = "fig_line_network.pdf", width = 5.5, height = 5.5)
plot(line, vertex.size = 25, vertex.color = "yellow", edge.color = "black")
hist(degree(line), col = "darkblue",
    xlab = "Node Degree", xlim = c(0,8), main = "",
    breaks = c(0,1,2,3,4,5,6,7,8))
dev.off()


# -------------------------
# clique (undirected network)
# -------------------------
# define graph object for undirected clique
# using links for pairs of nodes by number
```

```
a_clique <- graph.formula(1-2, 1-3, 1-4, 1-5, 1-6, 1-7,
                              2-3, 2-4, 2-5, 2-6, 2-7,
                                   3-4, 3-5, 3-6, 3-7,
                                        4-5, 4-6, 4-7,
                                             5-6, 5-7,
                                                  6-7)
# examine the graph object
print(str(a_clique))

# name the nodes (vertices)
V(a_clique)$name <- c("Ada", "Ala", "Ami", "Ana", "Ann", "Ara", "Ava")

# examine the degree of each node
print(degree(a_clique))

# determine the total number of links n(n-1)/2
print(sum(degree(a_clique))/2)

# create an adjacency matrix object for the clique
# this is a matrix of ones off-diagonal... fully connected
a_clique_mat <- get.adjacency(a_clique)
print(a_clique_mat)  # undirected networks are symmetric

# plot the network/graph to the console
plot(a_clique, vertex.size = 20, vertex.color = "yellow",
    edge.color = "black", layout = layout.circle)
# plot the network/graph and degree distribution to an external file
pdf(file = "fig_clique_network.pdf", width = 5.5, height = 5.5)
plot(a_clique, vertex.size = 25, vertex.color = "yellow",
    edge.color = "black", layout = layout.circle)
hist(degree(a_clique), col = "darkblue",
    xlab = "Node Degree", xlim = c(0,8), main = "",
    breaks = c(0,1,2,3,4,5,6,7,8))
dev.off()

# -------------------------------
# tree (directed network/digraph)
# -------------------------------
# define graph object for undirected tree network
# using links for pairs of nodes by number
tree <- graph.formula(1-+2, 1-+3, 2-+4, 2-+5, 3-+6, 6-+7, 6-+8)
# examine the graph object
print(str(tree))

# name the nodes (vertices)
V(tree)$name <- c("Art", "Bev", "Dan", "Fay", "Lee", "Mia", "Sal", "Van")

# examine the degree of each node
print(degree(tree))

# create an adjacency matrix object for the tree network
tree_mat <- get.adjacency(tree)
print(tree_mat)  # directed networks are not symmetric
```

```
# plot the network/graph to the console
plot(tree, vertex.size = 20, vertex.color = "yellow", edge.color = "black")

# plot the network/graph and degree distribution to an external file
# examine alternative layouts for plotting the tree
pdf(file = "fig_tree_network_four_ways.pdf", width = 5.5, height = 5.5)
par(mfrow = c(1,1))  # four plots on one page
plot(tree, vertex.size = 25, vertex.color = "yellow",
    layout = layout.fruchterman.reingold, edge.color = "black")
title("Fruchterman-Reingold Layout")
plot(tree, vertex.size = 25, vertex.color = "yellow",
    layout = layout.kamada.kawai, edge.color = "black")
title("Kamada-Kawai Layout")
plot(tree, vertex.size = 25, vertex.color = "yellow",
    layout = layout.circle, edge.color = "black")
title("Circle Layout")
plot(tree, vertex.size = 25, vertex.color = "yellow",
    layout = layout.reingold.tilford, edge.color = "black")
title("Reingold-Tilford Layout")
hist(degree(tree), col = "darkblue",
    xlab = "Node Degree", xlim = c(0,8), main = "",
    breaks = c(0,1,2,3,4,5,6,7,8))
dev.off()
```

Exhibit 6.3. *Visualizing Networks—Understanding Organizations (R)*

```
# Visualizing Networks---Understanding Organizations (R)

# bring in packages we rely upon for work in predictive analytics
library(igraph)  # network/graph methods
library(network)  # network representations
library(intergraph)  # for exchanges between igraph and network


# ------------------------------------------------------------
# Preliminary note about data preparation
#
# The first two records of the original link file
# contain vertices referenced by a zero index.
# For R network algorithms we need to ensure that
# the from- and to-node identifiers are
# treated as character strings, not integers.


# ------------------------------------------------------------
# Read in list of links... (from-node, to-node) pairs
# ------------------------------------------------------------
all_enron_links <- read.table('enron_email_links.txt', header = FALSE)
cat("\n\nNumber of Links on Input: ", nrow(all_enron_links))
# check the structure of the input data data frame
print(str(all_enron_links))

# convert the V1 and V2 to character strings
all_enron_links$V1 <- as.character(all_enron_links$V1)
all_enron_links$V2 <- as.character(all_enron_links$V2)

# ensure that no e-mail links are from an executive to himself/herself
# i.e. eliminate any nodes that are self-referring
enron_links <- subset(all_enron_links, subset = (V1 != V2))
cat("\n\nNumber of Valid Links: ", nrow(enron_links))

# create network object from the links
# multiple = TRUE allows for multiplex links/edges
# because it is possible to have two or more links
# between the same two nodes (multiple e-mail messages
# between the same two people)
enron_net <- network(as.matrix(enron_links),
    matrix.type = "edgelist", directed = TRUE, multiple = TRUE)

# create graph object with intergraph function asIgraph()
enron_graph <- asIgraph(enron_net)

# name the nodes noting that the first identifer on input was "0"
node_index <- as.numeric(V(enron_graph))
V(enron_graph)$name <- node_name <- as.character(V(enron_graph) - 1)
# node name lookup table
node_reference_table <- data.frame(node_index, node_name, stringsAsFactors = FALSE)
print(str(node_reference_table))
print(head(node_reference_table))
```

```
# consider the subgraph of all people that node "1"
# communicates with by e-mail (mail in or out)
# node "1" corresponds to node index 2 in R
ego_1_mail <- induced.subgraph(enron_graph,
    neighborhood(enron_graph, order = 1, nodes = 2)[[1]])
# examine alternative layouts for plotting the ego_1_mail
pdf(file = "fig_ego_1_mail_network_four_ways.pdf", width = 5.5, height = 5.5)
par(mfrow = c(1,1))  # four plots on one page
set.seed(9999)  # for reproducible results
plot(ego_1_mail, vertex.size = 10, vertex.color = "yellow",
    vertex.label = NA, edge.arrow.size = 0.25,
    layout = layout.fruchterman.reingold)
title("Fruchterman-Reingold Layout")
set.seed(9999)  # for reproducible results
plot(ego_1_mail, vertex.size = 10, vertex.color = "yellow",
    vertex.label = NA, edge.arrow.size = 0.25,
    layout = layout.kamada.kawai)
title("Kamada-Kawai Layout")
set.seed(9999)  # for reproducible results
plot(ego_1_mail, vertex.size = 10, vertex.color = "yellow",
    vertex.label = NA, edge.arrow.size = 0.25,
    layout = layout.circle)
title("Circle Layout")
set.seed(9999)  # for reproducible results
plot(ego_1_mail, vertex.size = 10, vertex.color = "yellow",
    vertex.label = NA, edge.arrow.size = 0.25,
    layout = layout.reingold.tilford)
title("Reingold-Tilford Layout")
dev.off()

set.seed(9999)  # for reproducible results
pdf(file = "fig_ego_1_mail_network.pdf", width = 8.5, height = 11)
plot(ego_1_mail, vertex.size = 15, vertex.color = "yellow",
    vertex.label.cex = 0.9, edge.arrow.size = 0.25,
    edge.color = "black", layout = layout.kamada.kawai)
dev.off()

# examine the degree of each node in the complete Enron e-mail network
# and add this measure (degree centrality) to the node reference table
node_reference_table$node_degree <- degree(enron_graph)
print(str(node_reference_table))
print(head(node_reference_table))

# sort the node reference table by degree and identify the indices
# of the most active nodes (those with the most links)
sorted_node_reference_table <-
    node_reference_table[sort.list(node_reference_table$node_degree,
        decreasing = TRUE),]
# check on the sort
print(head(sorted_node_reference_table))
print(tail(sorted_node_reference_table))
# select the top K executives... set K
K <- 50
```

```
# identify a subset of K Enron executives based on e-mail-activity
top_node_indices <- sorted_node_reference_table$node_index[1:K]
top_node_names <- sorted_node_reference_table$node_name[1:K]
print(top_node_indices)
print(top_node_names)

# define a top nodes reference table as subset of complete reference table
top_node_reference_table <- subset(node_reference_table,
    subset = (node_name %in% top_node_names))
print(str(top_node_reference_table))
print(head(top_node_reference_table))

# construct the subgraph of the top K executives
top_enron_graph <- induced.subgraph(enron_graph, top_node_indices)
# examine alternative layouts for plotting the top_enron_graph
pdf(file = "fig_top_enron_graph_four_ways.pdf", width = 5.5, height = 5.5)
par(mfrow = c(1,1))  # four plots on one page
set.seed(9999)  # for reproducible results
plot(top_enron_graph, vertex.size = 10, vertex.color = "yellow",
    vertex.label = NA, edge.arrow.size = 0.25,
    layout = layout.fruchterman.reingold)
title("Fruchterman-Reingold Layout")
set.seed(9999)  # for reproducible results
plot(top_enron_graph, vertex.size = 10, vertex.color = "yellow",
    vertex.label = NA, edge.arrow.size = 0.25,
    layout = layout.kamada.kawai)
title("Kamada-Kawai Layout")
set.seed(9999)  # for reproducible results
plot(top_enron_graph, vertex.size = 10, vertex.color = "yellow",
    vertex.label = NA, edge.arrow.size = 0.25,
    layout = layout.circle)
title("Circle Layout")
set.seed(9999)  # for reproducible results
plot(top_enron_graph, vertex.size = 10, vertex.color = "yellow",
    vertex.label = NA, edge.arrow.size = 0.25,
    layout = layout.reingold.tilford)
title("Reingold-Tilford Layout")
dev.off()
# let's use the Kamada-Kawai layout for the labeled plot
set.seed(9999)  # for reproducible results
pdf(file = "fig_top_enron_graph.pdf", width = 8.5, height = 11)
plot(top_enron_graph, vertex.size = 15, vertex.color = "yellow",
    vertex.label.cex = 0.9, edge.arrow.size = 0.25,
    edge.color = "darkgray", layout = layout.kamada.kawai)
dev.off()
# a clique is a subset of nodes that are fully connected
# (links between all pairs of nodes in the subset)
# perform a census of cliques in the top_enron_graph
table(sapply(cliques(top_enron_graph), length))  # shows two large cliques
# the two largest cliques have thirteen nodes/executives
# let's identify those cliques
two_cliques <-
    cliques(top_enron_graph)[sapply(cliques(top_enron_graph), length) == 13]
```

```
# show the new index values for the top cliques... note the overlap
print(two_cliques)
# finding our way to the executive core of the company
# note index numbers are reset by the induced.subgraph() function
# form a new subgraph from the union of the top two cliques
core_node_indices_new <- unique(unlist(two_cliques))
non_core_node_indices_new <- setdiff(1:K, core_node_indices_new)
set_node_colors <- rep("white", length = K)
set_node_colors[core_node_indices_new] <- "darkblue"
set_label_colors <- rep("black", length = K)
set_label_colors[core_node_indices_new] <- "white"
# again use the Kamada-Kawai layout for the labeled plot
# but this time we use white for non-core and blue for core nodes
set.seed(9999)  # for reproducible results
pdf(file = "fig_top_enron_graph_with_core.pdf", width = 8.5, height = 11)
plot(top_enron_graph, vertex.size = 15,
    vertex.color = set_node_colors,
    vertex.label.color = set_label_colors,
    vertex.label.cex = 0.9, edge.arrow.size = 0.25,
    edge.color = "darkgray", layout = layout.kamada.kawai)
dev.off()
# check on the tree/hierarchy to search for the source of power
set.seed(9999)  # for reproducible results
plot(top_enron_graph, vertex.size = 15, vertex.color = "white",
    vertex.label.cex = 0.9, edge.arrow.size = 0.25,
    layout = layout.reingold.tilford)  # node name 314?
# compute centrality idices for the top executive nodes
# begin by expressing the adjacency matrix in standard matrix form
top_enron_graph_mat <- as.matrix(get.adjacency(top_enron_graph))
top_node_reference_table$betweenness <-
    betweenness(top_enron_graph)  # betweenness centrality
top_node_reference_table$evcent <-
    evcent(top_enron_graph)$vector  # eigenvalue centrality
print(str(top_node_reference_table))
print(top_node_reference_table)  # data for top executive nodes


# Suggestions for the student:  Experiment with techniques for identifying
# core executive groupings and sources of power in the organization.
# Could it be that node 314, a node not in the core group,
# is the true source of power in the organization?
# What do indices of centrality suggest about possible sources
# of power, influence, or importance in the Enron network?
# Try other network visualizations for the Enron e-mail network.
# Note that nowhere in our analysis so far have we looked at the
# number of e-mails sent from one player/node to another.
# All we have are binary links. If there is at least one e-mail between
# a pair of nodes, we have drawn a link between those nodes.
# Perhaps that is why everything looks like spaghetti or a ball of string.
# There is much that can be done with the Enron e-mail corpus.
# We could work with the original Enron e-mail case data, assigning
# executive names (not just numbers) to the nodes.  We could
# explore methods of text analytics using the e-mail message text.
```

7

Understanding Communities

"She isn't my wife really. We just have some kids . . . No . . . No kids. Not even kids. Sometimes, though, it . . . it feels as if we had kids. She isn't beautiful, she's . . . easy to live with. No she isn't. That's why I don't live with her."

—DAVID HEMMINGS AS THOMAS IN *Blow-Up* (1966)

Part of my exercise regimen is to walk between five and six in the morning. Except for an old man I often see, I have the sidewalks to myself. As far as I can tell, he lives in a nearby park, or at least that is where he spends most of his time. When I pass him on the sidewalk, I wave, and he waves back. That is the extent of our social interaction. It is as though we are saying, "We are here sharing this space in the city, and that is okay."

Couples marry, and we trace genealogy. Contracts bind the party of the first part to the party of the second part, whatever those parties and parts happen to be. There are mergers and acquisitions, service providers and clients, students and teachers, friends and lovers. There are groups of people attending events, congregations, and conventions. Some connections are strong, others less so.

It is not always easy to understand why people do what they do, but we can at the very least observe their relationships with one another. Motives play out in communities of association, which we represent as social networks.

What is a social network? A social network is person nodes connected by links. What is a link? Whatever we want it to be. Sometimes, as with the old man and me, the link is geography accompanied by a gesture.

Social network analysis has been an active area of research in psychology, sociology, anthropology, and political science for many years. The invention of the sociogram and concepts of social structure may be traced back more than eighty years (Moreno 1934; Radcliffe-Brown 1940). Research topics include isolation and popularity, prestige, power, and influence, social cohesion, subgroups and cliques, status and roles within organizations, balance and reciprocity, marketplace relationships, and measures of centrality and connectedness.

In years past, social network data collection was a painstaking process. Not so today. Social media implemented through the web have rekindled interest in social network analysis. Technologies of the web provide a record of what people do, where they come from, where they go, with whom they communicate, and sometimes transcripts of what they say. There are online friends and followers, tweets and retweets, text messages, e-mail, and blog postings. The data are plentiful. The challenge is to find our way to useful data and make sense of them.

Recognizing growth in the use of electronic social networks and intelligent mobile devices, organizations see opportunities for communicating with and selling to friends of friends. Businesses, nonprofits, and governmental organizations (not to mention political campaigns) are interested in learning from the data of social media. Network data sets are large but accessible. The possibilities are many.

Social networks are useful in making predictions. We can expect that a our attitudes and behavior are affected by the people we know. People closest to us in "network space," as it were, may be most influential in affecting our attitudes and behavior.

To explore further the domain of web and network data science, we look for patterns or regularities in relationships. Relationships imply interactions between actors. The actor nodes in a social network can be people, organizations, firms, buyers and sellers in the marketplace, investors and entrepreneurs in the venture capital space, or nations in international commerce. The actors may be described in terms of their individual characteristics, the links by their strength and direction. Some links between actors are unidirectional, others bidirectional, corresponding to directed and undirected networks.

Various network measures come from mathematicians working with graph theory and from sociologists working with social networks. Network measures describe relationships among nodes and structural characteristics of networks.

Whether working with the web, a content domain, or social group, it may not be easy to determine which nodes and links to include in a network study. In other words, we may need to specify the *boundary* of a network. We may need to define arbitrarily where the network begins and ends—the population of interest.

We use mathematical models of networks in making predictions about network phenomena and in studying relationships across network measures. Three types of models are of special interest: random graph, preferential attachment, and small-world network models.

The first mathematical model of interest, developed by Paul Erdös and Alfred Rényi (1959, 1960, 1961), lays the foundation for all models to follow. A *random graph* is a set of nodes connected by links in a purely random fashion. Figure 7.1 uses a circular layout to show a random graph with fifty nodes and one hundred links. No discernable pattern is detected in this network. Links are randomly associated with nodes, which is to say that all pairs of nodes are equally likely to be linked.

A second mathematical model for networks is the *preferential attachment* model and follows from the work of Barabási and Albert (1999). This model begins as a random graph and adds new links in a manner that gives preference to nodes that are already well connected (nodes with higher degree centrality).

Figure 7.2 shows a preferential attachment network in a circular layout. Note that certain nodes have many more links than other nodes. In describing preferential attachment models, we can think of the saying, "The rich get richer, and the poor get poorer."

Networks arising from preferential attachment are sometimes called *scale-free* or *long-tail* networks. In the literature of network science, there have

Figure **7.1.** *A Random Graph*

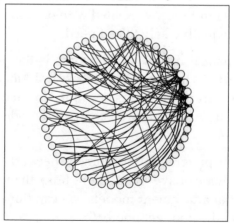

Figure **7.2.** *Network Resulting from Preferential Attachment*

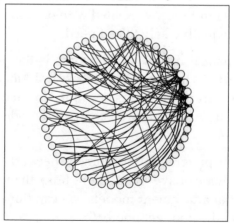

Figure 7.3. Building the Baseline for a Small World Network

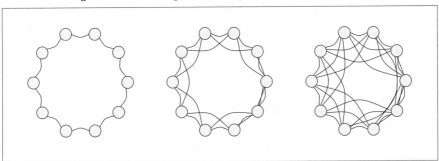

been many variations on the preferential attachment model (Albert and Barabási 2002). These models are especially useful for representing online networks.

A third mathematical model is especially important in representing social networks. This is the *small-world network* (Watts and Strogatz 1998). The small-world model defines a structure in which many nodes or actors are connected to nearby neighbors and some nodes are linked to nodes that are not nearby. Small-world networks have been studied extensively in social psychology, sociology, and network science (Milgram 1967; Travers and Milgram 1969; Watts 1999; Schnettler 2009).

To generate an undirected small-world network, we first specify the number of nodes and the size of the small-world neighborhood. Then we perturb the small world by rewiring a selected proportion of the nodes. The probability of random attachment or rewiring, like the size of the neighborhood, is a parameter defining the network.

Figure 7.3 shows baseline small-world networks prior to the rewiring of nodes. Using a circular layout, we see a ten-node network at the left with nodes associated only with adjacent nodes. Each node has two links. As neighborhoods grow in size, moving from left to right in the figure, the number of links between nodes grows. The network at the center has nodes linking to four other nodes, and the network at the right has nodes linking to six other nodes. The defining property of a small-world network is that many nodes are associated with nearby nodes.

Figure 7.4. *A Small-World Network*

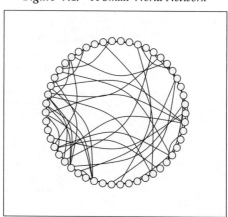

Moving from a baseline small-world network to an actual small-world network involves rewiring. We select links to rewire, detaching them from one of the nodes and reattaching to another node. The selection of links is conducted at random.

Again using a circular layout, we show a small-world network with fifty nodes and one hundred links in figure 7.4. Half of the links in this network are associated with nearest neighbors, the other half are associated with randomly selected nodes.

Figure 7.5 shows the degree distributions of networks generated from random graph, small-world, and preferential attachment models. Each of the models used to generate the figure contained about fifty nodes and one hundred links. Notice how small-world networks have a preponderance of low-degree nodes, compared with random graphs. Notice how the the preferential attachment model has a highly skewed or long-tail degree distribution.

We can use a mathematical model as a starting point in the statistical simulation of networks. They also have value in making statistical inferences about network parameters and predictions about the future state of networks.

Figure 7.5. *Degree Distributions for Network Models*

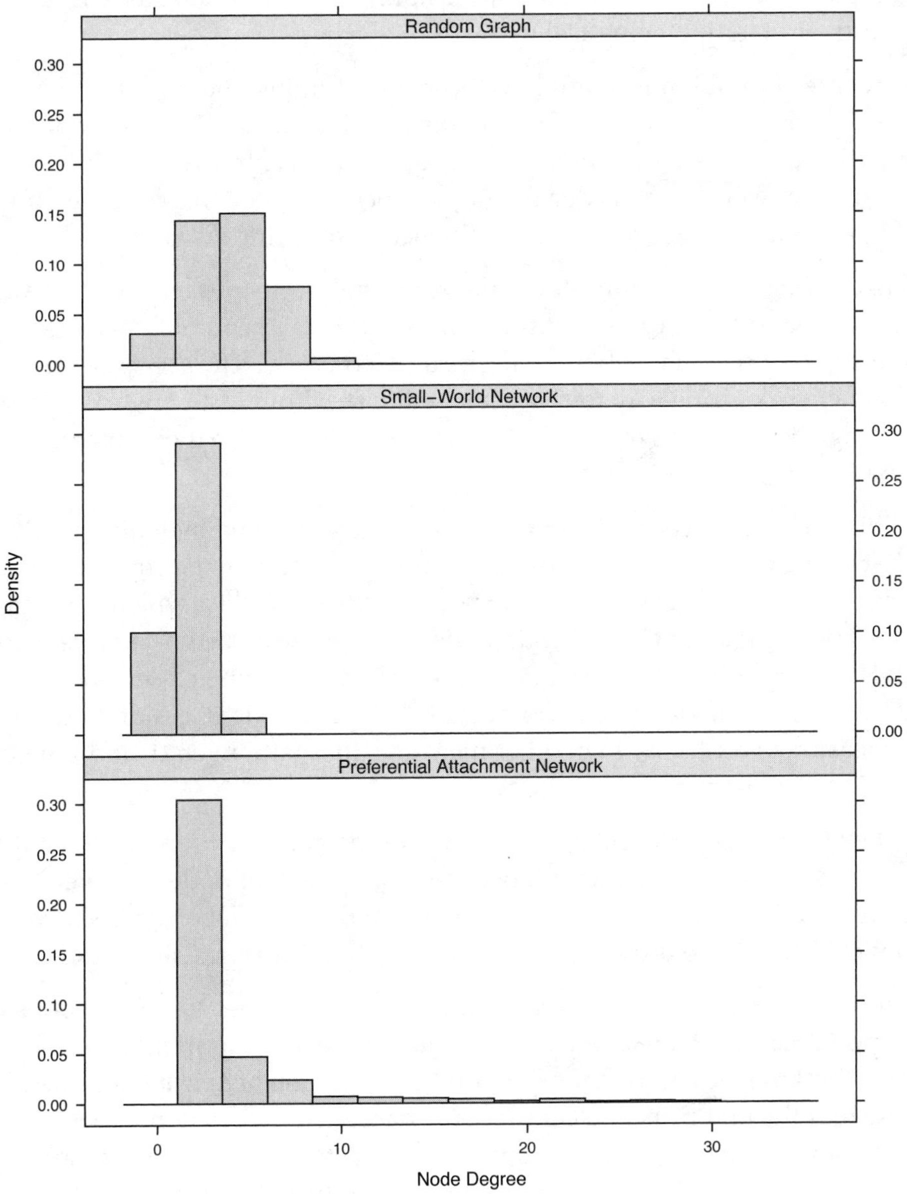

Before using a particular network model for inference or prediction, however, we may want to pose initial questions about the suitability of the model for the problem under investigation. Is it reasonable to believe that sample data come from a random graph, preferential attachment, small-world, or other mathematical model?

Measures of node importance are useful for finding sources of power, influence, or importance. We might want to characterize the importance of a politician to his party, thinking of measures of centrality as indicators of political power. We consider the importance of a page to a website, or the degree to which a journal article influences a scientific discipline.

We can begin by counting the links connected to the node. And for a directed network, we can count the number of links in and out—we measure the *degree* of each node, both *in-degree* and *out-degree*. The *degree distribution* of a network is one way to characterize the structure of a network, and the average degree or *degree centrality* is a way to summarize the connectedness of an entire network.

While the degree centrality of a node may be computed by looking only at that single node, other measures of centrality depend on the structure of the entire network. One nodal measure is *closeness centrality*, which characterizes how close a node is to all other nodes in the network, where closeness is the number of hops or links between nodes. Computing closeness centrality can be problematic in disconnected networks, so it is less often used as a measure of centrality than other measures (Borgatti, Everett, and Johnson 2013).

When measuring *betweenness centrality*, we consider the proportion of times a node lies on the shortest path between other pairs of nodes. Betweenness centrality reflects the degree to which a node is a broker or go-between node, affecting the traffic or flow of information across the network.

What if we consider a node as important if it is close to other nodes of importance, which are in turn close to other nodes of importance, and so on. This thinking moves us in the direction of the notion of *eigenvector centrality*. Much as the first principle component characterizes common variability in a set of variables, eigenvector centrality characterizes the degree to which a node is central to the set of nodes comprising the network.

Eigenvector centrality is a measure of overall connectedness or importance. A node acquires higher eigenvector centrality by being connected to other well-connected nodes. Eigenvector centrality depends on the entire network and is a good summary measure of node importance.

It should come as no surprise that various measures of node importance correlate highly. We can see this by randomly generating data from mathematical models for networks, computing alternative indices of importance or centrality for each node in each network, and then correlating the indices.

Figure 7.6 shows the average correlation heat map for a set of small-world networks with fifty nodes and one hundred links. Additional simulation studies using random graph and preferential attachment models would show even higher positive correlations among centrality measures. And simulation studies across networks generated from the three mathematical models (small-world, random graph, and preferential attachment models) would show that closeness centrality has lower correlations with degree centrality than either betweenness or eigenvector centrality.

Despite the fame and fortune that followed the publication of PageRank (Brin and Page 1998), Google's ranking of websites is fundamentally the idea that we should use the first eigenvector of an adjacency matrix to judge a node's importance. PageRank, like eigenvector centrality, is a network-wide measure. Getting a link from a web page of high importance or credibility is better than getting a link from a web page of low importance or credibility. PageRank for websites and eigenvector centrality for social networks are not new ideas. They flow from the work of mathematicians and statisticians in the early twentieth century.[1]

---

[1] Franceschet (2011) notes the mathematical genealogy of PageRank dating back to Markov in 1906 and the Perron-Frobenius theorem of 1912. Eigenvalue/eigenvector analysis may be employed to ranking and paired comparison problems, as well as website rankings. Keener (1993) reviewed the mathematics of ranking methods. Paired comparisons have been used in taste-testing experiments, psychophysical investigations, preference scaling, and studies of consistency across judges (inter-rater reliability), as well as in the ranking of players and teams, the design of sporting tournaments, and the seeding of teams in tournaments (Thompson 1975; Groeneveld 1990; Appleton 1995; Carlin 1996; West 2006). The literature on paired comparison and ranking methods, documented in Davidson and Farquhar (1976), follows from the work of Thurstone (1927), Guilford (1936), Kendall and Smith (1940), and Wei (1952). Alternatively, Bradley and Terry (1952) introduced traditional logistic or logit models for analyzing paired comparisons. Reviews of paired comparison methods have been provided by Torgerson (1958), David (1963), and Bradley (1976). Langville and Meyer (2006, 2012) review the linear algebra underlying PageRank and other ranking methods.

Figure 7.6. *Alternative Measures of Centrality are Positively Correlated*

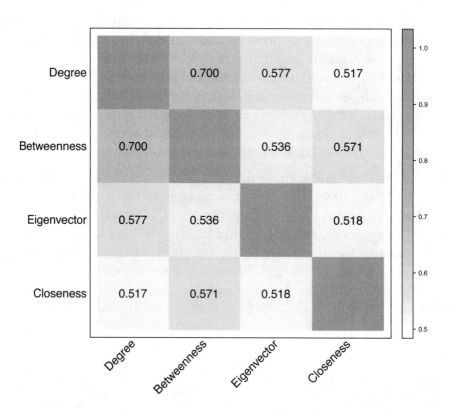

The importance of a web page is assessed by the number and quality of its links from other web pages. The importance of a person follows from the importance of associates, friends, or relatives. With measures of centrality, we have ways of finding the sources of credibility, power, or influence. These are associated with individual actors, the nodes of a social network. What about the links?

When looking at flow across a network or critical paths through a network, we focus on the links connecting nodes. To distinguish among links, we typically use betweenness. Links, like nodes, can lie on the path between pairs of nodes. And the more pairs of nodes potentially communicating across a link, the higher the betweenness of that link.

Various network-wide measures consider both nodes and links. One such measure is *density*. A completely unconnected or disconnected set of nodes has zero links, while a completely connected network of n nodes or clique has $\frac{n(n-1)}{2}$ links. If l is the number of links in a network, then network density is given by the formula $\frac{l}{n(n-1)/2}$.

Network density is easily calculated and takes values between zero and one. It is also easy to interpret as the proportion of actual links out of the set of all possible links. But density has limited value as a descriptive network measure across networks differing in size.

Social networks concern people, and people are constrained by time, distance, and communication media. There is a limit to the number of social connections or links people can maintain. As networks grow large, observed densities tend to decline.

Another network-wide measure is *transitivity*, which is the proportion of closed triads. A *triad* or triple is a set of three nodes, and a *closed triad* is a set of three nodes with links between each pair of nodes. We might think of a closed triad as a mini-clique. Transitivity, like other network measures, can be computed from the adjacency matrix.

There are many other measures of networks that we might consider. A comprehensive review of measures obtained from graphs and matrices is provided by Iacobucci in the volume by Wasserman and Faust (1994).

The measures we have been discussing relate to network topology or structure. They describe nodes relative to other nodes, links relative to other links, and, with density, links relative to possible links. They help us to identify cliques and core communities within a larger network.

Moving beyond topology or structure alone, we have what are known as metadata. There are node characteristics relating to player roles, demographics, attitudes, and behavior, and there link characteristics relating to width, bandwidth, length/distance, and cost of traversal. When we think about making predictions from social networks, we often think in terms of metadata.

Will a consumer buy a product? Will the citizen vote for a candidate? How much will a person spend on sporting events next year? Knowing the social network of the consumer, voter, or sports fan, can help to answer these questions.

Nearest-neighbor methods for predicting people's attitudes and behavior make no assumption about underlying processes or statistical distributions. These methods are data-driven rather than theory-dependent, and because of this they may be used for all kinds of data structures.

There are temporal, geographic, demographic, and psychographic nearest neighbors. Consider temporal nearest neighbors. Observations closer to one another in time are usually more highly correlated than observations distant in time. Nearest neighbors play out in autoregressive and moving average models for time series.

All other things being equal, observations closer together in space are more highly correlated than observations distant in space. We have geographic nearest neighbors. Someone living four miles from Dodger Stadium would be more likely a Dodgers fan than someone living in Tampa, Florida, say.

We can compute multivariate distance measures using sets of demographic or psychographic variables. We determine the distance between each pair of individuals in a study and use those distances to identify each person's five, ten, or fifteen nearest neighbors, for example. Who will get Mary's vote? Look at the votes of Mary's nearest neighbors to answer the question. See for whom they voted.

To employ nearest-neighbor methods, we need distances in time, physical space, demographics, psychographics, or network topology. For networks, we identify the neighborhood around each person and then identify the people in that neighborhood. Is Joe going to buy a multipurpose watch that displays physiological functions as well as time? Who are the people closest to Joe in his personal or ego-centric network? Are they buying the multipurpose watch?

Alternative versions of closeness may offer disparate nearest-neighbor predictions. The old man I see on my morning walks is close to me in geography and (I must acknowledge) close in age, but on other dimensions we have little in common. We do not frequent the same social circles.

For additional discussion of nearest-neighbor methods, see Duda, Hart, and Stork (2001), Shakhnarovich, Darrell, and Indyk (2006), and Hastie, Tibshirani, and Friedman (2009). Kolaczyk (2009) reviews nearest neighbors and other predictive modeling methods, including Markov random fields and kernel regression.

An introduction to social networks is provided by Kadushin (2012). Ackland (2013) applies social network concepts to web and social media networks. Reviews of network science are provided by Watts (2003), Lewis (2009), and Newman (2010). Works by Jackson (2008) and Easley and Kleinberg (2010) attempt to integrate network science and economic game theory.

Detection of clusters, information communities, core groups, and cliques is an important area of research in social network analysis. A recent review is provided by Zubcsek, Chowdhury, and Katona (2014).

Social network analyses often involve multidimensional scaling, hierarchical cluster analysis, log linear modeling, and a variety of specialized methods. Kolaczyk (2009) and Kolaczyk and Csárdi (2014) review statistical methods in network science. Wasserman and Faust (1994) and the contributors to Brandes and Erlebach (2005) review methods of social network analysis. Log-linear models have been used extensively in this area of research. For an overview of log linear models, see Bishop, Fienberg, and Holland (1975), Christensen (1997), and Fienberg (2007). Application of log-linear and logistic regression models to social networks is discussed in Wasserman and Iacobucci (1986) and Wasserman and Pattison (1996).

Stochastic blockmodels can be especially useful in the analysis of large networks. Individual nodes and their connections may be combined to form blocks within a network, and then these blocks may themselves be analyzed as networks. For reviews of blockmodels, see Holland et al. (1983) and Wasserman and Anderson (1987).

Do not underestimate the importance of analyst judgment in network science. The analyst chooses methods of sampling, modeling, and analysis. But more importantly, the analyst defines the network itself.

A link is a connection between nodes, but the analyst determines what it means to be connected. Consider the Enron e-mail network. Does one e-mail message between executives constitute a link? Do two, three, or four? The analyst sets the number of e-mail messages K between each pair of executives that constitutes a link. There are as many representations of the original Enron e-mail network as there are possible values of K. In sum, the structure of a network, the topology of links and nodes, much like the boundary of a network, is a matter of analyst judgment.

Network science and social networks provide interesting fodder for the popular press, as we can see with books by Buchanan (2002), Barabási (2003, 2010), Christakis and Fowler (2009), and Pentland (2014). We read that everything is connected to everything else, that order emerges out of chaos, that what appear to be chance events may not be chance events at all, and that collective intelligence is preeminent. While the notion of collective intelligence is strangely reminiscent of Jung (1968), and while it may be tempting to accept the new religion of network science, let us not forget that there is no collective, indeed there is no network and no science, without the actions of the one, and the one, and the one.

Exhibit 7.1 shows an R program for generating networks from random graph, small-world, and preferential attachment models. This program draws on R packages developed by Csardi (2014b), Kolaczyk and Csárdi (2014), and Sarkar (2014). Reviews of random network models have been provided by Albert and Barabási (2002), Cami and Deo (2008), Goldenberg et al. (2009), Kolaczyk (2009), and Toivonen et al. (2009).

When working with very large networks, processing times and memory requirements can be substantial. We must often sample from large networks to obtain networks of manageable size for modeling and analysis.

Sampling presents additional challenges. Shall we sample nodes or links (pairs of nodes), or some combination of nodes and links? A good sampling method should yield samples that have values of centrality and connectedness that are similar to the values in the complete network or population. Sampling and missing data issues have been reviewed by numerous researchers (Kossinets 2006; Leskovec and Faloutsos 2006; Handcock and Gile 2007; Huisman 2010; Smith and Moody 2013).

Exhibit 7.2 provides an R program that implements sampling methods described in Leskovec and Faloutsos (2006). We study these sampling methods using data from the Wikipedia Votes case (page 292). The study question is: "What sampling method should we use to obtain a representative sample of the complete network or population?" To assess the utility of alternative sampling methods in the example program, we use transitivity computed as the proportion of fully connected triples (triads). Transitivity, also known as the average clustering coefficient, is a measure of the connectedness of a network. For the complete Wikipedia Votes network, transitivity is 0.1646. We can think of this as the known value of a population parameter, and we want to identify an efficient network sampling method for estimating that parameter.

The example program shows how to implement three methods for obtaining a sample of N edges from the Wikipedia Votes network: random edge sampling, random node sampling, and random node-neighbor sampling. Random edge sampling, as the name would imply, selects edges at random from the network or population of edges. Using the complete data frame with each record representing an edge or connection between nodes, we select N edges at random. For random node sampling, we use the list of unique node numbers. We select nodes at random without replacement. This sampling is carried out using the combined set of from-node and to-node identifiers in the directed network. The preliminary sample data frame consists of all edges containing these nodes. Then we sample exactly N edges or rows from the preliminary sample data frame. Random node-neighbor sampling is a bit more complicated. We begin by selecting a node at random (referring to the from-node identifiers), together with all of its out-going neighbor nodes (using to-node identifiers). We continue selecting random from-node-identified nodes and their associated to-node-identified nodes until we have N edges in the sample.

Exhibit 7.1. *Network Models and Measures (R)*

```
# Network Models and Measures (R)

# install necessary packages

# bring packages into the workspace
library(igraph)  # network/graph models and methods
library(lattice)  # statistical graphics

# load correlation heat map utility
load(file = "correlation_heat_map.RData")

# note. evcent() often produces warning about lack of convergence
# we will ignore these warnings in this demonstration program
options(warn = -1)
# number of iterations for the statistical simulations
NITER <- 100

# user-defined function to compute centrality index correlations
get_centrality_matrix <- function(graph_object) {
    adjacency_mat <- as.matrix(get.adjacency(graph_object))
    node_degree <- degree(graph_object)
    node_betweenness <- betweenness(graph_object)
    node_closeness <- closeness.estimate(graph_object,
        mode = "all", cutoff = 0)
    node_evcent <- evcent(graph_object)$vector
    centrality <- cbind(node_degree, node_betweenness,
    node_closeness, node_evcent)
        colnames(centrality) <-  c("Degree", "Betweenness",
            "Closeness", "Eigenvector")
    return(cor(centrality))
    }
# ---------------------------------
# Random Graphs
# ---------------------------------
# show the plot of the first random graph model
set.seed(1)
# generate random graph with 50 nodes and 100 links/edges
random_graph <- erdos.renyi.game(n = 50, type = "gnm", p.or.m = 100)
pdf(file = "fig_network_random_graph.pdf", width = 5.5, height = 5.5)
plot(random_graph, vertex.size = 10, vertex.color = "yellow",
    vertex.label = NA, edge.arrow.size = 0.25, edge.color = "black",
    layout = layout.circle, edge.curved = TRUE)
dev.off()

# express adjacency matrix in standard matrix form
random_graph_mat <- as.matrix(get.adjacency(random_graph))
# verify that the network has one hundred links/edges
print(sum(degree(random_graph))/2)

aggregate_degree <- NULL  # initialize collection of node degree values
correlation_array <- array(NA, dim = c(4, 4, NITER))  # initialize array
```

```
for (i in 1:NITER) {
    set.seed(i)
    random_graph <- erdos.renyi.game(n = 50, type = "gnm", p.or.m = 100)
    aggregate_degree <- c(aggregate_degree,
        degree(random_graph))
    correlation_array[,,i] <- get_centrality_matrix(random_graph)
    }
average_correlation <- matrix(NA, nrow = 4, ncol = 4,
    dimnames = list(c("Degree", "Betweenness", "Closeness", "Eigenvector"),
        c("Degree", "Betweenness", "Closeness", "Eigenvector")))
for (i in 1:4)
    for(j in 1:4)
        average_correlation[i, j] <- mean(correlation_array[i, j, ])

pdf(file = "fig_network_random_graph_heat_map.pdf", width = 11,
  height = 8.5)
correlation_heat_map(cormat = average_correlation)
dev.off()


# create data frame for node degree distribution
math_model <- rep("Random Graph", rep = length(aggregate_degree))
random_graph_degree_data_frame <- data.frame(math_model, aggregate_degree)


# -------------------------------
# Small-World Networks
# -------------------------------
# example of a small-world network (no random links)
set.seed(1)
# one-dimensional small-world model with 10 nodes,
# links to additional adjacent nodes in a lattice
# (nei = 1 implies degree = 2 for all nodes prior to rewiring)
# rewiring probability of 0.00... no rewiring
small_world_network_prelim <- watts.strogatz.game(dim = 1, size = 10,
    nei = 1, p = 0.00, loops = FALSE, multiple = FALSE)
# remove any multiple links/edges
small_world_network_prelim <- simplify(small_world_network_prelim)
# express adjacency matrix in standard matrix form
# show that each node has four links
print(degree(small_world_network_prelim))
# verify that the network has one hundred links/edges
print(sum(degree(small_world_network_prelim))/2)
pdf(file = "fig_network_small_world_nei_1.pdf", width = 5.5, height = 5.5)
plot(small_world_network_prelim, vertex.size = 25, vertex.color = "yellow",
    vertex.label = NA, edge.arrow.size = 0.25, edge.color = "black",
    layout = layout.circle, edge.curved = TRUE)
dev.off()


# another of a small-world network (no random links)
set.seed(1)
# one-dimensional small-world model with 10 nodes,
# links to additional adjacent nodes in a lattice
# (nei = 2 implies degree = 2 for all nodes prior to rewiring)
# rewiring probability of 0.00... no rewiring
```

```
small_world_network_prelim <- watts.strogatz.game(dim = 1, size = 10,
    nei = 2, p = 0.00, loops = FALSE, multiple = FALSE)
# remove any multiple links/edges
small_world_network_prelim <- simplify(small_world_network_prelim)
# express adjacency matrix in standard matrix form
# show that each node has four links
print(degree(small_world_network_prelim))
# verify that the network has one hundred links/edges
print(sum(degree(small_world_network_prelim))/2)

pdf(file = "fig_network_small_world_nei_2.pdf", width = 5.5, height = 5.5)
plot(small_world_network_prelim, vertex.size = 25, vertex.color = "yellow",
    vertex.label = NA, edge.arrow.size = 0.25, edge.color = "black",
    layout = layout.circle, edge.curved = TRUE)
dev.off()

# yet another of a small-world network (no random links)
set.seed(1)
# one-dimensional small-world model with 10 nodes,
# links to additional adjacent nodes in a lattice
# (nei = 3 implies degree = 6 for all nodes prior to rewiring)
# rewiring probability of 0.00... no rewiring
small_world_network_prelim <- watts.strogatz.game(dim = 1, size = 10,
    nei = 3, p = 0.00, loops = FALSE, multiple = FALSE)
# remove any multiple links/edges
small_world_network_prelim <- simplify(small_world_network_prelim)
# express adjacency matrix in standard matrix form
# show that each node has four links
print(degree(small_world_network_prelim))
# verify that the network has one hundred links/edges
print(sum(degree(small_world_network_prelim))/2)
pdf(file = "fig_network_small_world_nei_3.pdf", width = 5.5, height = 5.5)
plot(small_world_network_prelim, vertex.size = 25, vertex.color = "yellow",
    vertex.label = NA, edge.arrow.size = 0.25, edge.color = "black",
    layout = layout.circle, edge.curved = TRUE)
dev.off()

# rewire a selected proportion of the links to get small world model
set.seed(1)
small_world_network <- watts.strogatz.game(dim = 1, size = 50, nei = 2,
    p = 0.2, loops = FALSE, multiple = FALSE)
# remove any multiple links/edges
small_world_network <- simplify(small_world_network)
# express adjacency matrix in standard matrix form
# show that each node has four links
print(degree(small_world_network))
# verify that the network has one hundred links/edges
print(sum(degree(small_world_network))/2)
pdf(file = "fig_network_small_world.pdf", width = 5.5, height = 5.5)
plot(small_world_network, vertex.size = 10, vertex.color = "yellow",
    vertex.label = NA, edge.arrow.size = 0.25, edge.color = "black",
    layout = layout.circle, edge.curved = TRUE)
dev.off()
```

```
aggregate_degree <- NULL  # initialize collection of node degree values
correlation_array <- array(NA, dim = c(4, 4, NITER))  # initialize array
for (i in 1:NITER) {
    set.seed(i)
    small_world_network <- watts.strogatz.game(dim = 1, size = 50, nei = 1,
        p = 0.2, loops = FALSE, multiple = FALSE)
    aggregate_degree <- c(aggregate_degree,
        degree(small_world_network))
    correlation_array[,,i] <- get_centrality_matrix(small_world_network)
    }
average_correlation <- matrix(NA, nrow = 4, ncol = 4,
    dimnames = list(c("Degree", "Betweenness", "Closeness", "Eigenvector"),
        c("Degree", "Betweenness", "Closeness", "Eigenvector")))
for (i in 1:4)
    for(j in 1:4)
        average_correlation[i, j] <- mean(correlation_array[i, j, ])

pdf(file = "fig_network_small_world_correlation_heat_map.pdf", width = 11,
  height = 8.5)
correlation_heat_map(cormat = average_correlation)
dev.off()

# create data frame for node degree distribution
math_model <- rep("Small-World Network", rep = length(aggregate_degree))
small_world_degree_data_frame <- data.frame(math_model, aggregate_degree)

# ---------------------------------
# Scale-Free Networks
# ---------------------------------
# show the plot of the first scale-free network model
set.seed(1)
# directed = FALSE to generate an undirected graph
# fifty nodes to be consistent with the models above
scale_free_network <- barabasi.game(n = 50, m = 2, directed = FALSE)

# remove any multiple links/edges
scale_free_network <- simplify(scale_free_network)

pdf(file = "fig_network_scale_free.pdf", width = 5.5, height = 5.5)
plot(scale_free_network, vertex.size = 10, vertex.color = "yellow",
    vertex.label = NA, edge.arrow.size = 0.25, edge.color = "black",
    layout = layout.circle, edge.curved = TRUE)
dev.off()

# express adjacency matrix in standard matrix form
scale_free_network_mat <- as.matrix(get.adjacency(scale_free_network))

# note that this model yields a graph with almost 100 links/edges
print(sum(degree(scale_free_network))/2)

aggregate_degree <- NULL  # initialize collection of node degree values
correlation_array <- array(NA, dim = c(4, 4, NITER))  # initialize array
```

```r
for (i in 1:NITER) {
    set.seed(i)
    scale_free_network <- barabasi.game(n = 50, m = 2, directed = FALSE)
    # remove any multiple links/edges
     scale_free_network <- simplify(scale_free_network)
    aggregate_degree <- c(aggregate_degree,
        degree(scale_free_network))
    correlation_array[,,i] <- get_centrality_matrix(scale_free_network)
    }
average_correlation <- matrix(NA, nrow = 4, ncol = 4,
    dimnames = list(c("Degree", "Betweenness", "Closeness", "Eigenvector"),
        c("Degree", "Betweenness", "Closeness", "Eigenvector")))
for (i in 1:4)
    for(j in 1:4)
        average_correlation[i, j] <- mean(correlation_array[i, j, ])

pdf(file = "fig_network_scale_free_correlation_heat_map.pdf", width = 11,
  height = 8.5)
correlation_heat_map(cormat = average_correlation)
dev.off()

# create data frame for node degree distribution
math_model <- rep("Preferential Attachment Network",
    rep = length(aggregate_degree))
scale_free_degree_data_frame <- data.frame(math_model, aggregate_degree)

# --------------------------------
# Compare Degree Distributions
# --------------------------------
plotting_data_frame <- rbind(scale_free_degree_data_frame,
    small_world_degree_data_frame,
    random_graph_degree_data_frame)

# use lattice graphics to compare degree distributions

pdf(file = "fig_network_model_degree_distributions.pdf", width = 8.5,
  height = 11)
lattice_object <- histogram(~aggregate_degree | math_model,
    plotting_data_frame, type = "density",
    xlab = "Node Degree", layout = c(1,3))
print(lattice_object)
dev.off()

# Suggestions for the student.
# Experiment with the three models, varying the numbers of nodes
# and methods for constructing links between nodes. Try additional
# measures of centrality to see how they relate to the four measures
# we explored in this program.  Explore summary network measures
# of centrality and connectedness to see how they relate across
# networks generated from random graph, preferential attachment,
# and small-world models.
```

Exhibit 7.2. *Methods of Sampling from Large Networks (R)*

```
# Methods of Sampling from Large Networks (R)

# install packages sna and network

# Background reference for sampling procedures:
# Leskovec, J. & Faloutsos, C. (2006). Sampling from large graphs.
# Proceedings of KDD '06. Available at
# <http://cs.stanford.edu/people/jure/pubs/sampling-kdd06.pdf>

# load packages into the workspace for this program
library("sna")  # social network analysis
library("network")  # network data methods

# user-defined function for computing network statistics
# this initial function computes network transitivity only
# additional code could be added to compute other network measures
# note that node-level measures may be converted to network measures
# by averaging across nodes
report.network.statistics <- function(selected.edges) {
    # selected.edges is a data frame of edges with two columns
    # corresponding to FromNodeId and ToNodeId for a directed graph
    # analysis of selected.edges can begin as follows
    # create a directed network/graph (digraph)
    selected.network <- network(as.matrix(selected.edges),
        matrix.type="edgelist",directed=TRUE)
    # convert to a selected.matrix/graph)
    selected.matrix <- as.matrix(selected.network)
    # transitivity (clustering coefficient) probability that
    # two nodes with a # common neighbor are also linked
    # (Consider three nodes A, B, and C.
    # If A is linked to C and B is linked to C,
    # what is the probability that A will be linked to B?)
    # If value above 0.50... indicates clustering of nodes.
    network.transitivity <-
        gtrans(selected.matrix, use.adjacency = FALSE)
    # report results for this run
    cat("\n\n","Network statistics for N = ",
      nrow(selected.edges),"\n   Transitivity = ",
        network.transitivity,sep="")
    }

# network data from the Wikipedia Votes case
# read in the data and set up a binary R data file
# wiki.edges = read.table("wiki_edges.txt",header=T)
# save(wiki.edges, file ="wiki_edges.Rdata")

# with the binary data file saved in your working directory
# you are ready to load it in and begin an analysis
load("wiki_edges.Rdata")  # brings in the data frame object wiki.edges

print(str(wiki.edges))  # shows 103,689 initial observations/edges
```

```
# check to see if there are any self-referring edges
self.referring.edges <- subset(wiki.edges,
    subset = (FromNodeId == ToNodeId))
print(nrow(self.referring.edges))  # shows that there are no such nodes

# this is a large network... so we will explore samples of the edges
# begin by identifying the number of unique nodes
unique.from.nodes <- unique(wiki.edges$FromNodeId)
print(length(unique.from.nodes))  # there are 6,110 unique from-nodes
unique.to.nodes <- unique(wiki.edges$ToNodeId)
print(length(unique.to.nodes))  # shows that there are 2,381 unique to-nodes
unique.nodes <- unique(c(unique.from.nodes,unique.to.nodes))
print(length(unique.nodes))  # there are 7,115 unique nodes total

# compute the transitivity for the complete network commented out here
# report.network.statistics(selected.edges = wiki.edges)
# RESULT: Transitivity of complete network: 0.1645504

# set the sample size N by setting a proportion of the edges to be sampled
P <- .1  # what proportion will work... .10, .20, .30, or higher?
N <- trunc(P * nrow(wiki.edges))

# specify the sampling method three methods are implemented here
# one is selected for each time the program is run
METHOD <- "RE"  # random edge sampling
# METHOD <- "RN"  # random node sampling
# METHOD <- "RNN"  # random node neighbor sampling

# seed the random number generator to obtain reporducible results
# if a loop is utilized to repeat sampling, ensure that the seed changes
# within each iteration of the loop
set.seed(9999)  # reporducible results are desired

# ------------------- random edge sampling -------------------
# Random edge (RE) sampling. Using the complete data frame with
# each record representing an edge or connection between nodes,
# we select N edges at random.
if(METHOD == "RE") {  # begin if-block for random edge sampling
    selected.indices <- sample(nrow(wiki.edges),N)
    selected.edges <- wiki.edges[selected.indices,]
    }  # end if-block for random edge sampling

# ------------------- random node sampling -------------------
# Random node (RN) sampling. Using the list of unique node numbers,
# we select nodes at random without replacement. This sampling is
# carried out using the combined set of FromNodeId and ToNodeId
# node identifiers. The preliminary sample data frame consists
# of all edges containing these nodes. Then we sample exactly N
# edges or rows from the preliminary sample data frame.
if(METHOD == "RN") {  # begin if-block for random node sampling
    selected.node.indices <-
        sample(length(unique.nodes),P*length(unique.nodes))
    selected.nodes <- unique.nodes[selected.node.indices]
```

```
    from.edge.samples <- subset(wiki.edges,
        subset = (FromNodeId %in% selected.nodes))
    to.edge.samples <- subset(wiki.edges,
        subset = (ToNodeId %in% selected.nodes))
    all.edge.samples <- rbind(from.edge.samples,to.edge.samples)
    selected.indices <- sample(nrow(all.edge.samples),N)
    selected.edges <- all.edge.samples[selected.indices,]
    }  # end if-block for random node sampling

# ------------------- random node-neighbor sampling ------------------
Random node neighbor (RNN) sampling. We begin by selecting a node at
# random (referring to the FromNodeId values), together with all
# of its out-going neighbors (ToNodeId values). We continue selecting
# random FromNodeId values and their associated ToNodeId values until
# we have N edges in the sample.
if(METHOD == "RNN") { # begin if-block for random node neighbor sampling
    # obtain from-node values in permuted order
    permuted.from.nodes <- sample(unique.from.nodes)
    selected.edges <- NULL  # initialize the sample data frame
    number.of.selected.edges.so.far <- 0  # initialize edge count
    # initialize index of permuted.from.nodes
    index.of.permuted.from.node <- 0
    while(number.of.selected.edges.so.far < N) {
        index.of.permuted.from.node <- index.of.permuted.from.node + 1
        this.selected.set.of.edges <- subset(wiki.edges,
            subset = (FromNodeId ==
                permuted.from.nodes[index.of.permuted.from.node]))
        selected.edges <- rbind(selected.edges,this.selected.set.of.edges)
        number.of.selected.edges.so.far <- nrow(selected.edges)
        }
    # just use the first N of the selected edges
    selected.edges <- selected.edges[1:N,]
    }  # end if-block for random node neighbor sampling

# report on network sampling results with the
# user-defined function report.network.statistics
cat("\n\n","Results for sampling method ", METHOD, sep="")
report.network.statistics(selected.edges)

# Suggestions for the student:
# Building on the example code, your job is to write and execute
# a program to answer a number of questions. How large of a sample
# is needed to obtain accurate estimates of transitivity?
# What type of sampling method (if any) provides the best
# estimate of transitivity for the complete network?
# Are there other network measures that might be more accurately
# estimated with network sampling?  Which network sampling method
# would you recommend for work in web and network data science?
# Go on to implement other network-wide measures, such as
# closeness, betweenness, and eigenvector centrality, and
# network density (number of observed links divided by
# the total possible links).
```

8

Measuring Sentiment

"I'm as mad as hell, and I'm not going to take this anymore!"

—PETER FINCH AS HOWARD BEALE IN *Network* (1976)

I have developed a reputation as an unconventional educator because I resist the notion of grading rubrics and step-by-step instructions for assignments. Crazy as it sounds, most of my assignments are mini case studies: *Here is a business question. Here are some messy data. Make sense out of them and write a report for management.*

I argue that this is the kind of problem we face every day in our work lives. Data are messy, often unstructured text. Managers do not lay out instructions about what to do with data or how to solve the problem. That is the data scientist's job.

Open-ended assignments give students a chance to be creative and to develop good judgment as data scientists. Students can be productive despite having large gaps in their understanding of a subject. Success depends on "tolerance of ambiguity."

What is tolerance of ambiguity? It is an attitude of acceptance. We accept the fact that we have uncertainties. While it is good to know many details, it is also okay not to know all the details. Things we do not understand today may become clearer later.

One of the things we need to understand better in data science is sentiment analysis. Tools for sentiment analysis or opinion mining, as it is sometimes called, are springing up everywhere on the web, and few understand what they mean.

What characterizes the mood of the public? Do voters like the candidate or not? Can well-placed messages alter candidate, product, or brand perceptions? These are the kinds of questions we might answer with a survey and primary research. But in today's information-rich world, we may have to answer these questions using secondary data sources. Sentiment analysis may be applied to much of the data we gather from the web. We can use it for news sites, blogs, and social media. Numerous studies of sentiment have been carried out on Twitter microblogs (tweets).

Sentiment analysis is a poorly understood measurement problem, with few reported measures having documented reliability and validity. Data scientists and modelers generally are not well schooled in measurement theory. There is great uncertainty about how to do sentiment analysis correctly, or whether there even is a correct way to do it. Sampling issues abound.

Tolerance of ambiguity may be good as an attitude for effective learning. But ambiguity is not so good when our job is to make predictions from the measures we are making. We want our measures to mean something. We want them to be as reliable and valid as possible. To do sentiment analysis correctly, we need to design text measures that work. The measures themselves are just the beginning. The real job of sentiment analysis is show that the measures can be used to make predictions.

Text feature selection and the definition of *text measures* are at the heart of sentiment analysis. Text measures are scores on attributes or features that describe text. Measurement, in its most basic sense, is the assignment of numbers to attributes according to rules. We can use text measures to assess personality, consumer preferences, and political opinions, just as we can use survey instruments. The difference between text measures and survey instruments is that text measures begin with unstructured or semistructured text as their input data, rather than questionnaire responses.

To demonstrate the process of sentiment analysis, we will draw on data from the IMDb movie reviews database. In addition to the IMDb data, I have written a few movie reviews. My reviews are shown in figures 8.1 and

8.2, I give each of the movies a rating from 1 to 10, with 1 being "horrible" and 10 being "fantastic."

Coming up with sentiment scores is just the beginning. The real job of sentiment analysis is to show that the scores mean something, that they can be used to make predictions. In this example we explore various methods for developing text measures of sentiment, including list-based measures, item-weighted measures, and models for text classification. And we employ a training-and-test regimen in evaluating the predictive performance of text measures and models.

One approach to sentiment analysis draws on positive and negative word sets (lexicons, dictionaries) that convey human emotion or feeling. These word sets are specific to the language being spoken and the context of application. To demonstrate list-based measures of sentiment, we can use a lexicon developed by Hu and Liu (2004). This lexicon includes lists of 2,006 positive and 4,783 negative opinion and sentiment words. Let us see how these lists match up with a movie reviews corpus from Mass et al. (2011).[1]

There are three data set types in the corpus provided by Mass et al. (2011). The first is a data set with review text but no ratings. We will use this first data set to identify word sets for sentiment text measures. The second data set contains movie reviews with known ratings. We use this as our training data for developing text measurement models. And the third data set, like the second, has known ratings. We use this for evaluating developed text measures. At the end of the process we will evaluate developed text measures and models on the ratings provided at the beginning of the chapter.

Figure 8.3 shows fifty words from the Hu and Liu (2004) lexicon that worked for the data set of movie reviews with no ratings. Review of the positive and negative word lists suggests that positive words may come from movie love stories and comedies, uplifting or "feel good" movies. Negative words, on

---

[1] The movie reviews data are available from the authors of Mass et al. (2011), Andrew L. Maas, Raymond E. Daly, Peter T. Pham, Dan Huang, Andrew Y. Ng, and Christopher Potts, who were kind enough to place these data in the public domain. The original source for these data was The Internet Movie Database (IMDb). One data set includes 50,000 movie reviews with no associated ratings. There are similarly sized training and test data sets, both with movie ratings attached. We are using a small sample from each of these data sets in this chapter. The movie reviews data may be downloaded from `http://ai.stanford.edu/~amaas/data/sentiment/`.

Figure 8.1. A Few Movie Reviews According to Tom

The Effect of Gamma Rays on Man-in-the-Moon Marigolds (1972)

Based on a Pulizer-Prize-winning play by Paul Zindel, with Paul Newman directing Joanne Woodward, this is one of the most uplifting movies you will ever see. It is a tribute to the human spirit, the will to look beyond the limitations of one's current circumstances and overcome adversity. Not a bad plug for education either. Watch it if you can find it.

My rating: 10

Blade Runner (1982)

Even better than Harrison Ford's *Raiders* movies. This one is a keeper. Replicants with feeling—that's a twist. The visual effects and fine photography draw you in. It's like you are really there in the metropolis of the future. Not a pretty picture of what's coming our way, but a great movie nonetheless.

My rating: 9

My Cousin Vinny (1992)

Joe Pesci and Marisa Tomei—now that's an odd couple. The movie builds on stereotypes of Brooklyn and Alabama, and it's hard to sympathize with the hapless cousin and his male friend. Nor is there much suspense because we know what's going to happen at the end. It has to work out. Tomei makes the hour or so go by just fine. Without the life she breathes into her fiancée role, this one would have been a waste.

My rating: 4

Mars Attacks (1996)

A mindless diversion to be sure. I have the DVD, and every six months or so I watch it for a laugh. Nicholson plays two roles: POTUS and a real estate developer. The exploding Martian heads are great. Not for country music fans, though.

My rating: 7

Figure 8.2. *A Few More Movie Reviews According to Tom*

Fight Club (1998)

Picked this up at a used video store. Thought it would be good. Boy, was I wrong. I guess I am just not much of a fan of violence as a way of life or entertainment. Had a hard time relating to any of the characters. I'll be taking this back for a trade-in when I get the chance.

My rating: 2

Miss Congeniality 2 (2006)

You have got to be kidding. The first one was a waste, and this one is worse. Do we really have to endure these insults to our intelligence every five years? I like Sandra Bullock, but why does she do this stuff?

My rating: 1

Find Me Guilty (2006)

I'm not really a Vin Diesel fan, but my son suggested that I watch this one. I must say, I was quite surprised at what they were able to do with a movie centered, as this was, around courtroom scenes. Vin was really good. The Sidney Lumet effect, I suppose. Too bad we lost him.

My rating: 7

Moneyball (2011)

You might think I would like this movie, given my love of sports and analytics, but I had a hard time caring about the Oakland Athletics or the main character. Not Brad Pitt again. That guy sure keeps busy. I suppose the movie is a fair rendering of a true story. After all, you can't have the As winning the World Series when they didn't. But I was left with an empty feeling at the end. I think the story would have been better told from the point of view of the nerdy analyst. Give Jonah Hill more time in front of the camera and see what he can do. Maybe you could even work in a love interest. Some of my students asked me what I thought of the movie. I said it was OK.

My rating: 4

Figure 8.3. *Fifty Words of Sentiment*

Positive Words

amazing	beautiful	classic	enjoy	enjoyed
entertaining	excellent	fans	favorite	fine
fun	humor	lead	liked	love
loved	modern	nice	perfect	pretty
recommend	strong	top	wonderful	worth

Negative Words

bad	boring	cheap	creepy	dark
dead	death	evil	hard	kill
killed	lack	lost	miss	murder
mystery	plot	poor	sad	scary
slow	terrible	waste	worst	wrong

the other hand, are more likely to be associated with horror, violence, and tragedy.[2]

Note that there is nothing inherently good about the positive words or inherently bad about the negative words. As with any words we might use in communication, it is context that gives them meaning. In fact, it would be easy for us to review a bad movie using words chosen from the positive list or a good movie using words chosen from the negative list.

Moving forward with these identified lists, suppose we count numbers of positive and negative words in each of the movie reviews, computing two list-based text measures. Let POSITIVE be the percentage of words in the review that match up with the list of twenty-five positive words, and let

---

[2] Notice that we are not using word stemming in creating word sets, as we have both "enjoy" and "enjoyed," as well as "love" and "loved," in the positive word set, and "kill" and "killed" in the negative word set. Stemming involves mapping words to a base word or stem. Stemming may be used in future studies with these data.

Table 8.1. *List-Based Sentiment Measures from Tom's Reviews*

Movie	Total Words	Positive Words	Negative Words	Text Measures POSITIVE	NEGATIVE	Rating	Thumbs Up/Down
Marigolds	26	0	1	0.00	3.85	10	UP
Blade Runner	21	2	0	9.52	0.00	9	UP
Vinny	29	1	2	3.45	6.90	4	DOWN
Mars Attacks	20	1	0	5.00	0.00	7	UP
Fight Club	18	0	2	0.00	11.11	2	DOWN
Congeniality	10	0	1	0.00	10.00	1	DOWN
Find Me Guilty	18	0	2	0.00	11.11	7	UP
Moneyball	36	2	1	5.56	2.78	4	DOWN

NEGATIVE be the percentage of words in the review that match up with the list of twenty-five negative words. Figure 8.4 shows four movie reviews and their associated POSITIVE and NEGATIVE scores.

The scatter plot in figure 8.5 shows how the two list-based measures play out across the set of 500 movie reviews. If POSITIVE and NEGATIVE were at opposite ends of a single underlying dimension, we would expect these text measures to correlate close to -1.0. In fact, when we compute the correlation between POSITIVE and NEGATIVE, we see that it is negative, but much closer to zero than -1.0.

How do POSITIVE and NEGATIVE scores play out as far as the eight movie reviews at the beginning of the chapter? Table 8.1 suggests that they may not be doing so well. The list-based measures are missing favorable ratings of *The Effect of Gamma Rays on Man-in-the-Moon Marigolds* and *Find Me Guilty*, as well as the unfavorable rating of *Moneyball*.

Perhaps POSITIVE and NEGATIVE can be combined in a way to yield effective predictions of movie ratings. Alternatively, we can go back to the list of fifty sentiment terms and use them in data-based models to predict movie ratings. To move forward with text measures and model development, let us look at the training set of movie reviews with known ratings.

We select 500 records from the training set of positive reviews (reviews with ratings between 7 and 10) and 500 records from the training set of negative reviews (reviews with ratings between 1 and 4). We combine these to form a training data set of 1,000 movie reviews. We employ the same procedure to create a test set of 1,000 movie reviews.

Figure 8.4. *List-Based Text Measures for Four Movie Reviews*

Ginger Snaps 2: Unleashed (2004)

I liked it a lot, in fact even more than the first movie. I loved the character of Ghost and all the comic book shots and her third person lines. Good ending. One thing they could have done was make the identity of the werewolf clearer. Also when the sister appeared it was kind of forced.. it didn't seem like she was a delusion

(20 analysis words, 2 from positive list, 0 from negative list)
List-based text measures: POSITIVE = 10, NEGATIVE = 0

Johnny Lingo (1969)

Beyond the tremendous and true romantic love Johnny Lingo proves for his dear Mahana, he gives a tremendous object lesson in how to properly treat others, and bring out the very best in them. If all husbands would treat their wives the way Johnny treated Mahana, there could be no evil in the world.

(20 analysis words, 1 from positive list, 1 from negative list)
List-based text measures: POSITIVE = 5, NEGATIVE = 5

Tomorrow Is Forever (1946)

The greatest and most poignant anguish we conscious beings experience is our recognition of the irretrievability of the past. All else we endure could be easily borne. The background strains of music as Orson Welles first recognizes Claudette Colbert haunts me still. She experienced a fragmentary nuance of remembrance that did not reach the level of her conscious recall.

(25 analysis words, 0 from positive list, 0 from negative list)
List-based text measures: POSITIVE = 0, NEGATIVE = 0

Malas temporadas (2005)

There are some exciting scenes in this movie but in general it is second-rate. The shoots are overextended, the characters are not life-like and some actors don't perform well either. I also didn't like multiple nationalist statements which have nothing to do with the plot. I guess the director intended to make his characters mysterious but instead they came out to be unnatural. We are supposed to see how different people successfully struggle with hard times in their lives. But two stories, the one of Carlos and that of Mikel, end up with nothing and the third, the story of Ana, makes a turn without any reason. The movie is very depressive but without any message that derives from it.

(40 analysis words, 0 from positive list, 2 from negative list)
List-based text measures: POSITIVE = 0, NEGATIVE = 5

Figure 8.5. *Scatter Plot of Text Measures of Positive and Negative Sentiment*

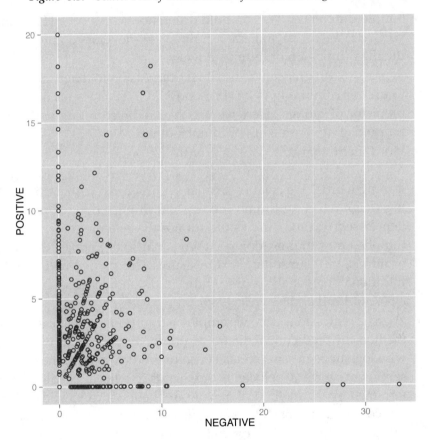

Working with alternative text measures and predictive modeling techniques, we employ a training-and-test regimen, developing measures and models on a training set and testing on a test set. Our goal is to predict whether a movie review is *thumbs-up* (a rating higher than 6) or *thumbs-down* (a rating less than 5). This is a text classification problem that may be addressed using various techniques. Here are six measures and modeling techniques we use for this example:

- Simple difference. We compute difference scores (POSITIVE minus NEGATIVE scores) and use a training-set-developed cutoff for predicting thumbs-up or thumbs-down.

- Regression difference. We use linear regression to determine weights to use for combining POSITIVE and NEGATIVE scores into a linear predictor of ratings. Here we use the predicted rating with a cutoff for predicting thumbs-up or thumbs-down.

- Word/item analysis. Working with the original set of fifty words, we use the training data to identify positive-leaning and negative-leaning words. We create a smaller set of words, with each word being weighted +1 for thumbs up or -1 for thumbs down. We define an item-based text measure as the sum of these item scores. Predicting thumbs-up or thumbs-down is then a matter of noting the sign (plus or minus) of the resulting text measure. This procedure is similar to traditional item analysis procedures in psychometrics. See Nunnally (1967) or Lord and Novick (1968).

- Logistic regression. Here we employ a traditional statistical modeling method for predicting a binary response. In particular, we use stepwise logistic regression to select useful predictors from the set of fifty sentiment words. Coefficients or weights in a linear predictor are determined by the method of maximum likelihood. Logistic regression is a common approach to binary classification problems. Discussion of logistic regression is provided in many sources (Ryan 2008; Fox and Weisberg 2011; Hosmer, Lemeshow, and Sturdivant 2013).

- Support vector machines. This machine learning algorithm has been shown to be an effective technique in text classification problems and, more generally, in problems with large numbers of explanatory variables, as we have here with the full set of fifty sentiment words. Most closely identified with Vladimir Vapnik (Boser, Guyon, and Vapnik

Table 8.2. *Accuracy of Text Classification for Movie Reviews (Thumbs-Up or Thumbs-Down)*

Text Measure/Model	Percentage of Reviews Correctly Classified	
	Training Set	Test Set
Simple difference	67.4	66.1
Regression difference	67.3	66.4
Word/item analysis	73.9	74.0
Logistic regression	75.2	72.6
Support vector machines	79.0	71.6
Random forests	82.2	74.0

1992; Vapnik 1998; Vapnik 2000), discussion of support vector machines may be found in Cristianini and Shawe-Taylor (2000), Izenman (2008), and Hastie, Tibshirani, and Friedman (2009). Tong and Koller (2001) discuss support vector machines for text classification.

- Random forests. This is a committee or ensemble method that uses thousands of tree-structured classifiers to arrive at a single prediction. The tree-structured classifiers themselves follow methods described in the work of Breiman et al. (1984). Review of tree-structured methods is provided by Izenman (2008) and Hastie, Tibshirani, and Friedman (2009). This is recursive partitioning on the training set to develop classification trees for predicting thumbs-up or thumbs-down. The set of explanatory variables is the full set of fifty sentiment words. Many such trees are constructed to form the random forest. Like support vector machines, this method has been shown to be very effective when working with large numbers of explanatory variables. It is based on bootstrap resampling techniques. Introduced by Breiman (2001a), useful introductions to this method may be found in Izenman (2008) and Hastie, Tibshirani, and Friedman (2009).

Testing various measurement and modeling techniques as we are doing here constitutes a first iteration of a benchmark experiment. How shall we evaluate the predictive accuracy in this study? For the movie reviews text classification problem, we use an index that is easy for managers to understand: the percentage of correct thumbs-up/thumbs-down predictions in the test set of movie reviews. In table 8.2 we report this statistic along with the percentage of correct predictions in the training set.

Table 8.3. *Random Forest Text Measurement Model Applied to Tom's Movie Reviews*

Movie	Rating	Actual Thumbs Up/Down	Predicted Thumbs Up/Down
Marigolds	10	UP	DOWN
Blade Runner	9	UP	UP
Vinny	4	DOWN	DOWN
Mars Attacks	7	UP	UP
Fight Club	2	DOWN	DOWN
Congeniality	1	DOWN	DOWN
Find Me Guilty	7	UP	DOWN
Moneyball	4	DOWN	UP

Word/item analysis following traditional psychometric methods and random forests do equally well in text classification in the test set, with random forests doing the best on the training data. Random forests have the added advantage of providing measures of explanatory variable importance (word importance in text classification), as shown in as a dot chart in figure 8.6.

To complete the work with the movie reviews, we select the best performing measurement model from the benchmark study, random forests, and use it to classify the movie reviews presented at the beginning of the chapter. Table 8.3 shows the results. Five of the eight reviews (62.5 percent) are correctly classified. Like the list-based measures NEGATIVE and POSITIVE, that we had reviewed earlier, the random forest method fails in its classification of the thumbs-up movies *The Effect of Gamma Rays on Man-in-the-Moon Marigolds* and *Find Me Guilty* and in its classification of the thumbs-down movie *Moneyball*.

If we were to build a simple tree classifier for the movie ratings data, it would look like the one in figure 8.7. The simple tree tells us that if we are to classify on the basis of one word and one word alone, that word would be "worst." People who use the word "worst" tend to give a movie thumbs-down, with "bad" and "waste" following closely as thumbs-down predictors. If a review has none of those three words, but instead has the word "amazing," the simple tree classifier would predict thumbs-up. Beyond that, we have to look at additional words, such as "plot," "favorite," "terrible," and "death."

Figure 8.6. *Word Importance in Classifying Movie Reviews as Thumbs-Up or Thumbs-Down*

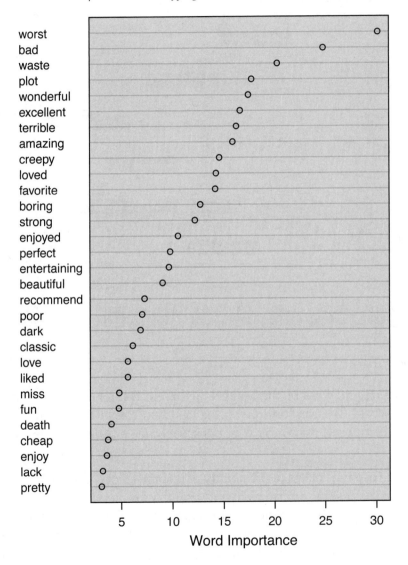

Figure 8.7. A Simple Tree Classifier for Thumbs-Up or Thumbs-Down

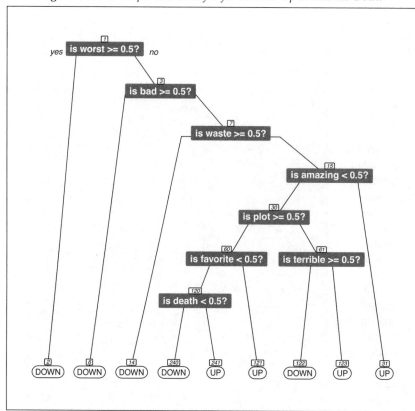

A simple tree may not be the best predictor, but it can help explain the way models work. The random forest, the best text classifier we found in this study is composed of hundreds of trees, each one a little different from the next.

The value of a model lies in the quality of its predictions. How well does the simple tree classifier do in predicting thumbs-up or thumbs-down for the movie reviews at the beginning of the chapter? Not very well. It makes the same mistakes as the list and random forest classifiers and one additional mistake—it classifies *Fight Club* as thumbs-up, when it should be thumbs-down. To see why it does this, we have only to trace the path taken down the tree. The review of *Fight Club* contains none of the words in the tree, so we end in the second terminal node from the right, a thumbs-up node.

Precursors to sentiment analysis may be found in content analysis, thematic, semantic, and network text analysis (Roberts 1997; Popping 2000; West 2001; Leetaru 2011; Krippendorff 2012). These methods have seen a wide range of applications within the social sciences, including analysis of political discourse. An early computer implementation of content analysis is found in the General Inquirer program (Stone et al. 1966; Stone 1997). Buvač and Stone (2001) describe a version of the program that provides text measures based upon word counts across numerous semantic categories.

Text measures flow from a measurement model (algorithms for scoring) and a dictionary, both defined by the researcher or analyst. A dictionary in this context is not a traditional dictionary; it is not an alphabetized list of words and their definitions. Rather, the dictionary used to construct text measures is a repository of word lists, such as synonyms and antonyms, positive and negative words, strong and weak sounding words, bipolar adjectives, parts of speech, and so on. The lists come from expert judgments about the meaning of words. A text measure assigns numbers to documents according to rules, with the rules being defined by the word lists, scoring algorithms, and modeling techniques in predictive analytics.

Among the more popular measurement schemes from the psychometric literature is Charles Osgood's semantic differential (Osgood, Suci, and Tannenbaum 1957; Osgood 1962). Exemplary bipolar dimensions include the positive–negative, strong–weak, and active–passive dimensions.

Text measurement holds promise as a technology for understanding consumer opinion and markets. Just as political researchers can learn from the words of the public, press, and politicians, business researchers can learn from the words of customers and competitors. There are customer service logs, telephone transcripts, and sales call reports, along with user group, listserv, and blog postings. And we have ubiquitous social media from which to build document collections for text and sentiment analysis.

Regarding Twitter-based text measures, there have been various attempts to predict the success of movies prior to their being distributed to theaters nationwide (Sharda and Delen 2006; Delen, Sharda, and Kumar 2007). Most telling is work completed at HP Labs that utilized chat on Twitter as a predictor of movie revenues (Asur and Huberman 2010). Bollen, Mao, and Zeng (2011) utilize Twitter sentiment analysis in predicting stock market movements. Taddy's (2013b, 2014) sentiment analysis work builds on the inverse regression methods of Cook (1998, 2007). Taddy (2013a) uses Twitter data to examine political sentiment.

We expect sentiment analysis to be an active area of research for many years. Some have voiced concerns about unidimensional measures of sentiment. There have been attempts to develop more extensive sentiment word sets, as well as multidimensional measures (Turney 2002; Asur and Huberman 2010). Recent developments in machine learning and quantitative linguistics point to sentiment measurement methods that employ natural language processing rather than relying on positive and negative word sets (Socher et al. 2011).

There are critics of social media and the Internet culture that emerges from what Geert Lovink (Lovink 2011) describes as the "Googlization of our lives." *Mute Magazine* and the edited volume *Proud to be Flesh* offers interesting perspectives on the interplay of art and technology (Slater et al. 2013). Additional criticism of social media comes from those with concerns about the effect of social media on privacy (Rosen 2001; Turow 2013).

Exhibit 8.1 lists the Python program for the movie reviews example. Exhibit 8.2 shows the corresponding R program. It shows the development of text measures of sentiment and models for predicting thumbs-up versus thumbs-down movie reviews based on the text from those reviews. Traditional methods and machine learning techniques are employed. The program draws on R packages provided by Feinerer (2014), Wickham (2014b), Wickham and Chang (2014), Sarkar and Andrews (2014), Kuhn (2014), Liaw and Wiener (2014), Therneau, Atkinson, and Ripley (2014), Meyer et al. (2014), and Milborrow (2014).

Exhibit 8.1. *Sentiment Analysis and Classification of Movie Ratings (Python)*

```
# Sentiment Analysis Using the Movie Ratings Data (Python)

# Note that results from this program may differ from the results
# documented in the book because algorithms for text parsing
# and text classification vary between Python and R.
# The objectives of the analysis and steps in completing the analysis
# are consistent with those in the book. And results, although
# not identical between Python and R, should be very similar.

# prepare for Python version 3x features and functions
from __future__ import division, print_function

# import packages for text processing and machine learning
import os  # operating system commands
import re  # regular expressions
import nltk  # draw on the Python natural language toolkit
import pandas as pd  # DataFrame structure and operations
import numpy as np  # arrays and numerical processing
import matplotlib.pyplot as plt  # 2D plotting
import statsmodels.api as sm  # logistic regression
import statsmodels.formula.api as smf  # R-like model specification
import patsy  # translate model specification into design matrices
from sklearn import svm  # support vector machines
from sklearn.ensemble import RandomForestClassifier  # random forests
# import user-defined module
from python_utilities import evaluate_classifier, get_text_measures,\
    get_summative_scores

# list files in directory omitting hidden files
def listdir_no_hidden(path):
    start_list = os.listdir(path)
    end_list = []
    for file in start_list:
        if (not file.startswith('.')):
            end_list.append(file)
    return(end_list)
# define list of codes to be dropped from document

# carriage-returns, line-feeds, tabs
codelist = ['\r', '\n', '\t']

# there are certain words we will ignore in subsequent
# text processing... these are called stop-words
# and they consist of prepositions, pronouns, and
# conjunctions, interrogatives, ...
# we begin with the list from the natural language toolkit
# examine this initial list of stopwords
nltk.download('stopwords')
# let's look at that list
print(nltk.corpus.stopwords.words('english'))
```

```
# previous analysis of a list of top terms showed a number of words, along
# with contractions and other word strings to drop from further analysis, we add
# these to the usual English stopwords to be dropped from a document collection
more_stop_words = ['cant','didnt','doesnt','dont','goes','isnt','hes',\
    'shes','thats','theres','theyre','wont','youll','youre','youve', 'br'\
    've', 're', 'vs']

some_proper_nouns_to_remove = ['dick','ginger','hollywood','jack',\
    'jill','john','karloff','kudrow','orson','peter','tcm','tom',\
    'toni','welles','william','wolheim','nikita']

# start with the initial list and add to it for movie text work
stoplist = nltk.corpus.stopwords.words('english') + more_stop_words +\
    some_proper_nouns_to_remove

# text parsing function for creating text documents
# there is more we could do for data preparation
# stemming... looking for contractions... possessives...
# but we will work with what we have in this parsing function
# if we want to do stemming at a later time, we can use
#     porter = nltk.PorterStemmer()
# in a construction like this
#     words_stemmed =  [porter.stem(word) for word in initial_words]

def text_parse(string):
    # replace non-alphanumeric with space
    temp_string = re.sub('[^a-zA-Z]', ' ', string)
    # replace codes with space
    for i in range(len(codelist)):
        stopstring = ' ' + codelist[i] + ' '
        temp_string = re.sub(stopstring, ' ', temp_string)
    # replace single-character words with space
    temp_string = re.sub('\s.\s', ' ', temp_string)
    # convert uppercase to lowercase
    temp_string = temp_string.lower()
    # replace selected character strings/stop-words with space
    for i in range(len(stoplist)):
        stopstring = ' ' + str(stoplist[i]) + ' '
        temp_string = re.sub(stopstring, ' ', temp_string)
    # replace multiple blank characters with one blank character
    temp_string = re.sub('\s+', ' ', temp_string)
    return(temp_string)

# read in positive and negative word lists from Hu and Liu (2004)
with open('Hu_Liu_positive_word_list.txt','rt') as f:
    positive_word_list = f.read().split()
with open('Hu_Liu_negative_word_list.txt','rt') as f:
    negative_word_list = f.read().split()

# define counts of positive, negative, and total words in text document
def count_positive(text):
    positive = [w for w in text.split() if w in positive_word_list]
    return(len(positive))
```

```
# define text measure for negative score as percentage of negative words
def count_negative(text):
    negative = [w for w in text.split() if w in negative_word_list]
    return(len(negative))

# count number of words
def count_total(text):
    total = [w for w in text.split()]
    return(len(total))

# define text measure for positive score as percentage of positive words
def score_positive(text):
    positive = [w for w in text.split() if w in positive_word_list]
    total = [w for w in text.split()]
    return 100 * len(positive)/len(total)


# define text measure for negative score as percentage of negative words
def score_negative(text):
    negative = [w for w in text.split() if w in negative_word_list]
    total = [w for w in text.split()]
    return 100 * len(negative)/len(total)

def compute_scores(corpus):
    # use the complete word lists for POSITIVE and NEGATIVE measures
    # to score all documents in a corpus or list of documents
    positive = []
    negative = []
    for document in corpus:
        positive.append(score_positive(document))
        negative.append(score_negative(document))
    return(positive, negative)

# we use movie ratings data from Mass et al. (2011)
# available at http://ai.stanford.edu/~amaas/data/sentiment/
# we set up a directory under our working directory structure
# /reviews/train/unsup/ for the unsupervised reviews
# /reviews/train/neg/ training set negative reviews
# /reviews/train/pos/ training set positive reviews
# /reviews/test/neg/ text set negative reviews
# /reviews/test/pos/ test set positive reviews
# /reviews/test/tom/ eight movie reviews from Tom
# function for creating corpus and aggregate document
# input is directory path for documents
# document parsing accomplished by text_parse function
# directory of parsed files set up for manual inspection

def corpus_creator (input_directory_path, output_directory_path):
    # identify the file names in unsup directory
    file_names = listdir_no_hidden(path = input_directory_path)
    # create list structure for storing parsed documents
    document_collection = []
```

```
    # initialize aggregate document for all documents in set
    aggregate_document = ''
    # create a directory for parsed files
    parsed_file_directory = output_directory_path
    os.mkdir(parsed_file_directory)
    # parse each file and write to directory of parsed files
    for filename in file_names:
        with open(os.path.join(input_directory_path, filename), 'r') as infile:
            this_document = text_parse(infile.read())
            aggregate_document = aggregate_document + this_document
            document_collection.append(this_document)
            outfile = parsed_file_directory + filename
            with open(outfile, 'wt') as f:
                f.write(str(this_document))
    aggregate_words = [w for w in aggregate_document.split()]
    aggregate_corpus = nltk.Text(aggregate_words)
    return(file_names, document_collection, aggregate_corpus)

# function for extracting rating from file name
# for file names of the form 'x_y.txt' where y is the rating
def get_rating(string):
    return(int(string.partition('.')[0].partition('_')[2]))

# dictionary for mapping of ratings to thumbsupdown
map_to_thumbsupdown = {1:'DOWN', 2:'DOWN', 3:'DOWN', 4:'DOWN',
    6:'UP', 7:'UP', 8:'UP', 9:'UP', 10:'UP'}

# begin working with the unsup corpus
unsup_file_names, unsup_corpus, unsup_aggregate_corpus = \
    corpus_creator(input_directory_path = 'reviews/train/unsup/',\
        output_directory_path = 'reviews/train/unsup_parsed/')

# examine frequency distribution of words in unsup corpus
unsup_freq = nltk.FreqDist(unsup_aggregate_corpus)
print('\nNumber of Unique Words in unsup corpus',len(unsup_freq.keys()))
print('\nTop Fifty Words in unsup Corpus:',unsup_freq.keys()[0:50])

# identify the most frequent unsup words from the positive word list
# here we use set intersection to find a list of the top 25 positive words
length_test = 0  # initialize test length
nkeys = 0  # slicing index for frequency table extent
while (length_test < 25):
    length_test =\
        len(set(unsup_freq.keys()[:nkeys]) & set(positive_word_list))
    nkeys = nkeys + 1
selected_positive_set =\
    set(unsup_freq.keys()[:nkeys]) & set(positive_word_list)
selected_positive_words = list(selected_positive_set)
selected_positive_words.sort()
print('\nSelected Positive Words:', selected_positive_words)

# identify the most frequent unsup words from the negative word list
# here we use set intersection to find a list of the top 25 negative words
```

```
length_test = 0  # initialize test length
nkeys = 0  # slicing index for frequency table extent
while (length_test < 25):
    length_test =\
        len(set(unsup_freq.keys()[:nkeys]) & set(negative_word_list))
    nkeys = nkeys + 1
selected_negative_set =\
    set(unsup_freq.keys()[:nkeys]) & set(negative_word_list)
# list is actually 26 items and contains both 'problem' and 'problems'
# so we will eliminate 'problems' from the selected negative words
selected_negative_set.remove('problems')
selected_negative_words = list(selected_negative_set)
selected_negative_words.sort()
print('\nSelected Negative Words:', selected_negative_words)

# use the complete word lists for POSITIVE and NEGATIVE measures/scores
positive, negative = compute_scores(unsup_corpus)

# create data frame to explore POSITIVE and NEGATIVE measures
unsup_data = {'file': unsup_file_names,\
    'POSITIVE': positive, 'NEGATIVE': negative}
unsup_data_frame = pd.DataFrame(unsup_data)

# summary of distributions of POSITIVE and NEGATIVE scores for unsup corpus
print(unsup_data_frame.describe())

print('\nCorrelation between POSITIVE and NEGATIVE',\
    round(unsup_data_frame['POSITIVE'].corr(unsup_data_frame['NEGATIVE']),3))

# scatter plot of POSITIVE and NEGATIVE scores for unsup corpus
ax = plt.axes()
ax.scatter(unsup_data_frame['NEGATIVE'], unsup_data_frame['POSITIVE'],\
    facecolors = 'none', edgecolors = 'blue')
ax.set_xlabel('NEGATIVE')
ax.set_ylabel('POSITIVE')
plt.savefig('fig_sentiment_text_measures_scatter_plot.pdf',
    bbox_inches = 'tight', dpi=None, facecolor='none', edgecolor='blue',
    orientation='portrait', papertype=None, format=None,
    transparent=True, pad_inches=0.25, frameon=None)

# work on the directory of training files-----------------------------------
# Perhaps POSITIVE and NEGATIVE can be combined in a way to yield effective
# predictions of movie ratings. Let us move to a set of movie reviews for
# supervised learning.  We select the 500 records from a set of positive
# reviews (ratings between 7 and 10) and 500 records from a set of negative
# reviews (ratings between 1 and 4). We begin with the training data.

# /reviews/train/pos/ training set positive reviews
train_pos_file_names, train_pos_corpus, train_pos_aggregate_corpus = \
    corpus_creator(input_directory_path = 'reviews/train/pos/',\
        output_directory_path = 'reviews/train/pos_parsed/')
# use the complete word lists for POSITIVE and NEGATIVE measures/scores
positive, negative = compute_scores(train_pos_corpus)
```

```
rating = []
for file_name in train_pos_file_names:
    rating.append(get_rating(str(file_name)))

# create data frame to explore POSITIVE and NEGATIVE measures
train_pos_data = {'train_test':['TRAIN'] * len(train_pos_file_names),\
    'pos_neg': ['POS'] * len(train_pos_file_names),\
    'file_name': train_pos_file_names,\
    'POSITIVE': positive, 'NEGATIVE': negative,\
    'rating': rating}
train_pos_data_frame = pd.DataFrame(train_pos_data)

# /reviews/train/neg/ training set negative reviews
train_neg_file_names, train_neg_corpus, train_neg_aggregate_corpus = \
    corpus_creator(input_directory_path = 'reviews/train/neg/',\
        output_directory_path = 'reviews/train/neg_parsed/')
# use the complete word lists for POSITIVE and NEGATIVE measures/scores
positive, negative = compute_scores(train_neg_corpus)
rating = []
for file_name in train_neg_file_names:
    rating.append(get_rating(str(file_name)))

# create data frame to explore POSITIVE and NEGATIVE measures
train_neg_data = {'train_test':['TRAIN'] * len(train_neg_file_names),\
    'pos_neg': ['NEG'] * len(train_neg_file_names),\
    'file_name': train_neg_file_names,\
    'POSITIVE': positive, 'NEGATIVE': negative,\
    'rating': rating}
train_neg_data_frame = pd.DataFrame(train_neg_data)

# merge the positive and negative training data frames
train_data_frame = pd.concat([train_pos_data_frame, train_neg_data_frame],\
    axis = 0, ignore_index = True)
# determining thumbs up or down based on rating
train_data_frame['thumbsupdown'] = \
    train_data_frame['rating'].map(map_to_thumbsupdown)
# compute simple measure of sentiment as POSITIVE - NEGATIVE
train_data_frame['simple'] = \
    train_data_frame['POSITIVE'] - train_data_frame['NEGATIVE']
# examine the data frame
print(pd.crosstab(train_data_frame['pos_neg'],\
    train_data_frame['thumbsupdown']))
print(train_data_frame.head())
print(train_data_frame.tail())
print(train_data_frame.describe())
ratings_grouped = train_data_frame['simple'].\
    groupby(train_data_frame['rating'])
print('\nTraining Data Simple Difference Means by Ratings:',\
    ratings_grouped.mean())
thumbs_grouped = \
    train_data_frame['simple'].groupby(train_data_frame['thumbsupdown'])
print('\nTraining Data Simple Difference Means by Thumbs UP/DOWN:',\
    thumbs_grouped.mean())
```

```python
# repeat methods for the test data ----------------------------
# /reviews/test/pos/ testing set positive reviews
test_pos_file_names, test_pos_corpus, test_pos_aggregate_corpus = \
    corpus_creator(input_directory_path = 'reviews/test/pos/',\
        output_directory_path = 'reviews/test/pos_parsed/')
# use the complete word lists for POSITIVE and NEGATIVE measures/scores
positive, negative = compute_scores(test_pos_corpus)
rating = []
for file_name in test_pos_file_names:
    rating.append(get_rating(str(file_name)))

# create data frame to explore POSITIVE and NEGATIVE measures
test_pos_data = {'train_test':['TEST'] * len(test_pos_file_names),\
    'pos_neg': ['POS'] * len(test_pos_file_names),\
    'file_name': test_pos_file_names,\
    'POSITIVE': positive, 'NEGATIVE': negative,\
    'rating': rating}
test_pos_data_frame = pd.DataFrame(test_pos_data)

# /reviews/test/neg/ testing set negative reviews
test_neg_file_names, test_neg_corpus, test_neg_aggregate_corpus = \
    corpus_creator(input_directory_path = 'reviews/test/neg/',\
        output_directory_path = 'reviews/test/neg_parsed/')
# use the complete word lists for POSITIVE and NEGATIVE measures/scores
positive, negative = compute_scores(test_neg_corpus)
rating = []
for file_name in test_neg_file_names:
    rating.append(get_rating(str(file_name)))

# create data frame to explore POSITIVE and NEGATIVE measures
test_neg_data = {'train_test':['TEST'] * len(test_neg_file_names),\
    'pos_neg': ['NEG'] * len(test_neg_file_names),\
    'file_name': test_neg_file_names,\
    'POSITIVE': positive, 'NEGATIVE': negative,\
    'rating': rating}
test_neg_data_frame = pd.DataFrame(test_neg_data)

# merge the positive and negative testing data frames
test_data_frame = pd.concat([test_pos_data_frame, test_neg_data_frame],\
    axis = 0, ignore_index = True)

# determining thumbs up or down based on rating
test_data_frame['thumbsupdown'] = \
    test_data_frame['rating'].map(map_to_thumbsupdown)
# compute simple measure of sentiment as POSITIVE - NEGATIVE
test_data_frame['simple'] = \
    test_data_frame['POSITIVE'] - test_data_frame['NEGATIVE']
# examine the data frame
print(pd.crosstab(test_data_frame['pos_neg'],\
    test_data_frame['thumbsupdown']))
print(test_data_frame.head())
print(test_data_frame.tail())
print(test_data_frame.describe())
```

```
ratings_grouped = test_data_frame['simple'].\
    groupby(test_data_frame['rating'])
print('\nTest Data Simple Difference Means by Ratings:',\
    ratings_grouped.mean())
thumbs_grouped = \
    test_data_frame['simple'].groupby(test_data_frame['thumbsupdown'])
print('\nTest Data Simple Difference Means by Thumbs UP/DOWN:',\
    thumbs_grouped.mean())


# repeat methods for the Tom's movie reviews ----------------------------
# /reviews/test/tom/ testing set directory path
test_tom_file_names, test_tom_corpus, test_tom_aggregate_corpus = \
    corpus_creator(input_directory_path = 'reviews/test/tom/',\
        output_directory_path = 'reviews/test/tom_parsed/')


# word counts for Tom's reviews
positive_words = []
negative_words = []
total_words = []
for file in test_tom_corpus:
    positive_words.append(count_positive(file))
    negative_words.append(count_negative(file))
    total_words.append(count_total(file))
# POSITIVE and NEGATIVE measures/scores for Tom's reviews
positive, negative = compute_scores(test_tom_corpus)
rating = []
for file_name in test_tom_file_names:
    rating.append(get_rating(str(file_name)))


# create data frame to check calculations of counts and scores
test_tom_data = {'train_test':['TOM'] * len(test_tom_file_names),\
    'pos_neg': ['POS', 'POS', 'NEG', 'POS', 'NEG', 'NEG', 'POS', 'NEG'],\
    'file_name': test_tom_file_names,\
    'movie': ['Marigolds',\
    'Blade Runner',\
    'Vinny',\
    'Mars Attacks',
    'Fight Club',\
    'Congeniality',\
    'Find Me Guilty',\
    'Moneyball'],\
    'positive_words' : positive_words,\
    'negative_words' : negative_words,\
    'total_words' : total_words,\
    'POSITIVE': positive, 'NEGATIVE': negative,\
    'rating': rating}
test_tom_data_frame = pd.DataFrame(test_tom_data)
# determing thumbs up or down based upon rating
test_tom_data_frame['thumbsupdown'] = \
    test_tom_data_frame['rating'].map(map_to_thumbsupdown)
# compute simple measure of sentiment as POSITIVE - NEGATIVE
test_tom_data_frame['simple'] = \
    test_tom_data_frame['POSITIVE'] - test_tom_data_frame['NEGATIVE']
```

```
# examine the data frame
print(test_tom_data_frame)
print(test_tom_data_frame.describe())
ratings_grouped = test_tom_data_frame['simple'].\
    groupby(test_tom_data_frame['rating'])
print('\nTom Simple Difference Means by Ratings:',ratings_grouped.mean())
thumbs_grouped = \
    test_tom_data_frame['simple'].groupby(test_tom_data_frame['thumbsupdown'])
print('\nTom Simple Difference Means by Thumbs UP/DOWN:',\
    thumbs_grouped.mean())

# develop predictive models using the training data
# -------------------------------------
# Simple difference method
# -------------------------------------
# use the median of the simple difference between POSITIVE and NEGATIVE
simple_cut_point = train_data_frame['simple'].median()

# algorithm for simple difference method based on training set median
def predict_simple(value):
    if (value > simple_cut_point):
        return('UP')
    else:
        return('DOWN')

train_data_frame['pred_simple'] = \
    train_data_frame['simple'].apply(lambda d: predict_simple(d))
print(train_data_frame.head())

print('\n Simple Difference Training Set Performance\n',\
    'Percentage of Reviews Correctly Classified:',\
    100 * round(evaluate_classifier(train_data_frame['pred_simple'],\
    train_data_frame['thumbsupdown'])[4], 3),'\n')

# evaluate simple difference method in the test set
# using algorithm developed with the training set
test_data_frame['pred_simple'] = \
    test_data_frame['simple'].apply(lambda d: predict_simple(d))
print(test_data_frame.head())
print('\n Simple Difference Test Set Performance\n',\
    'Percentage of Reviews Correctly Classified:',\
    100 * round(evaluate_classifier(test_data_frame['pred_simple'],\
    test_data_frame['thumbsupdown'])[4], 3), '\n')

# -------------------------------------
# Regression difference method
# -------------------------------------
# regression method for determining weights on POSITIVE AND NEGATIVE
# fit a regression model to the training data
regression_model = str('rating ~ POSITIVE + NEGATIVE')
# fit the model to the training set
train_regression_model_fit = smf.ols(regression_model,\
    data = train_data_frame).fit()
```

```
# summary of model fit to the training set
print(train_regression_model_fit.summary())

# because we are using predicted rating we use the midpoint
# rating of 5 as the cut-point for making thumbs up or down predictions
regression_cut_point = 5

# algorithm for simple difference method based on training set median
def predict_regression(value):
    if (value > regression_cut_point):
        return('UP')
    else:
        return('DOWN')

# training set predictions from the model fit to the training set
train_data_frame['pred_regression_rating'] =\
    train_regression_model_fit.fittedvalues

# predict thumbs up or down based upon the predicted rating
train_data_frame['pred_regression'] = \
    train_data_frame['pred_regression_rating'].\
        apply(lambda d: predict_regression(d))
print(train_data_frame.head())

print('\n Regression Difference Training Set Performance\n',\
    'Percentage of Reviews Correctly Classified:',\
    100 * round(evaluate_classifier(train_data_frame['pred_regression'],\
    train_data_frame['thumbsupdown'])[4], 3),'\n')

# evaluate regression difference method in the test set
# using algorithm developed with the training set
# predict thumbs up or down based upon the predicted rating

# test set predictions from the model fit to the training set
test_data_frame['pred_regression_rating'] =\
    train_regression_model_fit.predict(test_data_frame)

test_data_frame['pred_regression'] = \
    test_data_frame['pred_regression_rating'].\
        apply(lambda d: predict_regression(d))
print(test_data_frame.head())

print('\n Regression Difference Test Set Performance\n',\
    'Percentage of Reviews Correctly Classified:',\
    100 * round(evaluate_classifier(test_data_frame['pred_regression'],\
    test_data_frame['thumbsupdown'])[4], 3), '\n')

# ------------------------------------------------
# Compute text measures for each corpus
# ------------------------------------------------
# return to score the document collections with get_text_measures
# for each of the selected words from the sentiment lists
# these new variables will be given the names of the words
```

```
# to keep things simple.... there are 50 such variables/words
# identified from our analysis of the unsup corpus above
# start with the training document collection
working_corpus = train_pos_corpus + train_neg_corpus
add_corpus_data = get_text_measures(working_corpus)
add_corpus_data_frame = pd.DataFrame(add_corpus_data)
# merge the new text measures with the existing data frame
train_data_frame =\
    pd.concat([train_data_frame,add_corpus_data_frame],axis=1)
# examine the expanded training data frame
print('\n xtrain_data_frame (rows, cols):',train_data_frame.shape,'\n')
print(train_data_frame.describe())
print(train_data_frame.head())


# start with the test document collection
working_corpus = test_pos_corpus + test_neg_corpus
add_corpus_data = get_text_measures(working_corpus)
add_corpus_data_frame = pd.DataFrame(add_corpus_data)
# merge the new text measures with the existing data frame
test_data_frame = pd.concat([test_data_frame,add_corpus_data_frame],axis=1)
# examine the expanded testing data frame
print('\n xtest_data_frame (rows, cols):',test_data_frame.shape,'\n')
print(test_data_frame.describe())
print(test_data_frame.head())


# end with Tom's reviews as a document collection
working_corpus = test_tom_corpus
add_corpus_data = get_text_measures(working_corpus)
add_corpus_data_frame = pd.DataFrame(add_corpus_data)
# merge the new text measures with the existing data frame
tom_data_frame =\
    pd.concat([test_tom_data_frame,add_corpus_data_frame],axis=1)
# examine the expanded testing data frame
print('\n xtom_data_frame (rows, cols):',tom_data_frame.shape,'\n')
print(tom_data_frame.describe())
print(tom_data_frame.head())


# --------------------------------------------
# Word/item analysis method for training set
# --------------------------------------------
# item-rating correlations for all 50 words

item_list = selected_positive_words + selected_negative_words
item_rating_corr = []
for item in item_list:
    item_rating_corr.\
        append(train_data_frame['rating'].corr(train_data_frame[item]))
item_analysis_data_frame =\
    pd.DataFrame({'item': item_list, 'item_rating_corr': item_rating_corr})

# absolute value of item correlation with rating
item_analysis_data_frame['abs_item_rating_corr'] =\
    item_analysis_data_frame['item_rating_corr'].apply(lambda d: abs(d))
```

```
# look at sort by absolute value
print(item_analysis_data_frame.sort_index(by = ['abs_item_rating_corr'],\
    ascending = False))

# select subset of items with absolute correlations > 0.05
selected_item_analysis_data_frame =\
    item_analysis_data_frame\
        [item_analysis_data_frame['abs_item_rating_corr'] > 0.05]

# identify the positive items for word/item analysis measure
selected_positive_item_df =\
    selected_item_analysis_data_frame\
        [selected_item_analysis_data_frame['item_rating_corr'] > 0]
possible_positive_items = selected_positive_item_df['item']
print('Possible positive items:',possible_positive_items,'\n')
# note some surprises in the list of positive items
# select list consitent with initial list of positive words
selected_positive_items =\
    list(set(possible_positive_items) & set(positive_word_list))
print('Selected positive items:',selected_positive_items,'\n')

# identify the negative items for word/item analysis measure
selected_negative_item_df =\
    selected_item_analysis_data_frame\
        [selected_item_analysis_data_frame['item_rating_corr'] < 0]
possible_negative_items = selected_negative_item_df['item']
print('Possible negative items:',possible_negative_items,'\n')
# select list consitent with initial list of negative words
selected_negative_items =\
    list(set(possible_negative_items) & set(negative_word_list))
print('Selected negative items:',selected_negative_items,'\n')
# the word "funny" remains a mystery... kept in negative list for now

# selected positive and negative items entered into function
# for obtaining word/item analysis summative score in which
# postive items get +1 point and negative items get -1 point
# ... implemented in imported Python utility get_summative_scores
# start with the training set... identify a cut-off
working_corpus = train_pos_corpus + train_neg_corpus
add_corpus_data = get_summative_scores(working_corpus)
add_corpus_data_frame = pd.DataFrame(add_corpus_data)
# merge the new text measures with the existing data frame
train_data_frame = pd.concat([train_data_frame,add_corpus_data_frame],axis=1)
# examine the expanded training data frame and summative_scores
print('\n train_data_frame (rows, cols):',train_data_frame.shape,'\n')
print(train_data_frame['summative_score'].describe())
print('\nCorrelation of ratings and summative scores:'\
    ,round(train_data_frame['rating'].\
        corr(train_data_frame['summative_score'])),3))
ratings_grouped = train_data_frame['summative_score'].\
    groupby(train_data_frame['rating'])
print('\nTraining Data Summative Score Means by Ratings:',\
    ratings_grouped.mean())
```

```
thumbs_grouped = \
    train_data_frame['summative_score'].\
        groupby(train_data_frame['thumbsupdown'])
print('\nTraining Data Summative Score Means by Thumbs UP/DOWN:',\
    thumbs_grouped.mean())

# analyses suggest a simple positive/negative cut on summative scores
# algorithm for word/item method based on training set summative_scores
def predict_by_summative_score(value):
    if (value > 0):
        return('UP')
    else:
        return('DOWN')

# evaluate word/item analysis method on training set
train_data_frame['pred_summative_score'] = \
    train_data_frame['summative_score'].\
    apply(lambda d: predict_by_summative_score(d))

print('\n Word/item Analysis Training Set Performance\n',\
    'Percentage of Reviews Correctly Classified by Summative Scores:',\
    100 * round(evaluate_classifier(train_data_frame['pred_summative_score'],\
    train_data_frame['thumbsupdown'])[4], 3),'\n')

# compute summative scores on test data frame
working_corpus = test_pos_corpus + test_neg_corpus
add_corpus_data = get_summative_scores(working_corpus)
add_corpus_data_frame = pd.DataFrame(add_corpus_data)

# merge the new text measures with the existing data frame
test_data_frame = pd.concat([test_data_frame,add_corpus_data_frame],axis=1)

# evaluate word/item analysis method (summative score method) on test set
# using algorithm developed with the training set
test_data_frame['pred_summative_score'] = \
    test_data_frame['summative_score'].\
        apply(lambda d: predict_by_summative_score(d))

print('\n Word/item Analysis Test Set Performance\n',\
    'Percentage of Reviews Correctly Classified by Summative Scores:',\
    100 * round(evaluate_classifier(test_data_frame['pred_summative_score'],\
    test_data_frame['thumbsupdown'])[4], 3), '\n')

# ------------------------------------
# Logistic regression method
# ------------------------------------
# translate thumbsupdown into a binary indicator variable y
# here we let thumbs up have the higher value of 1

thumbsupdown_to_binary = {'UP':1,'DOWN':0}

train_data_frame['y'] =\
    train_data_frame['thumbsupdown'].map(thumbsupdown_to_binary)
```

```
# model specification in R-like formula syntax
text_classification_model = 'y ~ beautiful +\
    best + better + classic + enjoy + enough +\
    entertaining + excellent +\
    fans + fun + good + great + interesting + like +\
    love + nice + perfect + pretty + right +\
    top + well + won + wonderful + work + worth +\
    bad + boring + creepy + dark + dead+\
    death + evil + fear + funny + hard + kill +\
    killed + lack + lost + mystery +\
    plot + poor + problem + sad + scary +\
    slow + terrible + waste + worst + wrong'

# convert R-like formula into design matrix needed for statsmodels
y,x = patsy.dmatrices(text_classification_model,\
    train_data_frame, return_type = 'dataframe')

# define the logistic regression algorithm
my_logit_model = sm.Logit(y,x)
# fit the model to training set
my_logit_model_fit = my_logit_model.fit()
print(my_logit_model_fit.summary())

# predicted probability of thumbs up for training set
train_data_frame['pred_logit_prob'] =\
    my_logit_model_fit.predict(linear = False)

# map from probability to thumbsupdown with simple 0.5 cut-off
def prob_to_updown(x):
    if(x > 0.5):
        return('UP')
    else:
        return('DOWN')

train_data_frame['pred_logit'] =\
    train_data_frame['pred_logit_prob'].apply(lambda d: prob_to_updown(d))
print('\n Logistic Regression Training Set Performance\n',\
    'Percentage of Reviews Correctly Classified:',\
    100 * round(evaluate_classifier(train_data_frame['pred_logit'],\
    train_data_frame['thumbsupdown'])[4], 3),'\n')

# use the model developed on the training set to predict
# thumbs up or down reviews in the test set
# assume that y is not known... only x used from patsy
y,x = patsy.dmatrices(text_classification_model,\
        test_data_frame, return_type = 'dataframe')
y = []  # ignore known thumbs up/down from test set...
# we want to predict thumbs up/down from the model fit to
# the training set... my_logit_model_fit
test_data_frame['pred_logit_prob'] =\
    my_logit_model_fit.predict(exog = x, linear = False)
test_data_frame['pred_logit'] =\
    test_data_frame['pred_logit_prob'].apply(lambda d: prob_to_updown(d))
```

```python
print('\n Logistic Regression Test Set Performance\n',\
    'Percentage of Reviews Correctly Classified:',\
    100 * round(evaluate_classifier(test_data_frame['pred_logit'],\
    test_data_frame['thumbsupdown'])[4], 3),'\n')


# ----------------------------------------
# Support vector machines
# ----------------------------------------
# fit the model to the training set
y,x = patsy.dmatrices(text_classification_model,\
    train_data_frame, return_type = 'dataframe')
my_svm = svm.SVC()
my_svm_fit = my_svm.fit(x, np.ravel(y))
train_data_frame['pred_svm_binary'] = my_svm_fit.predict(x)
binary_to_thumbsupdown = {0: 'DOWN', 1: 'UP'}
train_data_frame['pred_svm'] =\
    train_data_frame['pred_svm_binary'].map(binary_to_thumbsupdown)
print('\n Support Vector Machine Training Set Performance\n',\
    'Percentage of Reviews Correctly Classified:',\
    100 * round(evaluate_classifier(train_data_frame['pred_svm'],\
    train_data_frame['thumbsupdown'])[4], 3),'\n')
# use the model developed on the training set to predict
# thumbs up or down reviews in the test set
# assume that y is not known... only x used from patsy
y,x = patsy.dmatrices(text_classification_model,\
        test_data_frame, return_type = 'dataframe')
y = []  # ignore known thumbs up/down from test set...
test_data_frame['pred_svm_binary'] = my_svm_fit.predict(x)
test_data_frame['pred_svm'] =\
    test_data_frame['pred_svm_binary'].map(binary_to_thumbsupdown)
print('\n Support Vector Machine Test Set Performance\n',\
    'Percentage of Reviews Correctly Classified:',\
    100 * round(evaluate_classifier(test_data_frame['pred_svm'],\
    test_data_frame['thumbsupdown'])[4], 3),'\n')


# ----------------------------------------
# Random forests
# ----------------------------------------
# fit random forest model to the training data
y,x = patsy.dmatrices(text_classification_model,\
    train_data_frame, return_type = 'dataframe')

# for reproducibility set random number seed with random_state
my_rf_model = RandomForestClassifier(n_estimators = 10, random_state = 9999)
my_rf_model_fit = my_rf_model.fit(x, np.ravel(y))
train_data_frame['pred_rf_binary'] = my_rf_model_fit.predict(x)
train_data_frame['pred_rf'] =\
    train_data_frame['pred_rf_binary'].map(binary_to_thumbsupdown)

print('\n Random Forest Training Set Performance\n',\
    'Percentage of Reviews Correctly Classified:',\
    100 * round(evaluate_classifier(train_data_frame['pred_rf'],\
    train_data_frame['thumbsupdown'])[4], 3),'\n')
```

```
# use the model developed on the training set to predict
# thumbs up or down reviews in the test set
# assume that y is not known... only x used from patsy
y,x = patsy.dmatrices(text_classification_model,\
        test_data_frame, return_type = 'dataframe')
y = []  # ignore known thumbs up/down from test set...
test_data_frame['pred_rf_binary'] = my_rf_model_fit.predict(x)
test_data_frame['pred_rf'] =\
    test_data_frame['pred_rf_binary'].map(binary_to_thumbsupdown)

print('\n Random Forest Test Set Performance\n',\
    'Percentage of Reviews Correctly Classified:',\
    100 * round(evaluate_classifier(test_data_frame['pred_rf'],\
    test_data_frame['thumbsupdown'])[4], 3),'\n')


# Suggestions for the student:
# Employ stemming prior to the creation of terms-by-document matrices.
# Try alternative positive and negative word sets for sentiment scoring.
# Try word sets that relate to a wider variety of emotional or opinion states.
# Better still, move beyond a bag-of-words approach to sentiment. Use
# the tools of natural language processing and define text features
# based upon combinations of words such as bigrams (pairs of words)
# and taking note of parts of speech.  Yet another approach would be
# to define ignore negative and positive word lists and work directly
# with identified text features that correlate with movie review ratings or
# do a good job of classifying reviews into positive and negative groups.
# Text features within text classification problems may be defined
# on term document frequency alone or on measures of term document
# frequency adjusted by term corpus frequency. Using alternative
# features and text measures as well as alternative classification methods,
# run a true benchmark within a loop, using hundreds or thousands of iterations.
# See if you can improve upon the performance of modeling methods by
# modifying the values of arguments to algorithms used here.
# Use various methods of classifier performance to evaluate classifiers.
# Try text classification for the movie reviews without using initial
# lists of positive an negative words. That is, identify text features
# for thumbs up/down text classification directly from the training set.
```

Exhibit 8.2. *Sentiment Analysis and Classification of Movie Ratings (R)*

```
# Sentiment Analysis Using the Movie Ratings Data (R)

# Note. Results from this program may differ from those published
#       in the book due to changes in the tm package.
#       The original analysis used the tm Dictionary() function,
#       which is no longer available in tm. This function has
#       been replaced by c(as.character()) to set the dictionary
#       as a character vector. Another necessary change concerns
#       the tolower() function, which must now be embedded within
#       the tm content_transformer() function.
#
# Despite changes in the tm functions, we have retained the
# earlier positive and negative word lists for scoring, as
# implemented in the code and utilities appendix under the file
# name <R_utility_program_5.R>, which is brought in by source().

# install these packages before bringing them in by library()
library(tm)  # text mining and document management
library(stringr)  # character manipulation with regular expressions
library(grid)  # grid graphics utilities
library(ggplot2)  # graphics
library(latticeExtra) # package used for text horizon plot
library(caret)  # for confusion matrix function
library(rpart)  # tree-structured modeling
library(e1071)  # support vector machines
library(randomForest)  # random forests
library(rpart.plot)  # plot tree-structured model information


# R preliminaries to get the user-defined utilities for plotting
# place the plotting code file <R_utility_program_3.R>
# in your working directory and execute it by
#     source("R_utility_program_3.R")
# Or if you have the R binary file in your working directory, use
#     load("mtpa_split_plotting_utilities.Rdata")
load("mtpa_split_plotting_utilities.Rdata")
# standardization needed for text measures
standardize <- function(x) {(x - mean(x)) / sd(x)}
# convert to bytecodes to avoid "invalid multibyte string" messages
bytecode.convert <- function(x) {iconv(enc2utf8(x), sub = "byte")}

# read in positive and negative word lists from Hu and Liu (2004)
positive.data.frame <- read.table(file = "Hu_Liu_positive_word_list.txt",
  header = FALSE, colClasses = c("character"), row.names = NULL,
  col.names = "positive.words")
positive.data.frame$positive.words <-
  bytecode.convert(positive.data.frame$positive.words)
negative.data.frame <- read.table(file = "Hu_Liu_negative_word_list.txt",
  header = FALSE, colClasses = c("character"), row.names = NULL,
  col.names = "negative.words")
negative.data.frame$negative.words <-
  bytecode.convert(negative.data.frame$negative.words)
```

```
# we use movie ratings data from Mass et al. (2011)
# available at http://ai.stanford.edu/~amaas/data/sentiment/
# we set up a directory under our working directory structure
# /reviews/train/unsup/ for the unsupervised reviews

directory.location <-
  paste(getwd(),"/reviews/train/unsup/",sep = "")
unsup.corpus <- Corpus(DirSource(directory.location, encoding = "UTF-8"),
  readerControl = list(language = "en_US"))
print(summary(unsup.corpus))

document.collection <- unsup.corpus

# strip white space from the documents in the collection
document.collection <- tm_map(document.collection, stripWhitespace)

# convert uppercase to lowercase in the document collection
document.collection <- tm_map(document.collection, content_transformer(tolower))

# remove numbers from the document collection
document.collection <- tm_map(document.collection, removeNumbers)

# remove punctuation from the document collection
document.collection <- tm_map(document.collection, removePunctuation)

# using a standard list, remove English stopwords from the document collection
document.collection <- tm_map(document.collection,
  removeWords, stopwords("english"))

# there is more we could do in terms of data preparation
# stemming... looking for contractions... possessives...
# previous analysis of a list of top terms showed a number of word
# contractions which we might like to drop from further analysis,
# recognizing them as stop words to be dropped from the document collection
initial.tdm <- TermDocumentMatrix(document.collection)
examine.tdm <- removeSparseTerms(initial.tdm, sparse = 0.96)
top.words <- Terms(examine.tdm)
print(top.words)

more.stop.words <- c("cant","didnt","doesnt","dont","goes","isnt","hes",
  "shes","thats","theres","theyre","wont","youll","youre","youve")
document.collection <- tm_map(document.collection,
  removeWords, more.stop.words)

some.proper.nouns.to.remove <-
  c("dick","ginger","hollywood","jack","jill","john","karloff",
    "kudrow","orson","peter","tcm","tom","toni","welles","william","wolheim")
document.collection <- tm_map(document.collection,
  removeWords, some.proper.nouns.to.remove)

# there is still more we could do in terms of data preparation
# but we will work with the bag of words we have for now
```

```
# the workhorse technique will be TermDocumentMatrix()
# for creating a terms-by-documents matrix across the document collection
# in previous text analytics with the taglines data we let the data
# guide us to the text measures... with sentiment analysis we have
# positive and negative dictionaries (to a large extent) defined in
# advance of looking at the data...
# positive.words and negative.words lists were read in earlier
# these come from the work of Hu and Liu (2004)
# positive.words = list of  positive words
# negative.words = list of  negative words
# we will start with these lists to build dictionaries
# that seem to make sense for movie reviews analysis
# Hu.Liu.positive.dictionary <- Dictionary(positive.data.frame$positive.words)
Hu.Liu.positive.dictionary <-
    c(as.character(positive.data.frame$positive.words))
reviews.tdm.Hu.Liu.positive <- TermDocumentMatrix(document.collection,
  list(dictionary = Hu.Liu.positive.dictionary))
examine.tdm <- removeSparseTerms(reviews.tdm.Hu.Liu.positive, 0.95)
top.words <- Terms(examine.tdm)
print(top.words)
Hu.Liu.frequent.positive <- findFreqTerms(reviews.tdm.Hu.Liu.positive, 25)
# this provides a list positive words occurring at least 25 times
# a review of this list suggests that all make sense (have content validity)
# test.positive.dictionary <- Dictionary(Hu.Liu.frequent.positive)
test.positive.dictionary <- c(as.character(Hu.Liu.frequent.positive))

# .... now for the negative words
# Hu.Liu.negative.dictionary <- Dictionary(negative.data.frame$negative.words)
Hu.Liu.negative.dictionary <-
    c(as.character(negative.data.frame$negative.words))
reviews.tdm.Hu.Liu.negative <- TermDocumentMatrix(document.collection,
  list(dictionary = Hu.Liu.negative.dictionary))
examine.tdm <- removeSparseTerms(reviews.tdm.Hu.Liu.negative, 0.97)
top.words <- Terms(examine.tdm)
print(top.words)
Hu.Liu.frequent.negative <- findFreqTerms(reviews.tdm.Hu.Liu.negative, 15)
# this provides a short list negative words occurring at least 15 times
# across the document collection... one of these words seems out of place
# as they could be thought of as positive: "funny"
test.negative <- setdiff(Hu.Liu.frequent.negative,c("funny"))
# test.negative.dictionary <- Dictionary(test.negative)
test.negative.dictionary <- c(as.character(test.negative))

# we need to evaluate the text measures we have defined
# for each of the documents count the total words
# and the number of words that match the positive and negative dictionaries
total.words <- integer(length(names(document.collection)))
positive.words <- integer(length(names(document.collection)))
negative.words <- integer(length(names(document.collection)))
other.words <- integer(length(names(document.collection)))

reviews.tdm <- TermDocumentMatrix(document.collection)
```

```
for(index.for.document in seq(along=names(document.collection))) {
  positive.words[index.for.document] <-
    sum(termFreq(document.collection[[index.for.document]],
    control = list(dictionary = test.positive.dictionary)))
  negative.words[index.for.document] <-
    sum(termFreq(document.collection[[index.for.document]],
    control = list(dictionary = test.negative.dictionary)))
  total.words[index.for.document] <-
    length(reviews.tdm[,index.for.document][["i"]])
  other.words[index.for.document] <- total.words[index.for.document] -
    positive.words[index.for.document] - negative.words[index.for.document]
  }
document <- names(document.collection)
text.measures.data.frame <- data.frame(document,total.words,
  positive.words, negative.words, other.words, stringsAsFactors = FALSE)
rownames(text.measures.data.frame) <- paste("D",as.character(0:499),sep="")
# compute text measures as percentages of words in each set
text.measures.data.frame$POSITIVE <-
  100 * text.measures.data.frame$positive.words /
  text.measures.data.frame$total.words
text.measures.data.frame$NEGATIVE <-
  100 * text.measures.data.frame$negative.words /
    text.measures.data.frame$total.words
# let us look at the resulting text measures we call POSITIVE and NEGATIVE
# to see if negative and positive dimensions appear to be on a common scale
# that is... is this a single dimension in the document space
# we use principal component biplots to explore text measures
# here we can use the technique to check on POSITIVE and NEGATIVE
principal.components.solution <-
  princomp(text.measures.data.frame[,c("POSITIVE","NEGATIVE")], cor = TRUE)
print(summary(principal.components.solution))
# biplot rendering of text measures and documents by year
pdf(file = "fig_sentiment_text_measures_biplot.pdf", width = 8.5, height = 11)
biplot(principal.components.solution, xlab = "First Pricipal Component",
  xlabs = rep("o", times = length(names(document.collection))),
  ylab = "Second Principal Component", expand = 0.7)
dev.off()
# results... the eigenvalues suggest that there are two underlying dimensions
# POSITIVE and NEGATIVE vectors rather than pointing in opposite directions
# they appear to be othogonal to one another... separate dimensions
# here we see the scatter plot for the two measures...
# if they were on the same dimension, they would be negatively correlated
# in fact they are correlated negatively but the correlation is very small
with(text.measures.data.frame, print(cor(POSITIVE, NEGATIVE)))
pdf(file = "fig_sentiment_text_measures_scatter_plot.pdf",
  width = 8.5, height = 8.5)
ggplot.object <- ggplot(data = text.measures.data.frame,
  aes(x = NEGATIVE, y = POSITIVE)) +
    geom_point(colour = "darkblue", shape = 1)
ggplot.print.with.margins(ggplot.object.name = ggplot.object,
  left.margin.pct=10, right.margin.pct=10,
  top.margin.pct=10,bottom.margin.pct=10)
dev.off()
```

```
# Perhaps POSITIVE and NEGATIVE can be combined in a way to yield effective
# predictions of movie ratings. Let us move to a set of movie reviews for
# supervised learning.  We select the 500 records from a set of positive
# reviews (ratings between 7 and 10) and 500 records from a set of negative
# reviews (ratings between 1 and 4).

# a set of 500 positive reviews... part of the training set
directory.location <-
  paste(getwd(),"/reviews/train/pos/",sep = "")
pos.train.corpus <- Corpus(DirSource(directory.location, encoding = "UTF-8"),
  readerControl = list(language = "en_US"))
print(summary(pos.train.corpus))
# a set of 500 negative reviews... part of the training set
directory.location <-
  paste(getwd(),"/reviews/train/neg/",sep = "")
neg.train.corpus <- Corpus(DirSource(directory.location, encoding = "UTF-8"),
  readerControl = list(language = "en_US"))
print(summary(neg.train.corpus))

# combine the positive and negative training sets
train.corpus <- c(pos.train.corpus, neg.train.corpus)

# strip white space from the documents in the collection
train.corpus <- tm_map(train.corpus, stripWhitespace)

# convert uppercase to lowercase in the document collection
train.corpus <- tm_map(train.corpus, content_transformer(tolower))

# remove numbers from the document collection
train.corpus <- tm_map(train.corpus, removeNumbers)

# remove punctuation from the document collection
train.corpus <- tm_map(train.corpus, removePunctuation)

# using a standard list, remove English stopwords from the document collection
train.corpus <- tm_map(train.corpus,
  removeWords, stopwords("english"))

# there is more we could do in terms of data preparation
# stemming... looking for contractions... possessives...
# previous analysis of a list of top terms showed a number of word
# contractions which we might like to drop from further analysis,
# recognizing them as stop words to be dropped from the document collection
initial.tdm <- TermDocumentMatrix(train.corpus)
examine.tdm <- removeSparseTerms(initial.tdm, sparse = 0.96)
top.words <- Terms(examine.tdm)
print(top.words)

more.stop.words <- c("cant","didnt","doesnt","dont","goes","isnt","hes",
  "shes","thats","theres","theyre","wont","youll","youre","youve")
train.corpus <- tm_map(train.corpus,
  removeWords, more.stop.words)
```

```
some.proper.nouns.to.remove <-
  c("dick","ginger","hollywood","jack","jill","john","karloff",
    "kudrow","orson","peter","tcm","tom","toni","welles","william","wolheim")
train.corpus <- tm_map(train.corpus,
  removeWords, some.proper.nouns.to.remove)

# compute list-based text measures for the training corpus
# for each of the documents count the total words
# and the number of words that match the positive and negative dictionaries
total.words <- integer(length(names(train.corpus)))
positive.words <- integer(length(names(train.corpus)))
negative.words <- integer(length(names(train.corpus)))
other.words <- integer(length(names(train.corpus)))

reviews.tdm <- TermDocumentMatrix(train.corpus)

for(index.for.document in seq(along=names(train.corpus))) {
  positive.words[index.for.document] <-
    sum(termFreq(train.corpus[[index.for.document]],
    control = list(dictionary = test.positive.dictionary)))
  negative.words[index.for.document] <-
    sum(termFreq(train.corpus[[index.for.document]],
    control = list(dictionary = test.negative.dictionary)))
  total.words[index.for.document] <-
    length(reviews.tdm[,index.for.document][["i"]])
  other.words[index.for.document] <- total.words[index.for.document] -
    positive.words[index.for.document] - negative.words[index.for.document]
  }

document <- names(train.corpus)
train.data.frame <- data.frame(document,total.words,
  positive.words, negative.words, other.words, stringsAsFactors = FALSE)
rownames(train.data.frame) <- paste("D",as.character(0:999),sep="")
# compute text measures as percentages of words in each set
train.data.frame$POSITIVE <-
  100 * train.data.frame$positive.words /
  train.data.frame$total.words
train.data.frame$NEGATIVE <-
  100 * train.data.frame$negative.words /
    train.data.frame$total.words

# rating is embedded in the document name... extract with regular expressions
for(index.for.document in seq(along = train.data.frame$document)) {
  first_split <- strsplit(train.data.frame$document[index.for.document],
    split = "[_]")
  second_split <- strsplit(first_split[[1]][2], split = "[.]")
  train.data.frame$rating[index.for.document] <- as.numeric(second_split[[1]][1])
  } # end of for-loop for defining ratings and thumbsupdown
train.data.frame$thumbsupdown <- ifelse((train.data.frame$rating > 5), 2, 1)
train.data.frame$thumbsupdown <-
  factor(train.data.frame$thumbsupdown, levels = c(1,2),
    labels = c("DOWN","UP"))
```

```
# a set of 500 positive reviews... part of the test set
directory.location <-
  paste(getwd(),"/reviews/test/pos/",sep = "")

pos.test.corpus <- Corpus(DirSource(directory.location, encoding = "UTF-8"),
  readerControl = list(language = "en_US"))
print(summary(pos.test.corpus))

# a set of 500 negative reviews... part of the test set
directory.location <-
  paste(getwd(),"/reviews/test/neg/",sep = "")

neg.test.corpus <- Corpus(DirSource(directory.location, encoding = "UTF-8"),
  readerControl = list(language = "en_US"))
print(summary(neg.test.corpus))

# combine the positive and negative testing sets
test.corpus <- c(pos.test.corpus, neg.test.corpus)

# strip white space from the documents in the collection
test.corpus <- tm_map(test.corpus, stripWhitespace)

# convert uppercase to lowercase in the document collection
test.corpus <- tm_map(test.corpus, content_transformer(tolower))

# remove numbers from the document collection
test.corpus <- tm_map(test.corpus, removeNumbers)

# remove punctuation from the document collection
test.corpus <- tm_map(test.corpus, removePunctuation)

# using a standard list, remove English stopwords from the document collection
test.corpus <- tm_map(test.corpus,
  removeWords, stopwords("english"))

# there is more we could do in terms of data preparation
# stemming... looking for contractions... possessives...
# previous analysis of a list of top terms showed a number of word
# contractions which we might like to drop from further analysis,
# recognizing them as stop words to be dropped from the document collection
initial.tdm <- TermDocumentMatrix(test.corpus)
examine.tdm <- removeSparseTerms(initial.tdm, sparse = 0.96)
top.words <- Terms(examine.tdm)
print(top.words)
more.stop.words <- c("cant","didnt","doesnt","dont","goes","isnt","hes",
  "shes","thats","theres","theyre","wont","youll","youre","youve")
test.corpus <- tm_map(test.corpus,
  removeWords, more.stop.words)
some.proper.nouns.to.remove <-
  c("dick","ginger","hollywood","jack","jill","john","karloff",
    "kudrow","orson","peter","tcm","tom","toni","welles","william","wolheim")
test.corpus <- tm_map(test.corpus,
  removeWords, some.proper.nouns.to.remove)
```

```
# compute list-based text measures for the testing corpus
# for each of the documents count the total words
# and the number of words that match the positive and negative dictionaries
total.words <- integer(length(names(test.corpus)))
positive.words <- integer(length(names(test.corpus)))
negative.words <- integer(length(names(test.corpus)))
other.words <- integer(length(names(test.corpus)))

reviews.tdm <- TermDocumentMatrix(test.corpus)

for(index.for.document in seq(along=names(test.corpus))) {
  positive.words[index.for.document] <-
    sum(termFreq(test.corpus[[index.for.document]],
    control = list(dictionary = test.positive.dictionary)))
  negative.words[index.for.document] <-
    sum(termFreq(test.corpus[[index.for.document]],
    control = list(dictionary = test.negative.dictionary)))
  total.words[index.for.document] <-
    length(reviews.tdm[,index.for.document][["i"]])
  other.words[index.for.document] <- total.words[index.for.document] -
    positive.words[index.for.document] - negative.words[index.for.document]
  }
document <- names(test.corpus)
test.data.frame <- data.frame(document,total.words,
  positive.words, negative.words, other.words, stringsAsFactors = FALSE)
rownames(test.data.frame) <- paste("D",as.character(0:999),sep="")

# compute text measures as percentages of words in each set
test.data.frame$POSITIVE <-
  100 * test.data.frame$positive.words /
  test.data.frame$total.words
 test.data.frame$NEGATIVE <-
  100 * test.data.frame$negative.words /
    test.data.frame$total.words

# rating is embedded in the document name... extract with regular expressions
for(index.for.document in seq(along = test.data.frame$document)) {
  first_split <- strsplit(test.data.frame$document[index.for.document],
    split = "[_]")
  second_split <- strsplit(first_split[[1]][2], split = "[.]")
  test.data.frame$rating[index.for.document] <- as.numeric(second_split[[1]][1])
  } # end of for-loop for defining

test.data.frame$thumbsupdown <- ifelse((test.data.frame$rating > 5), 2, 1)
test.data.frame$thumbsupdown <-
  factor(test.data.frame$thumbsupdown, levels = c(1,2),
    labels = c("DOWN","UP"))

# a set of 4 positive and 4 negative reviews... testing set of Tom's reviews
directory.location <-
  paste(getwd(),"/reviews/test/tom/",sep = "")
```

```r
tom.corpus <- Corpus(DirSource(directory.location, encoding = "UTF-8"),
  readerControl = list(language = "en_US"))
print(summary(tom.corpus))

# strip white space from the documents in the collection
tom.corpus <- tm_map(tom.corpus, stripWhitespace)

# convert uppercase to lowercase in the document collection
tom.corpus <- tm_map(tom.corpus, content_transformer(tolower))

# remove numbers from the document collection
tom.corpus <- tm_map(tom.corpus, removeNumbers)

# remove punctuation from the document collection
tom.corpus <- tm_map(tom.corpus, removePunctuation)

# using a standard list, remove English stopwords from the document collection
tom.corpus <- tm_map(tom.corpus,
  removeWords, stopwords("english"))

# there is more we could do in terms of data preparation
# stemming... looking for contractions... possessives...
# previous analysis of a list of top terms showed a number of word
# contractions which we might like to drop from further analysis,
# recognizing them as stop words to be dropped from the document collection

initial.tdm <- TermDocumentMatrix(tom.corpus)
examine.tdm <- removeSparseTerms(initial.tdm, sparse = 0.96)
top.words <- Terms(examine.tdm)
print(top.words)

more.stop.words <- c("cant","didnt","doesnt","dont","goes","isnt","hes",
  "shes","thats","theres","theyre","wont","youll","youre","youve")
tom.corpus <- tm_map(tom.corpus,
  removeWords, more.stop.words)

some.proper.nouns.to.remove <-
  c("dick","ginger","hollywood","jack","jill","john","karloff",
    "kudrow","orson","peter","tcm","tom","toni","welles","william","wolheim")
tom.corpus <- tm_map(tom.corpus,
  removeWords, some.proper.nouns.to.remove)

# compute list-based text measures for tom corpus
# for each of the documents count the total words
# and the number of words that match the positive and negative dictionaries

total.words <- integer(length(names(tom.corpus)))
positive.words <- integer(length(names(tom.corpus)))
negative.words <- integer(length(names(tom.corpus)))
other.words <- integer(length(names(tom.corpus)))

reviews.tdm <- TermDocumentMatrix(tom.corpus)
```

```r
for(index.for.document in seq(along=names(tom.corpus))) {
  positive.words[index.for.document] <-
    sum(termFreq(tom.corpus[[index.for.document]],
    control = list(dictionary = test.positive.dictionary)))
  negative.words[index.for.document] <-
    sum(termFreq(tom.corpus[[index.for.document]],
    control = list(dictionary = test.negative.dictionary)))
  total.words[index.for.document] <-
    length(reviews.tdm[,index.for.document][["i"]])
  other.words[index.for.document] <- total.words[index.for.document] -
    positive.words[index.for.document] - negative.words[index.for.document]
  }

document <- names(tom.corpus)
tom.data.frame <- data.frame(document,total.words,
  positive.words, negative.words, other.words, stringsAsFactors = FALSE)
rownames(tom.data.frame) <- paste("D",as.character(0:7),sep="")

# compute text measures as percentages of words in each set
tom.data.frame$POSITIVE <-
  100 * tom.data.frame$positive.words /
  tom.data.frame$total.words

tom.data.frame$NEGATIVE <-
  100 * tom.data.frame$negative.words /
  tom.data.frame$total.words

# rating is embedded in the document name... extract with regular expressions

for(index.for.document in seq(along = tom.data.frame$document)) {
  first_split <- strsplit(tom.data.frame$document[index.for.document],
    split = "[_]")
  second_split <- strsplit(first_split[[1]][2], split = "[.]")
  tom.data.frame$rating[index.for.document] <- as.numeric(second_split[[1]][1])
  } # end of for-loop for defining

tom.data.frame$thumbsupdown <- ifelse((tom.data.frame$rating > 5), 2, 1)
tom.data.frame$thumbsupdown <-
  factor(tom.data.frame$thumbsupdown, levels = c(1,2),
    labels = c("DOWN","UP"))

tom.movies <- data.frame(movies =
  c("The Effect of Gamma Rays on Man-in-the-Moon Marigolds",
    "Blade Runner","My Cousin Vinny","Mars Attacks",
    "Fight Club","Miss Congeniality 2","Find Me Guilty","Moneyball"))

# check out the measures on Tom's ratings
tom.data.frame.review <-
  cbind(tom.movies,tom.data.frame[,names(tom.data.frame)[2:9]])
print(tom.data.frame.review)

# develop predictive models using the training data
```

```
# ------------------------------------
# Simple difference method
# ------------------------------------
train.data.frame$simple <-
      train.data.frame$POSITIVE - train.data.frame$NEGATIVE
# check out simple difference method... is there a correlation with ratings?
with(train.data.frame, print(cor(simple, rating)))
# we use the training data to define an optimal cutoff...
# trees can help with finding the optimal split point for simple.difference
try.tree <- rpart(thumbsupdown ~ simple, data = train.data.frame)
print(try.tree)  # note that the first split value
# an earlier analysis had this value as -0.7969266
# create a user-defined function for the simple difference method
predict.simple <- function(x) {
  if (x >= -0.7969266) return("UP")
  if (x < -0.7969266) return("DOWN")
  }
# evaluate predictive accuracy in the training data
train.data.frame$pred.simple <- character(nrow(train.data.frame))
for (index.for.review in seq(along = train.data.frame$pred.simple)) {
  train.data.frame$pred.simple[index.for.review] <-
    predict.simple(train.data.frame$simple[index.for.review])
  }
train.data.frame$pred.simple <-
  factor(train.data.frame$pred.simple)
train.pred.simple.performance <-
  confusionMatrix(data = train.data.frame$pred.simple,
  reference = train.data.frame$thumbsupdown, positive = "UP")
# report full set of statistics relating to predictive accuracy
print(train.pred.simple.performance)
cat("\n\nTraining set percentage correctly predicted by",
  " simple difference method = ",
  sprintf("%1.1f",train.pred.simple.performance$overall[1]*100)," Percent",sep="")
# evaluate predictive accuracy in the test data
# SIMPLE DIFFERENCE METHOD
test.data.frame$simple <-
      test.data.frame$POSITIVE - train.data.frame$NEGATIVE
test.data.frame$pred.simple <- character(nrow(test.data.frame))
for (index.for.review in seq(along = test.data.frame$pred.simple)) {
  test.data.frame$pred.simple[index.for.review] <-
    predict.simple(test.data.frame$simple[index.for.review])
  }

test.data.frame$pred.simple <-
  factor(test.data.frame$pred.simple)
test.pred.simple.performance <-
  confusionMatrix(data = test.data.frame$pred.simple,
  reference = test.data.frame$thumbsupdown, positive = "UP")
# report full set of statistics relating to predictive accuracy
print(test.pred.simple.performance)
cat("\n\nTest set percentage correctly predicted = ",
  sprintf("%1.1f",test.pred.simple.performance$overall[1]*100),"
    Percent",sep="")
```

```
# ----------------------------------------
# Regression difference method
# ----------------------------------------
# regression method for determining weights on POSITIVE AND NEGATIVE
# fit a regression model to the training data
regression.model <- lm(rating ~ POSITIVE + NEGATIVE, data = train.data.frame)
print(regression.model)  # provides 5.5386 + 0.2962(POSITIVE) -0.3089(NEGATIVE)
train.data.frame$regression <-
  predict(regression.model, newdata = train.data.frame)
# determine the cutoff for regression.difference
  try.tree <- rpart(thumbsupdown ~ regression, data = train.data.frame)
print(try.tree)  # note that the first split is at 5.264625
# create a user-defined function for the simple difference method
predict.regression <- function(x) {
  if (x >= 5.264625) return("UP")
  if (x < 5.264625) return("DOWN")
  }
train.data.frame$pred.regression <-  character(nrow(train.data.frame))
for (index.for.review in seq(along = train.data.frame$pred.simple)) {
  train.data.frame$pred.regression[index.for.review] <-
    predict.regression(train.data.frame$regression[index.for.review])
  }
train.data.frame$pred.regression <-
  factor(train.data.frame$pred.regression)
train.pred.regression.performance <-
  confusionMatrix(data = train.data.frame$pred.regression,
  reference = train.data.frame$thumbsupdown, positive = "UP")
# report full set of statistics relating to predictive accuracy
print(train.pred.regression.performance)  # result 67.3 percent
cat("\n\nTraining set percentage correctly predicted by regression = ",
  sprintf("%1.1f",train.pred.regression.performance$overall[1]*100),
    " Percent",sep="")
# regression method for determining weights on POSITIVE AND NEGATIVE
# for the test set we use the model developed on the training set
test.data.frame$regression <-
  predict(regression.model, newdata = test.data.frame)

test.data.frame$pred.regression <-  character(nrow(test.data.frame))
for (index.for.review in seq(along = test.data.frame$pred.simple)) {
  test.data.frame$pred.regression[index.for.review] <-
    predict.regression(test.data.frame$regression[index.for.review])
  }

test.data.frame$pred.regression <-
  factor(test.data.frame$pred.regression)
test.pred.regression.performance <-
  confusionMatrix(data = test.data.frame$pred.regression,
  reference = test.data.frame$thumbsupdown, positive = "UP")
# report full set of statistics relating to predictive accuracy
print(test.pred.regression.performance)  # result 67.3 percent
cat("\n\nTest set percentage correctly predicted = ",
  sprintf("%1.1f",test.pred.regression.performance$overall[1]*100),
    " Percent",sep="")
```

```
# ----------------------------------------------
# Word/item analysis method for train.corpus
# ----------------------------------------------
# return to the training corpus to develop simple counts
# for each of the words in the sentiment list
# these new variables will be given the names of the words
# to keep things simple.... there are 50 such variables/words
# identified from an earlier analysis, as published in the book
working.corpus <- train.corpus
# run common code from utilities for scoring the working corpus
# this common code uses 25 positive and 25 negative words
# identified in an earlier analysis of these data
source("R_utility_program_5.R")
add.data.frame <- data.frame(amazing,beautiful,classic,enjoy,
   enjoyed,entertaining,excellent,fans,favorite,fine,fun,humor,
   lead,liked,love,loved,modern,nice,perfect,pretty,
   recommend,strong,top,wonderful,worth,bad,boring,cheap,creepy,dark,dead,
   death,evil,hard,kill,killed,lack,lost,miss,murder,mystery,plot,poor,
   sad,scary,slow,terrible,waste,worst,wrong)
train.data.frame <- cbind(train.data.frame,add.data.frame)
# ----------------------------------------------
# Word/item analysis method for test.corpus
# ----------------------------------------------
# return to the testing corpus to develop simple counts
# for each of the words in the sentiment list
# these new variables will be given the names of the words
# to keep things simple.... there are 50 such variables/words
working.corpus <- test.corpus
# run common code from utilities for scoring the working corpus
source("R_utility_program_5.R")
add.data.frame <- data.frame(amazing,beautiful,classic,enjoy,
   enjoyed,entertaining,excellent,fans,favorite,fine,fun,humor,
   lead,liked,love,loved,modern,nice,perfect,pretty,
   recommend,strong,top,wonderful,worth,bad,boring,cheap,creepy,dark,dead,
   death,evil,hard,kill,killed,lack,lost,miss,murder,mystery,plot,poor,
   sad,scary,slow,terrible,waste,worst,wrong)
 test.data.frame <- cbind(test.data.frame,add.data.frame)
# ----------------------------------------------
# Word/item analysis method for tom.corpus
# ----------------------------------------------
# return to the toming corpus to develop simple counts
# for each of the words in the sentiment list
# these new variables will be given the names of the words
# to keep things simple.... there are 50 such variables/words
working.corpus <- tom.corpus
# run common code from utilities for scoring the working corpus
source("R_utility_program_5.R")
add.data.frame <- data.frame(amazing,beautiful,classic,enjoy,
   enjoyed,entertaining,excellent,fans,favorite,fine,fun,humor,
   lead,liked,love,loved,modern,nice,perfect,pretty,
   recommend,strong,top,wonderful,worth,bad,boring,cheap,creepy,dark,dead,
   death,evil,hard,kill,killed,lack,lost,miss,murder,mystery,plot,poor,
   sad,scary,slow,terrible,waste,worst,wrong)
```

```
tom.data.frame <- cbind(tom.data.frame,add.data.frame)

# use phi coefficient... correlation with rating as index of item value
# again we draw upon the earlier positive and negative lists
phi <- numeric(50)
item <- c("amazing","beautiful","classic","enjoy",
  "enjoyed","entertaining","excellent","fans","favorite","fine","fun","humor",
  "lead","liked","love","loved","modern","nice","perfect","pretty",
  "recommend","strong","top","wonderful","worth",
  "bad","boring","cheap","creepy","dark","dead",
  "death","evil","hard","kill","killed","lack",
  "lost","miss","murder","mystery","plot","poor",
  "sad","scary","slow","terrible","waste","worst","wrong")
item.analysis.data.frame <- data.frame(item,phi)
item.place <- 14:63
for (index.for.column in 1:50) {
  item.analysis.data.frame$phi[index.for.column] <-
    cor(train.data.frame[, item.place[index.for.column]],train.data.frame[,8])
  }

# sort by absolute value of the phi coefficient with the rating
item.analysis.data.frame$absphi <- abs(item.analysis.data.frame$phi)
item.analysis.data.frame <-
  item.analysis.data.frame[sort.list(item.analysis.data.frame$absphi,
    decreasing = TRUE),]

# subset of words with phi coefficients greater than 0.05 in absolute value
selected.items.data.frame <-
  subset(item.analysis.data.frame, subset = (absphi > 0.05))

# use the sign of the phi coefficient as the item weight
selected.positive.data.frame <-
  subset(selected.items.data.frame, subset = (phi > 0.0))
selected.positive.words <- as.character(selected.positive.data.frame$item)

selected.negative.data.frame <-
  subset(selected.items.data.frame, subset = (phi < 0.0))
selected.negative.words <- as.character(selected.negative.data.frame$item)

# these lists define new dictionaries for scoring

reviews.tdm <- TermDocumentMatrix(train.corpus)

temp.positive.score <- integer(length(names(train.corpus)))
temp.negative.score <- integer(length(names(train.corpus)))
for(index.for.document in seq(along=names(train.corpus))) {
  temp.positive.score[index.for.document] <-
    sum(termFreq(train.corpus[[index.for.document]],
    control = list(dictionary = selected.positive.words)))
  temp.negative.score[index.for.document] <-
    sum(termFreq(train.corpus[[index.for.document]],
    control = list(dictionary = selected.negative.words)))
  }
```

```
train.data.frame$item.analysis.score <-
  temp.positive.score - temp.negative.score

# use the training set and tree-structured modeling to determine the cutoff
  try.tree<-rpart(thumbsupdown ~ item.analysis.score, data = train.data.frame)
print(try.tree)  # note that the first split is at -0.5
# create a user-defined function for the simple difference method
predict.item.analysis <- function(x) {
  if (x >= -0.5) return("UP")
  if (x < -0.5) return("DOWN")
  }
train.data.frame$pred.item.analysis <-  character(nrow(train.data.frame))
for (index.for.review in seq(along = train.data.frame$pred.simple)) {
  train.data.frame$pred.item.analysis[index.for.review] <-
  predict.item.analysis(train.data.frame$item.analysis.score[index.for.review])
  }
train.data.frame$pred.item.analysis <-
  factor(train.data.frame$pred.item.analysis)
train.pred.item.analysis.performance <-
  confusionMatrix(data = train.data.frame$pred.item.analysis,
  reference = train.data.frame$thumbsupdown, positive = "UP")
# report full set of statistics relating to predictive accuracy
print(train.pred.item.analysis.performance)  # result 73.9 percent
cat("\n\nTraining set percentage correctly predicted by item analysis = ",
  sprintf("%1.1f",train.pred.item.analysis.performance$overall[1]*100),
    " Percent",sep="")

# use item analysis method of scoring with the test set

reviews.tdm <- TermDocumentMatrix(test.corpus)

temp.positive.score <- integer(length(names(test.corpus)))
temp.negative.score <- integer(length(names(test.corpus)))
for(index.for.document in seq(along=names(test.corpus))) {
  temp.positive.score[index.for.document] <-
    sum(termFreq(test.corpus[[index.for.document]],
    control = list(dictionary = selected.positive.words)))
  temp.negative.score[index.for.document] <-
    sum(termFreq(test.corpus[[index.for.document]],
    control = list(dictionary = selected.negative.words)))
  }

test.data.frame$item.analysis.score <-
  temp.positive.score - temp.negative.score

test.data.frame$pred.item.analysis <-  character(nrow(test.data.frame))
for (index.for.review in seq(along = test.data.frame$pred.simple)) {
  test.data.frame$pred.item.analysis[index.for.review] <-
  predict.item.analysis(test.data.frame$item.analysis.score[index.for.review])
  }
test.data.frame$pred.item.analysis <-
  factor(test.data.frame$pred.item.analysis)
```

```
test.pred.item.analysis.performance <-
  confusionMatrix(data = test.data.frame$pred.item.analysis,
  reference = test.data.frame$thumbsupdown, positive = "UP")

# report full set of statistics relating to predictive accuracy
print(test.pred.item.analysis.performance)  # result 74 percent

cat("\n\nTest set percentage correctly predicted by item analysis = ",
  sprintf("%1.1f",test.pred.item.analysis.performance$overall[1]*100),
    " Percent",sep="")

# ----------------------------------------
# Logistic regression method
# ----------------------------------------
text.classification.model <- {thumbsupdown ~ amazing + beautiful +
  classic + enjoy + enjoyed +
  entertaining + excellent +
  fans + favorite + fine + fun + humor + lead + liked +
  love + loved + modern + nice + perfect + pretty +
  recommend + strong + top + wonderful + worth +
  bad + boring + cheap + creepy + dark + dead +
  death + evil + hard + kill +
  killed + lack + lost + miss + murder + mystery +
  plot + poor + sad + scary +
  slow + terrible + waste + worst + wrong}

# full logistic regression model
logistic.regression.fit <- glm(text.classification.model,
  family=binomial(link=logit), data = train.data.frame)
print(summary(logistic.regression.fit))

# obtain predicted probability values for training set
logistic.regression.pred.prob <-
  as.numeric(predict(logistic.regression.fit, newdata = train.data.frame,
  type="response"))

train.data.frame$pred.logistic.regression <-
  ifelse((logistic.regression.pred.prob > 0.5),2,1)

train.data.frame$pred.logistic.regression <-
  factor(train.data.frame$pred.logistic.regression, levels = c(1,2),
    labels = c("DOWN","UP"))

train.pred.logistic.regression.performance <-
  confusionMatrix(data = train.data.frame$pred.logistic.regression,
  reference = train.data.frame$thumbsupdown, positive = "UP")

# report full set of statistics relating to predictive accuracy
print(train.pred.logistic.regression.performance)  # result 75.2 percent
cat("\n\nTraining set percentage correct by logistic regression = ",
  sprintf("%1.1f",train.pred.logistic.regression.performance$overall[1]*100),
    " Percent",sep="")
```

```
# now we use the model developed on the training set with the test set
# obtain predicted probability values for test set
logistic.regression.pred.prob <-
  as.numeric(predict(logistic.regression.fit, newdata = test.data.frame,
  type="response"))

test.data.frame$pred.logistic.regression <-
  ifelse((logistic.regression.pred.prob > 0.5),2,1)

test.data.frame$pred.logistic.regression <-
  factor(test.data.frame$pred.logistic.regression, levels = c(1,2),
    labels = c("DOWN","UP"))

test.pred.logistic.regression.performance <-
  confusionMatrix(data = test.data.frame$pred.logistic.regression,
  reference = test.data.frame$thumbsupdown, positive = "UP")

# report full set of statistics relating to predictive accuracy
print(test.pred.logistic.regression.performance)  # result 72.6 percent

cat("\n\nTest set percentage correctly predicted by logistic regression = ",
  sprintf("%1.1f",test.pred.logistic.regression.performance$overall[1]*100),
    " Percent",sep="")
# -------------------------------------
# Support vector machines
# -------------------------------------
# determine tuning parameters prior to fitting model
train.tune <- tune(svm, text.classification.model, data = train.data.frame,
                ranges = list(gamma = 2^(-8:1), cost = 2^(0:4)),
                tunecontrol = tune.control(sampling = "fix"))
# display the tuning results (in text format)
print(train.tune)
# fit the support vector machine to the training data using tuning parameters
train.data.frame.svm <- svm(text.classification.model, data = train.data.frame,
  cost=4, gamma=0.00390625, probability = TRUE)
train.data.frame$pred.svm <- predict(train.data.frame.svm, type="class",
newdata=train.data.frame)
train.pred.svm.performance <-
  confusionMatrix(data = train.data.frame$pred.svm,
  reference = train.data.frame$thumbsupdown, positive = "UP")

# report full set of statistics relating to predictive accuracy
print(train.pred.svm.performance)  # result 79.0 percent

cat("\n\nTraining set percentage correctly predicted by SVM = ",
  sprintf("%1.1f",train.pred.svm.performance$overall[1]*100),
    " Percent",sep="")

# use the support vector machine model identified in the training set
# to do text classification on the test set
test.data.frame$pred.svm <- predict(train.data.frame.svm, type="class",
newdata=test.data.frame)
```

```
test.pred.svm.performance <-
  confusionMatrix(data = test.data.frame$pred.svm,
  reference = test.data.frame$thumbsupdown, positive = "UP")

# report full set of statistics relating to predictive accuracy
print(test.pred.svm.performance)  # result 71.6 percent

cat("\n\nTest set percentage correctly predicted by SVM = ",
  sprintf("%1.1f",test.pred.svm.performance$overall[1]*100),
    " Percent",sep="")

# ---------------------------------------
# Random forests
# ---------------------------------------
# fit random forest model to the training data
set.seed (9999)  # for reproducibility
train.data.frame.rf <- randomForest(text.classification.model,
  data=train.data.frame, mtry=3, importance=TRUE, na.action=na.omit)

# review the random forest solution
print(train.data.frame.rf)

# check importance of the individual explanatory variables
pdf(file = "fig_sentiment_random_forest_importance.pdf",
width = 11, height = 8.5)
varImpPlot(train.data.frame.rf, main = "")
dev.off()

train.data.frame$pred.rf <- predict(train.data.frame.rf, type="class",
  newdata = train.data.frame)

train.pred.rf.performance <-
  confusionMatrix(data = train.data.frame$pred.rf,
  reference = train.data.frame$thumbsupdown, positive = "UP")

# report full set of statistics relating to predictive accuracy
print(train.pred.rf.performance)  # result 82.2 percent

cat("\n\nTraining set percentage correctly predicted by random forests = ",
  sprintf("%1.1f",train.pred.rf.performance$overall[1]*100),
    " Percent",sep="")

# use the model fit to the training data to predict the the test data
test.data.frame$pred.rf <- predict(train.data.frame.rf, type="class",
  newdata = test.data.frame)

test.pred.rf.performance <-
  confusionMatrix(data = test.data.frame$pred.rf,
  reference = test.data.frame$thumbsupdown, positive = "UP")

# report full set of statistics relating to predictive accuracy
print(test.pred.rf.performance)  # result 74.0 percent
```

```
cat("\n\nTest set percentage correctly predicted by random forests = ",
  sprintf("%1.1f",test.pred.rf.performance$overall[1]*100),
    " Percent",sep="")

# measurement model performance summary

methods <- c("Simple difference","Regression difference",
  "Word/item analysis","Logistic regression",
  "Support vector machines","Random forests")

methods.performance.data.frame <- data.frame(methods)

methods.performance.data.frame$training <-
  c(train.pred.simple.performance$overall[1]*100,
    train.pred.regression.performance$overall[1]*100,
    train.pred.item.analysis.performance$overall[1]*100,
    train.pred.logistic.regression.performance$overall[1]*100,
    train.pred.svm.performance$overall[1]*100,
    train.pred.rf.performance$overall[1]*100)

methods.performance.data.frame$test <-
  c(test.pred.simple.performance$overall[1]*100,
    test.pred.regression.performance$overall[1]*100,
    test.pred.item.analysis.performance$overall[1]*100,
    test.pred.logistic.regression.performance$overall[1]*100,
    test.pred.svm.performance$overall[1]*100,
    test.pred.rf.performance$overall[1]*100)

# random forest predictions for Tom's movie reviews

tom.data.frame$pred.rf <- predict(train.data.frame.rf, type="class",
  newdata = tom.data.frame)

print(tom.data.frame[,c("thumbsupdown","pred.rf")])

tom.pred.rf.performance <-
  confusionMatrix(data = tom.data.frame$pred.rf,
  reference = tom.data.frame$thumbsupdown, positive = "UP")

# report full set of statistics relating to predictive accuracy

print(tom.pred.rf.performance)  # result 74.0 percent

cat("\n\nTraining set percentage correctly predicted by random forests = ",
  sprintf("%1.1f",tom.pred.rf.performance$overall[1]*100),
    "Percent",sep="")

# building a simple tree to classify reviews

simple.tree <- rpart(text.classification.model,
  data=train.data.frame,)
```

```
# plot the regression tree result from rpart

pdf(file = "fig_sentiment_simple_tree_classifier.pdf", width = 8.5, height = 8.5)
prp(simple.tree, main="",
  digits = 3,  # digits to display in terminal nodes
  nn = TRUE,  # display the node numbers
  fallen.leaves = TRUE,  # put the leaves on the bottom of the page
  branch = 0.5,  # change angle of branch lines
  branch.lwd = 2,  # width of branch lines
  faclen = 0,  # do not abbreviate factor levels
  trace = 1,  # print the automatically calculated cex
  shadow.col = 0,  # no shadows under the leaves
  branch.lty = 1,  # draw branches using dotted lines
  split.cex = 1.2,  # make the split text larger than the node text
  split.prefix = "is ",  # put "is" before split text
  split.suffix = "?",  # put "?" after split text
  split.box.col = "blue",  # lightgray split boxes (default is white)
  split.col = "white",  # color of text in split box
  split.border.col = "blue",  # darkgray border on split boxes
  split.round = .25)  # round the split box corners a tad
dev.off()

# simple tree predictions for Tom's movie reviews
tom.data.frame$pred.simple.tree <- predict(simple.tree, type="class",
  newdata = tom.data.frame)

print(tom.data.frame[,c("thumbsupdown","pred.rf","pred.simple.tree")])
```

9

Discovering Common Themes

Curly: "Do you know what the secret of life is?"

[Curly holds up one finger]

Curly: "This."

Mitch: "Your finger?"

Curly: "One thing. Just one thing.
You stick to that and the rest don't mean shit."

Mitch: "But, what is the 'one thing' ?"

Curly: "That's what you have to find out."

—JACK PALANCE AS CURLY AND BILLY CRYSTAL AS MITCH ROBBINS
IN *City Slickers* (1991)

I choose the quiet live. I have little political ambition or desire for celebrity. The extent of my political involvement is to vote in national elections. But even a quiet person can be concerned about worldwide events and what seems to be never-ending conflict among nations.

Maybe there are things we can learn from history to understand better the nature of foreign affairs. Perhaps we can find common themes—ideas that define who we are as a nation and why we do the things we do.

Common historical themes may be ferreted out by choosing a set of documents relating to our questions. Using unsupervised text analytics, we approach our study with no preconceptions. We let the data guide us to themes. In this chapter we illustrate the process using multidimensional scaling and cluster analysis on text data obtained from the web.

To explore common political themes, we selected State of the Union addresses delivered by the President of the United States (POTUS), as described in the *POTUS Speeches* case study (page 295). Our focus is recent political history, so we select documents following the Eisenhower Presidency. Our time frame is 1960–2014, during which fifty-two State of the Union speeches were given. These comprise a document collection or text corpus for analysis.

With text analytics it is often judgement rather than algorithms that makes an analysis successful. At each stage in the process, we decide what to keep and what to throw away.

There are standard *stop words* in English. These are pronouns and prepositions that add little value in identifying common themes or distinguishing among documents. We identify these using data from the Natural Language Took Kit (Bird, Klein, and Loper 2009).

Reviewing the POTUS data, we note that there are other common words in the State of the Union addresses that will have little value in identifying common themes or distinguishing among speeches. The speeches are delivered to the United States Congress, so words addressing Congress and its representatives add little value, as do words such as "United States" and "America." These words are added to our stop words list.

Parsing the initial text documents involves dropping punctuation and converting uppercase to lowercase letters, and then dropping the stop words. The result is a document collection or corpus of fifty-two speeches, twenty-five from Democratic Presidents and twenty-seven from Republican Presidents.

We form a terms-by-documents matrix for the corpus. Here again, we use our judgement. We must determine the number of words to include—that is, the number of rows in the matrix. Across the parsed fifty-two documents, there are 10,938 distinct words after the stop words have been omitted. We need to determine the number of words to include in the study. Shall we use the top 50, 100, 150, or 200 words in defining the matrix? Our answer to this question will affect all subsequent analyses requiring the terms-by-documents matrix as input.

The objective is to convert the POTUS text into a concise representation using features or words that are particularly useful in discriminating across documents. One way to do this is to find words that occur frequently in one document but infrequently in the document collection as a whole. That is, we can use a matrix with adjusted term frequencies in the cells. The usual choice is to employ the term frequency times the inverse of the document frequency (across the entire corpus). This technique is identified with the acronym *TF-IDF*. We will use this method going forward.

We bring the TF-IDF matrix for the top 200 words into distance metrics, finding the distance between all pairs of speeches. The resulting distance matrix is input into a multidimensional scaling algorithm. We choose a two-dimensional solution for this problem, obtaining the plot of the Presidents' speeches in figure 9.1. We see that there is a natural grouping of Presidents' speeches in space. Many Kennedy and Johnson speeches are close to one another. Obama's speeches are seen in close proximity to one another, and so on.

The multidimensional scaling solution gives us confidence that the TF-IDF matrix provides useful features for identifying the Presidents' speeches. But we are not at a stopping point yet. Recall our initial objective was to find common themes. For this we turn to cluster analysis. In particular, we seek a clustering of words rather than speeches.

We utilize statistical criteria to determine the number of clusters, identifying two clusters of words. Using distances to the center of each cluster, we identify eight words most closely associated with the first cluster and eight words most closely associated with the second.

The first cluster, which we call "Stay Strong," has these words closest to its center: "since," "keep," "far," "day," "strong," "protect," "high," and "important." The second cluster, which we call "Help Nation," has these words closest to its center: "make," "time," "help," "nation," "years," "work," "every," and "tonight." Alternatively, we could think of foreign defense and protection versus domestic policy focused on helping others and pro-

Figure 9.1. *Mapping the Presidents from Their Own Words*

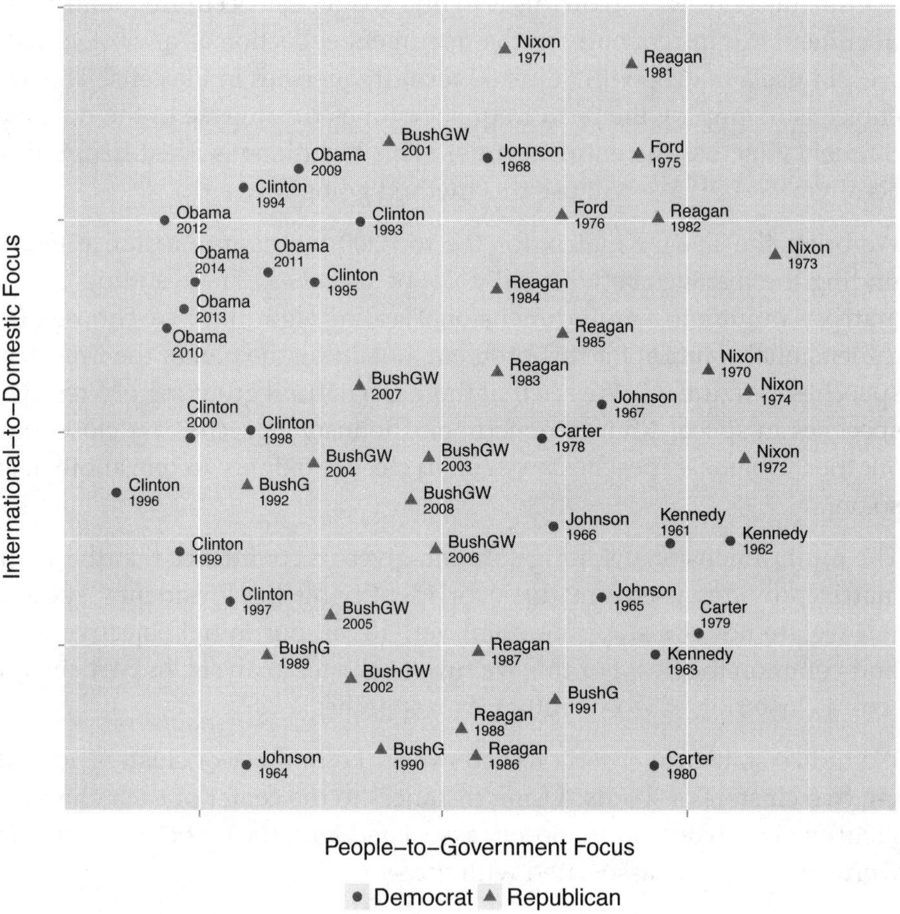

International–to–Domestic Focus

Nixon
1971

Reagan
1981

BushGW
2001

Johnson
1968

Ford
1975

Obama
2009

Clinton
1994

Obama
2012

Clinton
1993

Ford
1976

Reagan
1982

Nixon
1973

Obama
2014

Obama
2011

Obama
2013

Clinton
1995

Reagan
1984

Reagan
1985

Nixon
1970

Nixon
1974

Obama
2010

BushGW
2007

Reagan
1983

Johnson
1967

Clinton
2000

Clinton
1998

Carter
1978

Nixon
1972

Clinton
1996

BushG
1992

BushGW
2004

BushGW
2003

BushGW
2008

Johnson
1966

Kennedy
1961

Kennedy
1962

Clinton
1999

BushGW
2006

Clinton
1997

BushGW
2005

Johnson
1965

Carter
1979

BushG
1989

BushGW
2002

Reagan
1987

Kennedy
1963

BushG
1991

Johnson
1964

BushG
1990

Reagan
1988

Reagan
1986

Carter
1980

People–to–Government Focus

● Democrat ▲ Republican

viding employment. We could use these word lists to develop text measures for the individual speeches or for the Presidents. We could also observe the direction of these themes across the fifty-two years from Kennedy to Obama.

This example illustrates methods for working with large quantities of text and finding common themes. The methods apply to all types of text and content areas.

It would be a mistake to think that algorithms alone tell the story of text. Success in finding common themes is as much dependent on analyst judgment as on algorithms. We are the expert speakers of natural language, and we should not expect computers to replace us any time soon.

Political text analysis is a rich area of research, exemplified by Roderick P. Hart's (2000a, 2000b, 2001) work with political discourse across thirteen U. S. presidential campaigns from 1948 to 1996. Hart developed general text measures summarizing the tone of documents along dimensions called certainty, optimism, activity, realism, and commonality (shared values). Each measure involved counting words from selected word lists. Some words received positive weights, others negative weights. The optimism measure, for example, assigned positive weights to words relating to praise, satisfaction, and inspiration, while recording negative weights for words relating to blame, hardship, and denial. The realism measure drew on the concept of familiarity.

A text measure in its own right, familiarity was computed with reference to a dictionary of forty-four words that are regarded as the most common words in the English language. Reviewing the voices of three main groups (politicians, press, and public), Hart observed increasing complexity (lower familiarity scores) over time. Across the entire period of the study, the voice of the press was decidedly negative in tone (low in optimism), compared to politicians and the public.

Normalcy in speech, if not in policy, is rewarded in American politics. For political candidates, centrist speakers, those who speak in common parlance, are more successful than non-centrist candidates.

Modeling techniques for unsupervised text analytics include multivariate methods such as multidimensional scaling and cluster analysis, which are based upon dissimilarity or distance measures, and principal components analysis, which works from covariance or correlation matrices. Dissimilarity and distance measures and cluster analysis are discussed by Meyer (2014a, 2014b), Kaufman and Rousseeuw (1990), and Izenman (2008), with R code provided by Maechler (2014a). For multidimensional scaling, see Davison (1992), Cox and Cox (1994), Everitt and Rabe-Hesketh (1997), Izenman (2008) and Borg and Groenen (2010). Principal components are reviewed in various multivariate texts (Manly 1994; Sharma 1996; Gnanadesikan 1997; Johnson and Wichern 1998; Everitt and Dunn 2001; Seber 2000; Izenman 2008), with biplots discussed by Gabriel (1971) and Gower and Hand (1996). Latent semantic analysis and latent Dirichlet allocation offer additional tools for identifying common themes in text (Landauer, Foltz, and Laham 1998; Blei, Ng, and Jordan 2003; Murphy 2012; Ingersoll, Morton, and Farris 2013; Landauer, McNamara, Dennis, and Kintsch 2014).

Exhibit 9.1 shows the text analysis of *POTUS Speeches* data using Python. The program draws on numerous packages for text parsing, natural language processing, and multivariate methods.

Utilities for plotting word clouds and non-overlapping text in scatter plots are available in R (Fellows 2014a; Fellows 2014b). We use these utilities to provide pictures of text from the State of the Union addresses. These are shown in figures 9.2 through 9.11 (pages 178 through 182). Exhibit 9.2 lists the code used to produce these word clouds.

In many contexts, we regard word clouds as a mere diversion, quantitative art of sorts. But with the *POTUS Speeches*, word clouds help us to identify the meaning of dimensions in multidimensional scaling. In other words, we can use word clouds for the Presidents as a guide in labeling the axes of the map.

To label the horizontal axis, we look to word cloud differences between Clinton on the left-hand side and Nixon on the right-hand side. The vertical axis is more difficult to discern as it is not easily identified with individual Presidents. Some Presidents run the gamut from top to bottom. Notice how Presidents Johnson and George W. Bush are not identified with any area of the map. For the vertical axis, we can consider differences between Nixon at the top and Kennedy at the bottom or from Obama at the top to George Bush at the bottom. While the labeling of dimensions relies on researcher subjectivity, as much of text analytics in general, our subjective senses are more fully informed by word clouds.

The upshot of the process for *POTUS Speeches* has us labeling the horizontal dimension "People-to-Government Focus" and the vertical dimension "International-to-Domestic Focus." These axes do not line up with what we ordinarily consider to be liberal versus conservative political philosophies. Kennedy and Carter, usually regarded as liberals, are on right-hand side of the map along with Nixon, Ford, and Reagan.

More telling perhaps than an election night choropleth map of blue and red states, a map of the Presidents in their own words reveals common themes across fifty-two years of United States politics. That is the multidimensional scaling solution we provide in figure 9.1. Exhibit 9.2 lists the code used to produce this multidimensional scaling solution and the text map.

We start with text, a document collection. Then we identify salient words or features in text, assigning numbers to those features—a process some describe as *text annotation*. We analyze the numbers using techniques of predictive analytics, including multivariate methods. Finally, we provide reports of our findings in words and pictures. Text analytics takes us from words to numbers (text measures), and back to words. And sometimes we convert text measures into pictures—a process we call *text mapping*.

A curious reader might ask, "What does all this political research have to do with web and network data science?" Our answer is that it has a lot to do with web and network data science. The *POTUS Speeches* were a convenient public-domain source of text to demonstrate a process that we use extensively in working with web data.

We can employ the methods of this chapter to analyze any document collection, including web pages, blogs, microblogs, wikis, search queries and listings, research reports, and customer reviews. Methods of text analytics may also be employed in the analysis of qualitative research data from focus groups and online interviews, as described in appendix B. The data of the web are in large measure text, so it is most useful to understand how to analyze text and how to identify common themes.

Figure 9.2. Word Cloud for John F. Kennedy Speeches

Figure 9.3. Word Cloud for Lyndon B. Johnson Speeches

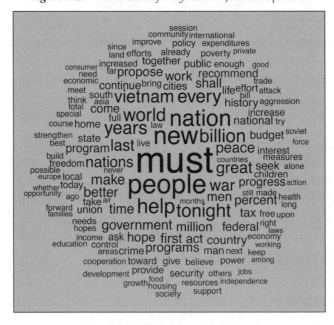

Figure 9.4. Word Cloud for Richard M. Nixon Speeches

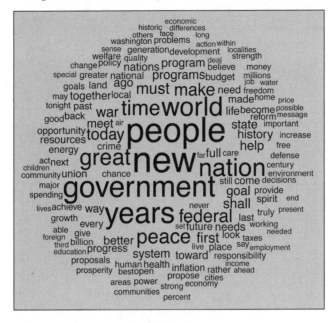

Figure 9.5. Word Cloud for Gerald R. Ford Speeches

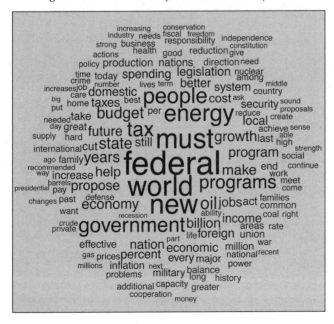

Figure 9.6. Word Cloud for Jimmy Carter Speeches

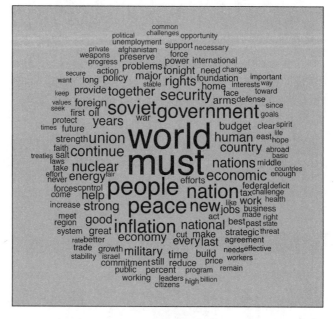

Figure 9.7. Word Cloud for Ronald Reagan Speeches

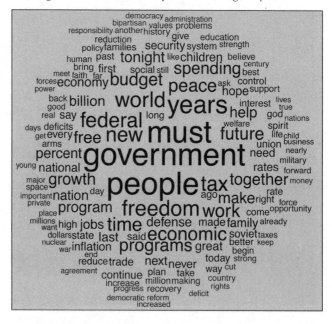

Figure 9.8. *Word Cloud for George Bush Speeches*

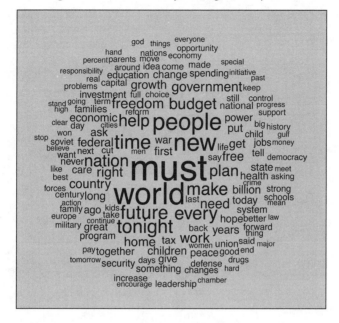

Figure 9.9. *Word Cloud for William J. Clinton Speeches*

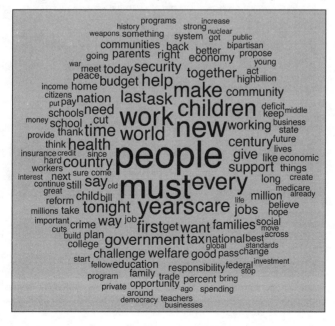

Figure 9.10. Word Cloud for George W. Bush Speeches

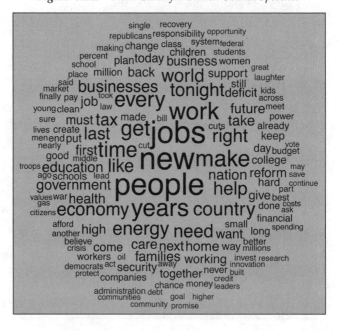

Figure 9.11. Word Cloud for Barack Obama Speeches

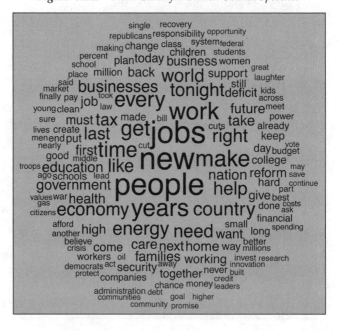

Exhibit 9.1. *Discovering Common Themes: POTUS Speeches (Python)*

```
# Discovering Common Themes: POTUS Speeches (Python)

# prepare for Python version 3x features and functions
from __future__ import division, print_function

# import packages for text processing and multivariate analysis
import os  # operating system functions
from fnmatch import fnmatch   # character string matching
import re  # regular expressions
import nltk # draw on the Python natural language toolkit
import pandas as pd  # DataFrame structure and operations
import numpy as np # arrays and numerical processing
import scipy
import matplotlib.pyplot as plt  # 2D plotting

# terms-by-documents matrix
from sklearn.feature_extraction.text import CountVectorizer
from sklearn.feature_extraction.text import TfidfTransformer

# alternative distance metrics for multidimensional scaling
from sklearn.metrics import euclidean_distances
from sklearn.metrics.pairwise import linear_kernel as cosine_distances
from sklearn.metrics.pairwise import manhattan_distances as manhattan_distances
from sklearn.metrics import silhouette_score as silhouette_score

from sklearn import manifold  # multidimensional scaling
from sklearn.cluster import KMeans # cluster analysis by partitioning

# function for walking and printing directory structure
def list_all(current_directory):
    for root, dirs, files in os.walk(current_directory):
        level = root.replace(current_directory, '').count(os.sep)
        indent = ' ' * 4 * (level)
        print('{}{}/'.format(indent, os.path.basename(root)))
        subindent = ' ' * 4 * (level + 1)
        for f in files:
            print('{}{}'.format(subindent, f))

# define list of codes to be dropped from documents
# carriage-returns, line-feeds, tabs
codelist = ['\r', '\n', '\t']

# we will drop the standard stop words
standard_stopwords = nltk.corpus.stopwords.words('english')
print(map(lambda t: t.encode('ascii'), standard_stopwords))
# special words associated with the occasion to be dropped from analysis
# along with the usual English stopwords
more_stopwords = ['speaker','president', 'mr', 'ms', 'mrs', 'th', 'house',\
    'representative', 'representatives', 'senate', 'senator',\
    'americans', 'american', 'america', \
    'one', 'two', 'three', 'four', 'five', 'six', 'seven', 'eight', 'nine',\
```

```
        'united', 'states', 'us', 'we', 'applause', 'ladies', 'gentlemen',\
        'congress', 'country' 'weve', 'youve', 'member', 'members',\
        'with', 'without', 'also', 'yet', 'half', 'also', 'many', 'see', 'said'\
        'years', 'year', 'even', 'ever', 'use', 'well', 'much', 'know', 'let', 'less']
print(map(lambda t: t.encode('ascii'), more_stopwords))

# start with the initial list and add to it for POTUS text work
stoplist = standard_stopwords + more_stopwords
print(map(lambda t: t.encode('ascii'), stoplist))

# employ regular expressions to parse documents
# here we are working with characters and character sequences
def text_parse(string):
    # replace non-alphanumeric with space
    temp_string = re.sub('[^a-zA-Z]', ' ', string)
    # replace codelist codes with space
    for i in range(len(codelist)):
        stopstring = ' ' + codelist[i] + ' '
        temp_string = re.sub(stopstring, ' ', temp_string)
    # replace single-character words with space
    temp_string = re.sub('\s.\s', ' ', temp_string)
    # convert uppercase to lowercase
    temp_string = temp_string.lower()
    # replace selected character strings/stop-words with space
    for i in range(len(stoplist)):
        stopstring = ' ' + str(stoplist[i]) + ' '
        temp_string = re.sub(stopstring, ' ', temp_string)
    # replace multiple blank characters with one blank character
    temp_string = re.sub('\s+', ' ', temp_string)
    return(temp_string)

# word stemming... looking for contractions... possessives...
# if we want to do stemming at a later time, we could use
#     porter = nltk.PorterStemmer()
# in a construction like this
#     words_stemmed =  [porter.stem(word) for word in initial_words]
# we examined stemming but found the results to be undesirable

# data for POTUS example are in one directory called POTUS
# all oral State of the Union addresses after Dwight D. Eisenhower
# 52 files (25 Democratic speeches, 27 Republican speeches)
# there is no scraping requirement for working with these text data
# but there are some comments with the text files, such as [Applause]
# because the speeches were delivered orally. Such comments need to
# be deleted from the files before additional text processing is done

# examine the directory structure to ensure POTUS is present
current_directory = os.getcwd()
list_all(current_directory)
# identify text file names of files/documents... should be 52 files
POTUS_file_names =\
    [name for name in os.listdir('POTUS') if fnmatch(name, '*.txt')]
print('\n\nNumber of files/documents:',len(POTUS_file_names))
```

```
# extract metadata from file names for labels on plots/reports
party_label = []  # initialize list
pres_label = []  # initialize list
year_label = []  # initialize list
for file_name in POTUS_file_names:
    file_name_less_extension = file_name.split('.')[0]
    party_label.append(file_name_less_extension.split('_')[0])
    pres_label.append(file_name_less_extension.split('_')[2])
    year_label.append(file_name_less_extension.split('_')[3])

# for numeric year in plots
year = []
for y in year_label: year.append(int(y))

# make working directory for a text corpus of parsed documents
# used for review of files and subsequent processing from files
os.mkdir('WORK_POTUS')

noutfiles = 0  # intialize count of working files for corpus
working_corpus = []  # initialize corpus with simple list structure
# work on files one at a time parsing and saving to new directory
for input_file_name in POTUS_file_names:
    # read in file
    this_file_name = input_file_name
    with open('POTUS/' + this_file_name, 'rt') as finput:
        text = finput.read()  # text string
        clean_text = text_parse(text)
        # output file name will be the same as input file name
        # but we store the file in the new directory WORK_POTUS
        output_file_name = "WORK_POTUS/" + input_file_name
        with open(output_file_name, 'wt') as foutput:
            foutput.write(str(clean_text))
            noutfiles = noutfiles + 1
            working_corpus.append(clean_text)  # build list-structured corpus
print('\nParsing complete: ', str(noutfiles) + ' files written to WORK_POTUS')
print('\nworking_corpus list: ', str(len(working_corpus)) + ' items')

# terms-by-documents matrix method to be employed
# with various numbers of top words
# check on top 200 words
tdm_method = CountVectorizer(max_features = 200, binary = True)
# employ simple term frequency
tdm_method = CountVectorizer(max_features = 200, binary = True)
examine_POTUS_tdm = tdm_method.fit(working_corpus)
top_200_words = examine_POTUS_tdm.get_feature_names()
# get clean printing of the top words
print('\nTop 200 words in POTUS corpus\n')
print(map(lambda t: t.encode('ascii'), top_200_words))  # print sans unicode

# check on top 100 words
tdm_method = CountVectorizer(max_features = 100, binary = True)
examine_POTUS_tdm = tdm_method.fit(working_corpus)
```

```
top_100_words = examine_POTUS_tdm.get_feature_names()
# get clean printing of the top words
print('\nTop 100 words in POTUS corpus\n')
print(map(lambda t: t.encode('ascii'), top_100_words))  # print sans unicode


# check on top 50 words
tdm_method = CountVectorizer(max_features = 50, binary = True)
examine_POTUS_tdm = tdm_method.fit(working_corpus)
top_100_words = examine_POTUS_tdm.get_feature_names()
# get clean printing of the top words
print('\nTop 100 words in POTUS corpus\n')
print(map(lambda t: t.encode('ascii'), top_100_words))  # print sans unicode


# -----------------------------------
# Full solution with all words TF-IDF
# -----------------------------------
# use the term frequency inverse document frequency matrix
# . . . begin by computing simple term frequency
count_vectorizer = CountVectorizer(min_df = 1)
term_freq_matrix = count_vectorizer.fit_transform(working_corpus)
# print vocabulary without unicode symbols... big vocabulary commented out
# print(map(lambda t: t.encode('ascii'), count_vectorizer.vocabulary_))
# . . . then transform to term frequency times inverse document frequency
# To get TF-IDF weighted word vectors tf times idf
tfidf = TfidfTransformer()  # accept default settings
tfidf.fit(term_freq_matrix)
tfidf_matrix = tfidf.transform(term_freq_matrix)
print('Shape of term frequency/inverse document frequency matrix',\
    str(tfidf_matrix.shape))


# -----------------------------
# Try top 200 words TF-IDF
# -----------------------------
# use the term frequency inverse document frequency matrix
# . . . begin by computing simple term frequency
count_vectorizer = CountVectorizer(min_df = 1, max_features = 200)
term_freq_matrix = count_vectorizer.fit_transform(working_corpus)
# print the feature names without unicode symbols...
top_200_words = count_vectorizer.get_feature_names()
print(map(lambda t: t.encode('ascii'), top_200_words))


# . . . then transform to term frequency times inverse document frequency
# To get TF-IDF weighted word vectors tf times idf
tfidf = TfidfTransformer()  # accept default settings
tfidf.fit(term_freq_matrix)
tfidf_matrix = tfidf.transform(term_freq_matrix)
print('Shape of term frequency/inverse document frequency matrix',\
    str(tfidf_matrix.shape))
# for input to subsequent analyses we need to choose
# either the standard terms x documents frequency matrix
# or the term frequency/inverse document frequency matrix
# we choose the latter with 200 words
```

```
# -----------------------------
# Multidimensional Scaling
# -----------------------------
# dissimilarity measures and multidimensional scaling
# consider alternative pairwise distance metrics from sklearn modules
# euclidean_distances, cosine_distances, manhattan_distances (city-block)
# note that different metrics provide different solutions
POTUS_distance_matrix = euclidean_distances(tfidf_matrix)
# POTUS_distance_matrix = manhattan_distances(tfidf_matrix)
# POTUS_distance_matrix = cosine_distances(tfidf_matrix)

mds_method = manifold.MDS(n_components = 2, random_state = 9999,\
    dissimilarity = 'precomputed')
mds_fit = mds_method.fit(POTUS_distance_matrix)
mds_coordinates = mds_method.fit_transform(POTUS_distance_matrix)

# plot mds solution in two dimensions using party labels
# defined by multidimensional scaling
plt.figure()
plt.scatter(mds_coordinates[:,0],mds_coordinates[:,1],\
    facecolors = 'none', edgecolors = 'none')  # points in white (invisible)
labels = party_label
for label, x, y in zip(labels, mds_coordinates[:,0], mds_coordinates[:,1]):
    plt.annotate(label, (x,y), xycoords = 'data')
plt.xlabel('First Dimension')
plt.ylabel('Second Dimension')
plt.show()
plt.savefig('fig_text_mds_POTUS_party.pdf',
    bbox_inches = 'tight', dpi=None, facecolor='w', edgecolor='b',
    orientation='landscape', papertype=None, format=None,
    transparent=True, pad_inches=0.25, frameon=None)

# plot mds solution in two dimensions using President names as labels
# defined by multidimensional scaling
plt.figure()
plt.scatter(mds_coordinates[:,0],mds_coordinates[:,1],\
    facecolors = 'none', edgecolors = 'none')  # points in white (invisible)
labels = pres_label
for label, x, y in zip(labels, mds_coordinates[:,0], mds_coordinates[:,1]):
    plt.annotate(label, (x,y), xycoords = 'data')
plt.xlabel('First Dimension')
plt.ylabel('Second Dimension')
plt.show()
plt.savefig('fig_text_mds_POTUS_pres.pdf',
    bbox_inches = 'tight', dpi=None, facecolor='w', edgecolor='b',
    orientation='landscape', papertype=None, format=None,
    transparent=True, pad_inches=0.25, frameon=None)

# plot mds solution in two dimensions using years as labels
# defined by multidimensional scaling
plt.figure()
plt.scatter(mds_coordinates[:,0],mds_coordinates[:,1],\
    facecolors = 'none', edgecolors = 'none')  # points in white (invisible)
```

```
labels = year_label
for label, x, y in zip(labels, mds_coordinates[:,0], mds_coordinates[:,1]):
    plt.annotate(label, (x,y), xycoords = 'data')
plt.xlabel('First Dimension')
plt.ylabel('Second Dimension')
plt.show()
plt.savefig('fig_text_mds_POTUS_year.pdf',
    bbox_inches = 'tight', dpi=None, facecolor='w', edgecolor='b',
    orientation='landscape', papertype=None, format=None,
    transparent=True, pad_inches=0.25, frameon=None)

# ------------------------------
# Cluster Analysis
# ------------------------------
# investigate alternative numbers of clusters using silhouette score
silhouette_value = []
k = range(2,10)
for i in k:
    kmeans_model = KMeans(n_clusters = i, random_state = 9999).\
        fit(np.transpose(tfidf_matrix))
    labels = kmeans_model.labels_
    silhouette_value.append(silhouette_score(np.transpose(tfidf_matrix),\
        labels, metric = 'euclidean'))
# highest silhouette score is for two clusters

# classification of words into groups for further analysis
# use transpose of the terms-by-document matrix and cluster analysis
# try two clusters/groups of words

clustering_method = KMeans(n_clusters = 2, random_state = 9999)
clustering_solution = clustering_method.fit(np.transpose(tfidf_matrix))
cluster_membership = clustering_method.predict(np.transpose(tfidf_matrix))
word_distance_to_center = clustering_method.transform(np.transpose(tfidf_matrix))

# top words data frame for reporting k-means clustering results
top_words_data = {'word': top_200_words, 'cluster': cluster_membership,\
    'dist_to_0': word_distance_to_center[0:,0],\
    'dist_to_1': word_distance_to_center[0:,1]}
distance_name_list = ['dist_to_0','dist_to_1']
top_words_data_frame = pd.DataFrame(top_words_data)
for cluster in range(2):
    words_in_cluster =\
        top_words_data_frame[top_words_data_frame['cluster'] == cluster]
    sorted_data_frame =\
        top_words_data_frame.sort_index(by = distance_name_list[cluster],\
        ascending = True)
    print('\n Top Words in Cluster :',cluster,'------------------------------')
    print(sorted_data_frame[:8])  # top 8 words in each cluster

# ------------------------------------------------
# Cluster Analysis Results Suggest Common Themes
# ------------------------------------------------
```

```
# Top Words in Cluster : 0 -------------------------------
#     cluster  dist_to_0  dist_to_1        word
#         0    0.150253   0.862167         since
#         0    0.154762   0.789895         keep
#         0    0.160067   0.862156         far
#         0    0.160653   0.801106         day
#         0    0.168222   0.808046         strong
#         0    0.169240   0.843582         protect
#         0    0.171769   0.777582         high
#         0    0.173527   0.848002         important

# Top Words in Cluster : 1 -------------------------------
#     cluster  dist_to_0  dist_to_1        word
#         1    0.602360   0.369731         make
#         1    0.575850   0.425409         time
#         1    0.623849   0.446006         help
#         1    0.727506   0.446941         nation
#         1    0.875372   0.454196         years
#         1    0.645027   0.508289         work
#         1    0.679918   0.508682         every
#         1    0.650285   0.532537         tonight

# a two-cluster solution seems to make sense with words
# toward the center of each cluster fitting together
# let's use pairs of top words from each cluster to name the clusters
# cluster index 0: "Stay Strong"
# cluster index 1: "Help Nation"

# name the clusters in the top words data frame
cluster_to_name = {0:'Stay Strong',1:'Help Nation'}
top_words_data_frame['cluster_name'] =\
    top_words_data_frame['cluster'].map(cluster_to_name)


# ------------------------------------------------
# Output Results of MDS and Cluster Analysis
# ------------------------------------------------

# write multidimensional scaling solution to comma-delimited text file
mds_data = {'party': party_label, 'pres': pres_label, 'year': year_label,\
    'first_dimension': list(mds_coordinates[:,0]),\
    'second_dimension': list(mds_coordinates[:,1])}
mds_data_frame = pd.DataFrame(mds_data)
mds_data_frame.to_csv('POTUS_mds.csv')

# write cluster analysis solution to comma-delimited text file
top_words_data_frame.to_csv('POTUS_top_words_clustering.csv')


# ------------------------------------------------
# Create Aggregate Text Files for Presidents
# ------------------------------------------------
# aggregate State of the Union addresses for each of
# ten presidents formed by simple concatenation
# working from the WORK_POTUS directory of parsed text
```

```
Kennedy = ''  # initialize aggregate text string
Johnson = ''
Nixon = ''
Ford = ''
Carter = ''
Reagan = ''
BushG = ''
Clinton = ''
BushGW = ''
Obama = ''

for input_file_name in POTUS_file_names:
    # read in file
    this_file_name = input_file_name
    with open('WORK_POTUS/' + this_file_name, 'rt') as finput:
        text = finput.read()  # text string
        file_name_less_extension = input_file_name.split('.')[0]
        pres = file_name_less_extension.split('_')[2]  # President
        # append to the appropriate President string with ' ' separator
        if pres == 'Kennedy':
            Kennedy = Kennedy + ' ' + text
        elif pres == 'Johnson':
            Johnson = Johnson + ' ' + text
        elif pres == 'Nixon':
            Nixon = Nixon + ' ' + text
        elif pres == 'Ford':
            Ford = Ford + ' ' + text
        elif pres == 'Carter':
            Carter = Carter + ' ' + text
        elif pres == 'Reagan':
            Reagan = Reagan + ' ' + text
        elif pres == 'BushG':
            BushG = BushG + ' ' + text
        elif pres == 'Clinton':
            Clinton = Clinton + ' ' + text
        elif pres == 'BushGW':
            BushGW = BushGW + ' ' + text
        elif pres == 'Obama':
            Obama = Obama + ' ' + text
        else:
            print('\n\nError in processing file:',this_file_name,'\n\n')

# store the strings as text files in a new directory ALL_POTUS
os.mkdir('ALL_POTUS')
with open('ALL_POTUS/Kennedy.txt', 'wt') as foutput:
    foutput.write(str(Kennedy))
with open('ALL_POTUS/Johnson.txt', 'wt') as foutput:
    foutput.write(str(Johnson))
with open('ALL_POTUS/Nixon.txt', 'wt') as foutput:
    foutput.write(str(Nixon))
with open('ALL_POTUS/Ford.txt', 'wt') as foutput:
    foutput.write(str(Ford))
```

```
with open('ALL_POTUS/Carter.txt', 'wt') as foutput:
    foutput.write(str(Carter))
with open('ALL_POTUS/Reagan.txt', 'wt') as foutput:
    foutput.write(str(Reagan))
with open('ALL_POTUS/BushG.txt', 'wt') as foutput:
    foutput.write(str(BushG))
with open('ALL_POTUS/Clinton.txt', 'wt') as foutput:
    foutput.write(str(Clinton))
with open('ALL_POTUS/BushGW.txt', 'wt') as foutput:
    foutput.write(str(BushGW))
with open('ALL_POTUS/Obama.txt', 'wt') as foutput:
    foutput.write(str(Obama))

# Suggestions for the student: Use word clusters to define text
# measures that vary across addresses, Presidents, and years.
# Try word stemming prior to the definition of a
# terms-by-documents matrix. Try longer lists of words
# for the identified clusters. Try alternative numbers
# of top words or alternative numbers of clusters.
# Repeat the multidimensional scaling and cluster analysis
# using the aggregate text files for the ten Presidents.
# Try other methods for identifying common themes, such as
# latent semantic analysis or latent Dirichlet allocation.
```

Exhibit 9.2. *Making Word Clouds: POTUS Speeches (R)*

```r
# Making Word Clouds: POTUS Speeches (R)
# install R wordcloud package
library(wordcloud)


# -----------------------------------
# word cloud for John F. Kennedy
# -----------------------------------
Kennedy.text <- scan("ALL_POTUS/Kennedy.txt", what = "char", sep = "\n")
# replace uppercase with lowercase letters
Kennedy.text <- tolower(Kennedy.text)
# strip out all non-letters and return vector
Kennedy.text.preword.vector <- unlist(strsplit(Kennedy.text, "\\W"))
# drop all empty words
Kennedy.text.vector <-
 Kennedy.text.preword.vector[which(nchar(Kennedy.text.preword.vector) > 0)]
pdf(file = "fig_text_wordcloud_of_Kennedy_speeches.pdf", width = 8.5, height = 8.5)
set.seed(1234)
wordcloud(Kennedy.text.vector, min.freq = 5,
  max.words = 150,
  random.order=FALSE,
  random.color=FALSE,
  rot.per=0.0,
  colors="black",
  ordered.colors=FALSE,
  use.r.layout=FALSE,
  fixed.asp=TRUE)
dev.off()


# -----------------------------------
# word cloud for Lyndon B. Johnson
# -----------------------------------
Johnson.text <- scan("ALL_POTUS/Johnson.txt", what = "char", sep = "\n")
# replace uppercase with lowercase letters
Johnson.text <- tolower(Johnson.text)
# strip out all non-letters and return vector
Johnson.text.preword.vector <- unlist(strsplit(Johnson.text, "\\W"))
# drop all empty words
Johnson.text.vector <-
 Johnson.text.preword.vector[which(nchar(Johnson.text.preword.vector) > 0)]
pdf(file = "fig_text_wordcloud_of_Johnson_speeches.pdf", width = 8.5, height = 8.5)
set.seed(1234)
wordcloud(Johnson.text.vector, min.freq = 5,
  max.words = 150,
  random.order=FALSE,
  random.color=FALSE,
  rot.per=0.0,
  colors="black",
  ordered.colors=FALSE,
  use.r.layout=FALSE,
  fixed.asp=TRUE)
dev.off()
```

```
# -----------------------------------
# word cloud for Richard M Nixon
# -----------------------------------
Nixon.text <- scan("ALL_POTUS/Nixon.txt", what = "char", sep = "\n")
# replace uppercase with lowercase letters
Nixon.text <- tolower(Nixon.text)
# strip out all non-letters and return vector
Nixon.text.preword.vector <- unlist(strsplit(Nixon.text, "\\W"))
# drop all empty words
Nixon.text.vector <-
 Nixon.text.preword.vector[which(nchar(Nixon.text.preword.vector) > 0)]
pdf(file = "fig_text_wordcloud_of_Nixon_speeches.pdf", width = 8.5, height = 8.5)
set.seed(1234)
wordcloud(Nixon.text.vector, min.freq = 5,
  max.words = 150,
  random.order=FALSE,
  random.color=FALSE,
  rot.per=0.0,
  colors="black",
  ordered.colors=FALSE,
  use.r.layout=FALSE,
  fixed.asp=TRUE)
dev.off()


# -----------------------------------
# word cloud for Gerald R. Ford
# -----------------------------------
Ford.text <- scan("ALL_POTUS/Ford.txt", what = "char", sep = "\n")
# replace uppercase with lowercase letters
Ford.text <- tolower(Ford.text)
# strip out all non-letters and return vector
Ford.text.preword.vector <- unlist(strsplit(Ford.text, "\\W"))
# drop all empty words
Ford.text.vector <-
 Ford.text.preword.vector[which(nchar(Ford.text.preword.vector) > 0)]
pdf(file = "fig_text_wordcloud_of_Ford_speeches.pdf", width = 8.5, height = 8.5)
set.seed(1234)
wordcloud(Ford.text.vector, min.freq = 5,
  max.words = 150,
  random.order=FALSE,
  random.color=FALSE,
  rot.per=0.0,
  colors="black",
  ordered.colors=FALSE,
  use.r.layout=FALSE,
  fixed.asp=TRUE)
dev.off()


# -----------------------------------
# word cloud for Jimmy Carter
# -----------------------------------
Carter.text <- scan("ALL_POTUS/Carter.txt", what = "char", sep = "\n")
# replace uppercase with lowercase letters
```

```
Carter.text <- tolower(Carter.text)
# strip out all non-letters and return vector
Carter.text.preword.vector <- unlist(strsplit(Carter.text, "\\W"))
# drop all empty words
Carter.text.vector <-
 Carter.text.preword.vector[which(nchar(Carter.text.preword.vector) > 0)]
pdf(file = "fig_text_wordcloud_of_Carter_speeches.pdf", width = 8.5, height = 8.5)
set.seed(1234)
wordcloud(Carter.text.vector, min.freq = 5,
  max.words = 150,
  random.order=FALSE,
  random.color=FALSE,
  rot.per=0.0,
  colors="black",
  ordered.colors=FALSE,
  use.r.layout=FALSE,
  fixed.asp=TRUE)
dev.off()


# -----------------------------------
# word cloud for Ronald Reagan
# -----------------------------------
Reagan.text <- scan("ALL_POTUS/Reagan.txt", what = "char", sep = "\n")
# replace uppercase with lowercase letters
Reagan.text <- tolower(Reagan.text)
# strip out all non-letters and return vector
Reagan.text.preword.vector <- unlist(strsplit(Reagan.text, "\\W"))
# drop all empty words
Reagan.text.vector <-
 Reagan.text.preword.vector[which(nchar(Reagan.text.preword.vector) > 0)]
pdf(file = "fig_text_wordcloud_of_Reagan_speeches.pdf", width = 8.5, height = 8.5)
set.seed(1234)
wordcloud(Reagan.text.vector, min.freq = 5,
  max.words = 150,
  random.order=FALSE,
  random.color=FALSE,
  rot.per=0.0,
  colors="black",
  ordered.colors=FALSE,
  use.r.layout=FALSE,
  fixed.asp=TRUE)
dev.off()


# -----------------------------------
# word cloud for George Bush
# -----------------------------------
BushG.text <- scan("ALL_POTUS/BushG.txt", what = "char", sep = "\n")
# replace uppercase with lowercase letters
BushG.text <- tolower(BushG.text)
# strip out all non-letters and return vector
BushG.text.preword.vector <- unlist(strsplit(BushG.text, "\\W"))
# drop all empty words
```

```
BushG.text.vector <-
 BushG.text.preword.vector[which(nchar(BushG.text.preword.vector) > 0)]
pdf(file = "fig_text_wordcloud_of_BushG_speeches.pdf", width = 8.5, height = 8.5)
set.seed(1234)
wordcloud(BushG.text.vector, min.freq = 5,
  max.words = 150,
  random.order=FALSE,
  random.color=FALSE,
  rot.per=0.0,
  colors="black",
  ordered.colors=FALSE,
  use.r.layout=FALSE,
  fixed.asp=TRUE)
dev.off()


# -----------------------------------
# word cloud for William J. Clinton
# -----------------------------------
Clinton.text <- scan("ALL_POTUS/Clinton.txt", what = "char", sep = "\n")
# replace uppercase with lowercase letters
Clinton.text <- tolower(Clinton.text)
# strip out all non-letters and return vector
Clinton.text.preword.vector <- unlist(strsplit(Clinton.text, "\\W"))
# drop all empty words
Clinton.text.vector <-
 Clinton.text.preword.vector[which(nchar(Clinton.text.preword.vector) > 0)]
pdf(file = "fig_text_wordcloud_of_Clinton_speeches.pdf", width = 8.5, height = 8.5)
set.seed(1234)
wordcloud(Clinton.text.vector, min.freq = 5,
  max.words = 150,
  random.order=FALSE,
  random.color=FALSE,
  rot.per=0.0,
  colors="black",
  ordered.colors=FALSE,
  use.r.layout=FALSE,
  fixed.asp=TRUE)
dev.off()


# -----------------------------------
# word cloud for George W. Bush
# -----------------------------------
BushGW.text <- scan("ALL_POTUS/BushGW.txt", what = "char", sep = "\n")
# replace uppercase with lowercase letters
BushGW.text <- tolower(BushGW.text)
# strip out all non-letters and return vector
BushGW.text.preword.vector <- unlist(strsplit(BushGW.text, "\\W"))
# drop all empty words
BushGW.text.vector <-
 BushGW.text.preword.vector[which(nchar(BushGW.text.preword.vector) > 0)]
pdf(file = "fig_text_wordcloud_of_BushGW_speeches.pdf", width = 8.5, height = 8.5)
set.seed(1234)
```

```
wordcloud(BushGW.text.vector, min.freq = 5,
  max.words = 150,
  random.order=FALSE,
  random.color=FALSE,
  rot.per=0.0,
  colors="black",
  ordered.colors=FALSE,
  use.r.layout=FALSE,
  fixed.asp=TRUE)
dev.off()

# -----------------------------------
# word cloud for Barack Obama
# -----------------------------------
Obama.text <- scan("ALL_POTUS/Obama.txt", what = "char", sep = "\n")
# replace uppercase with lowercase letters
Obama.text <- tolower(Obama.text)
# strip out all non-letters and return vector
Obama.text.preword.vector <- unlist(strsplit(Obama.text, "\\W"))
# drop all empty words
Obama.text.vector <-
 Obama.text.preword.vector[which(nchar(Obama.text.preword.vector) > 0)]
pdf(file = "fig_text_wordcloud_of_Obama_speeches.pdf", width = 8.5, height = 8.5)
set.seed(1234)
wordcloud(Obama.text.vector, min.freq = 5,
  max.words = 150,
  random.order=FALSE,
  random.color=FALSE,
  rot.per=0.0,
  colors="black",
  ordered.colors=FALSE,
  use.r.layout=FALSE,
  fixed.asp=TRUE)
dev.off()
```

Exhibit 9.3. *From Text Measures to Text Maps: POTUS Speeches (R)*

```
# From Text Measures to Text Maps: POTUS Speeches (R)

# install R wordcloud package

library(ggplot2)  # grammar of graphics plotting system

# --------------------
# Presidents Studied
# --------------------
# John F. Kennedy
# Lyndon B. Johnson
# Richard M Nixon
# Gerald R. Ford
# Jimmy Carter
# Ronald Reagan
# George Bush
# William J. Clinton
# George W. Bush
# Barack Obama
# --------------------

# read in results from multidimensional scaling analysis
mds_data_frame <- read.csv("POTUS_mds.csv", stringsAsFactors = FALSE)
print(str(mds_data_frame))

mds_data_frame$year_label <- as.character(mds_data_frame$year)
mds_data_frame$party_label <- rep("", length = nrow(mds_data_frame))
for (i in seq(along = mds_data_frame$party)) {
    if(mds_data_frame$party[i] == "D")
        mds_data_frame$party_label[i] <- "Democrat"
    if(mds_data_frame$party[i] == "R")
        mds_data_frame$party_label[i] <- "Republican"
    }

# direct manipulation of text position to avoid overlapping text
mds_data_frame$pres_x <- mds_data_frame$first_dimension + 0.025
mds_data_frame$pres_y <- mds_data_frame$second_dimension + 0.015
mds_data_frame$year_x <- mds_data_frame$first_dimension + 0.025
mds_data_frame$year_y <- mds_data_frame$second_dimension - 0.015

# Carter 1979 up and to the left
mds_data_frame$pres_x[2] <- mds_data_frame$first_dimension[2] + 0.025 -0.0225
mds_data_frame$pres_y[2] <- mds_data_frame$second_dimension[2] + 0.015 + 0.035
mds_data_frame$year_x[2] <- mds_data_frame$first_dimension[2] + 0.025 -0.0225
mds_data_frame$year_y[2] <- mds_data_frame$second_dimension[2] - 0.015 + 0.035

# Clinton 1994 move down
mds_data_frame$pres_x[5] <- mds_data_frame$first_dimension[5] + 0.025
mds_data_frame$pres_y[5] <- mds_data_frame$second_dimension[5] + 0.015 - 0.0125
mds_data_frame$year_x[5] <- mds_data_frame$first_dimension[5] + 0.025
mds_data_frame$year_y[5] <- mds_data_frame$second_dimension[5] - 0.015 - 0.0125
```

```
# Clinton 2000 up and to the left
mds_data_frame$pres_x[11] <- mds_data_frame$first_dimension[11] + 0.025 -0.0325
mds_data_frame$pres_y[11] <- mds_data_frame$second_dimension[11] + 0.015 + 0.045
mds_data_frame$year_x[11] <- mds_data_frame$first_dimension[11] + 0.025 -0.0325
mds_data_frame$year_y[11] <- mds_data_frame$second_dimension[11] - 0.015 + 0.045

# Kennedy 1961 move up and to the left
mds_data_frame$pres_x[17] <- mds_data_frame$first_dimension[17] + 0.025 -0.0455
mds_data_frame$pres_y[17] <- mds_data_frame$second_dimension[17] + 0.015 + 0.04
mds_data_frame$year_x[17] <- mds_data_frame$first_dimension[17] + 0.025 - 0.0455
mds_data_frame$year_y[17] <- mds_data_frame$second_dimension[17] - 0.015 + 0.04

# Kennedy 1963 move down
mds_data_frame$pres_x[19] <- mds_data_frame$first_dimension[19] + 0.025
mds_data_frame$pres_y[19] <- mds_data_frame$second_dimension[19] + 0.015 - 0.0125
mds_data_frame$year_x[19] <- mds_data_frame$first_dimension[19] + 0.025
mds_data_frame$year_y[19] <- mds_data_frame$second_dimension[19] - 0.015 - 0.0125

# Obama 2009 move up
mds_data_frame$pres_x[20] <- mds_data_frame$first_dimension[20] + 0.025
mds_data_frame$pres_y[20] <- mds_data_frame$second_dimension[20] + 0.015 + 0.0125
mds_data_frame$year_x[20] <- mds_data_frame$first_dimension[20] + 0.025
mds_data_frame$year_y[20] <- mds_data_frame$second_dimension[20] - 0.015 + 0.0125

# Obama 2010 move down and to the left
mds_data_frame$pres_x[21] <- mds_data_frame$first_dimension[21] + 0.025 -0.0125
mds_data_frame$pres_y[21] <- mds_data_frame$second_dimension[21] + 0.015 - 0.03
mds_data_frame$year_x[21] <- mds_data_frame$first_dimension[21] + 0.025 -0.0125
mds_data_frame$year_y[21] <- mds_data_frame$second_dimension[21] - 0.015 - 0.03

# Obama 2011 move up and to the left
mds_data_frame$pres_x[22] <- mds_data_frame$first_dimension[22] + 0.025 -0.0125
mds_data_frame$pres_y[22] <- mds_data_frame$second_dimension[22] + 0.015 + 0.0355
mds_data_frame$year_x[22] <- mds_data_frame$first_dimension[22] + 0.025 -0.0125
mds_data_frame$year_y[22] <- mds_data_frame$second_dimension[22] - 0.015 + 0.0355

# Obama 2014 move up and to the left
mds_data_frame$pres_x[25] <- mds_data_frame$first_dimension[25] + 0.025 -0.0255
mds_data_frame$pres_y[25] <- mds_data_frame$second_dimension[25] + 0.015 + 0.0415
mds_data_frame$year_x[25] <- mds_data_frame$first_dimension[25] + 0.025 - 0.0255
mds_data_frame$year_y[25] <- mds_data_frame$second_dimension[25] - 0.015 + 0.0415

# BushG 1990 move down
mds_data_frame$pres_x[27] <- mds_data_frame$first_dimension[27] + 0.025
mds_data_frame$pres_y[27] <- mds_data_frame$second_dimension[27] + 0.015 - 0.0125
mds_data_frame$year_x[27] <- mds_data_frame$first_dimension[27] + 0.025
mds_data_frame$year_y[27] <- mds_data_frame$second_dimension[27] - 0.015 - 0.0125

# BushG 1992 move down
mds_data_frame$pres_x[29] <- mds_data_frame$first_dimension[29] + 0.025
mds_data_frame$pres_y[29] <- mds_data_frame$second_dimension[29] + 0.015 - 0.0125
mds_data_frame$year_x[29] <- mds_data_frame$first_dimension[29] + 0.025
mds_data_frame$year_y[29] <- mds_data_frame$second_dimension[29] - 0.015 - 0.0125
```

```
# Nixon 1970 move up
mds_data_frame$pres_x[40] <- mds_data_frame$first_dimension[40] + 0.025
mds_data_frame$pres_y[40] <- mds_data_frame$second_dimension[40] + 0.015 + 0.0125
mds_data_frame$year_x[40] <- mds_data_frame$first_dimension[40] + 0.025
mds_data_frame$year_y[40] <- mds_data_frame$second_dimension[40] - 0.015 + 0.0125

# Reagan 1988 move up
mds_data_frame$pres_x[52] <- mds_data_frame$first_dimension[52] + 0.025
mds_data_frame$pres_y[52] <- mds_data_frame$second_dimension[52] + 0.015 + 0.0125
mds_data_frame$year_x[52] <- mds_data_frame$first_dimension[52] + 0.025
mds_data_frame$year_y[52] <- mds_data_frame$second_dimension[52] - 0.015 + 0.0125

pdf(file = "fig_mds_solution_POTUS_Speeches.pdf", width = 8.5, height = 8.5)
ggplot_object <- ggplot(data = mds_data_frame,
    aes(x = first_dimension, y = second_dimension,
        shape = party_label, colour = party_label)) +
    geom_point(size = 3) +
    scale_colour_manual(values =
        c(Democrat = "darkblue", Republican = "darkred")) +
    geom_text(aes(x = pres_x , y = pres_y, label = pres),
        size = 4, hjust = 0, colour = "black") +
    geom_text(aes(x = year_x , y = year_y, label = year_label),
        size = 3, hjust = 0, colour = "black") +
    xlim(-0.70, 0.85) +
    xlab("People-to-Government Focus") +
    ylab("International-to-Domestic Focus") +
    theme(axis.title.x = element_text(size = 15, colour = "black")) +
    theme(axis.title.y = element_text(size = 15, colour = "black")) +
    theme(axis.text.x = element_text(colour = "white")) +
    theme(axis.text.y = element_text(colour = "white")) +
    theme(legend.position = "bottom") +
    theme(legend.title = element_text(size = 0.000001)) +
    theme(legend.text = element_text(size = 15))
print(ggplot_object)
dev.off()
```

10

Making Recommendations

"Who are those guys?"

—PAUL NEWMAN AS BUTCH CASSIDY IN
Butch Cassidy and the Sundance Kid (1969)

Being a frequent customer of Amazon.com, I receive many suggestions for book purchases. In May 2013, before the release of the initial edition of my first *Modeling Techniques* book, Amazon sent me an e-mail list of ten recommended books, and the book I was writing at the time was at the top of the list. My publisher, being a good publisher, made the book available for pre-order, and Amazon was asking if I wanted to order it.

Who are those guys who seem to know so much about me? Amazon for books, Netflix movies, Pandora music, among others—they seem to know a lot about what I like without my telling them very much.

Recommender systems build on *sparse matrices*. A well known recommender system problem, the Netflix $1 million prize competition (May 29, 2006 though September 21, 2009), presented Netflix data for 17,770 movies and 480,189 customers. Any given customer rents at most a couple hundred movies, so if we were to place a one in each cell of a customers-by-movies matrix for rentals and a zero otherwise, the resulting matrix would have 99 percent zeroes—it would be 99 percent sparse.

Algorithms for recommender systems, or recommender engines as they are sometimes called, must be capable of dealing with sparse matrices. There are many possible algorithms. One class of recommender systems is *content-based* recommender systems. These draw on customer personal characteristics, past orders, and revealed preferences. Content-based systems may also rely on the characteristics of products.

Another class of recommender systems is *collaborative filtering* (also called social or group filtering), which builds on the premise that customers with similar ratings for one set of products will have similar ratings for other sets of products. To predict what music Janiya is going to like, find Janiya's nearest neighbors and see what music they like.

Suppose Brit has chosen ten movies and a movie service provider wants to suggest a new movie for Brit. One way to find that movie is to identify other customers who have chosen many of the same movies as Brit. Then, searching across those nearest-neighbor customers, the service provider detects movies they have in common. Movies most in common across these customers provide a basis for recommendations to Brit.

Interestingly, a competition offered by Kaggle asked analysts to develop a recommendation engine for users of R packages. Conway and White (2012) show how to build a nearest-neighbor recommender system using data from this competition. We note that data from the Kaggle R competition could have been supplemented with information from the R programming environment itself, with its "depends" and "reverse depends" links. We can represent any programming environment as a network, with one package, function, or module calling another. And using the network topology, we can identify nearest neighbors or nodes in close proximity to any given node.

Association rules modeling (also called affinity or market basket analysis) is another method for building recommender systems. Association rules modeling asks, *What goes with what? What products are ordered or purchased together? What activities go together? What website areas are viewed together?*

Association rules modeling is contingency table analysis on a very large scale, and our job is to determine which contingency tables to look at. A key challenge in association rule modeling is the sheer number of rules that are generated.

An *item set* is a collection of items selected from all items in the store. The size of an item set is the number of items in that set. Item sets may be composed of two items, three items, and so on. The number of distinct item sets is very large, even for the Microsoft case.

An *association rule* is a division of each item set into two subsets with one subset, the *antecedent*, thought of as preceding the other subset, the *consequent*. There are more association rules than there are item sets.[1] The Apriori algorithm of Agraval et al. (1996) deals with the large-number-of-rules problem by using selection criteria that reflect the potential utility of association rules.

The first criterion relates to the *support* or prevalence of an item set. Each item set is evaluated to determine the proportion of times it occurs in the store data set. If this proportion exceeds a minimum support threshold or criterion, then it is passed along to the next phase of analysis. A support criterion of 0.01 implies that one in every one hundred market baskets must contain the item set. A support criterion of 0.001 implies that one in every one thousand market baskets must contain the item set.

The second criterion relates to the *confidence* or predictability of an association rule. This is computed as the support of the item set divided by the support of the subset of items in the antecedent. This is an estimate of the conditional probability of the consequent, given the antecedent. In the selection of association rules, we set the confidence criterion much higher than the support criterion.[2] Support and confidence criteria are arbitrarily set by the researcher.

---

[1] For the Microsoft data set, with its $K = 294$ website areas and its corresponding binary data matrix, the number of distinct item sets will be

$$2^K = 2^{294} \approx 3.182868 \times 10^{88}$$

[2] With item subsets identified as A and B, there are two possible association rules: $(A \Rightarrow B)$ and $(B \Rightarrow A)$. We need to consider only one of these rules because we favor rules with higher confidence. Consider our confidence in the rule $(A \Rightarrow B)$. This is the conditional probability of B given A. Similarly, confidence in rule $(B \Rightarrow A)$ is the conditional probability of A given B:

$$P(B|A) = \frac{P(AB)}{P(A)} \qquad P(A|B) = \frac{P(AB)}{P(B)}$$

It follows that, if the item subset A has more support than the item subset B, then $P(A) > P(B)$ and $P(A|B) > P(B|A)$. Given our preference for rules of higher confidence, then, it is clear that the item subset with higher support will take the role of the consequent.

To demonstrate an association-rules-based system for making recommendations, we draw on the *Anonymous Microsoft Web Data* case (page 296). The case data include user identifiers and web area identifiers. Visitor choices reveal visitor preferences about the Microsoft site.

Consider the visitors-by-areas matrix for these data. If we were to place a one in each cell of a visitors-by-areas matrix for visits and a zero otherwise, the resulting matrix would have mostly zeroes—it would be a sparse binary matrix. Association rules algorithms are designed to work with such matrices.

Settings for support and confidence vary from one problem to another. For the Microsoft data set, we set the initial support criterion to 0.025 and make a plot showing website area frequencies for individual areas meeting this criterion. See figure 10.1 and the directory names and descriptions in table 10.1.

A set of 268 association rules may be obtained by setting thresholds for support and confidence of 0.01 and 0.025, respectively. Figure 10.2 provides a scatter plot of these rules with support on the horizontal axis and confidence on the vertical axis. Color coding of the points relates to *lift*, a measure of relative predictive confidence.[3]

Figure 10.3 provides another view of the identified association rules: a matrix bubble chart. An item in the antecedent subset or left-hand side (LHS) of an association rule provides the label at the top of the matrix, and an item in the consequent or right-hand side (RHS) of an association rule provides the label at the right of the matrix. Support relates to the size of each bubble and lift is reflected in color intensity, as shown in the color gradient legend.

---

[3] As with support and confidence, probability formulas define lift. Think of lift as the confidence we have in predicting the consequent B with the rule $(A \Rightarrow B)$, divided by the confidence we would have in predicting B without the rule. Without knowledge of A and the association rule $(A \Rightarrow B)$, our confidence in observing the item subset B is $P(B)$. With knowledge of A and the association rule $(A \Rightarrow B)$, our confidence in observing item subset B is $P(B|A)$, as we have defined earlier. The ratio of these quantities is lift:

$$\frac{P(B|A)}{P(B)} = \left(\frac{P(AB)}{P(A)}\right)\left(\frac{1}{P(B)}\right) = \frac{P(AB)}{P(A)P(B)}$$

Looking at the numerator and denominator of the ratio on the far right-hand side of this equation, we note that these are equivalent under an independence assumption. That is, if there is no relationship between item subsets A and B, then the joint probability of A and B is the product of their individual probabilities: $P(AB) = P(A)P(B)$. So lift is a measure of the degree to which item subsets in an association rule depart from independence.

Table 10.1. *Most Frequently Visited Website Areas and Descriptions*

Directory Name	Description
/gallery	Web Site Builders Gallery
/games	Games
/ie	Internet Explorer
/ie_intl	International Internet Explorer Content
/isapi	Internet Server Application Programming Interface
/kb	Knowledge Base
/msdn	Developer Network
/msdownload	Free Downloads
/ntserver	Windows NT Server
/office	Microsoft Office Information
/products	Products (a generic category)
/regwiz	Registration Wizard
/sbnmember	SiteBuilder Network Membership
/search	Microsoft.com Search
/sitebuilder	Internet Site Construction for Developers
/support	Support Desktop
/windows	Windows Family of Operating Systems
/windows95	Windows 95
/windowssupport	Windows95 Support
/word	Microsoft Word News
/workshop	Developer Workshop

Figure 10.1. *Most Frequently Visited Website Areas*

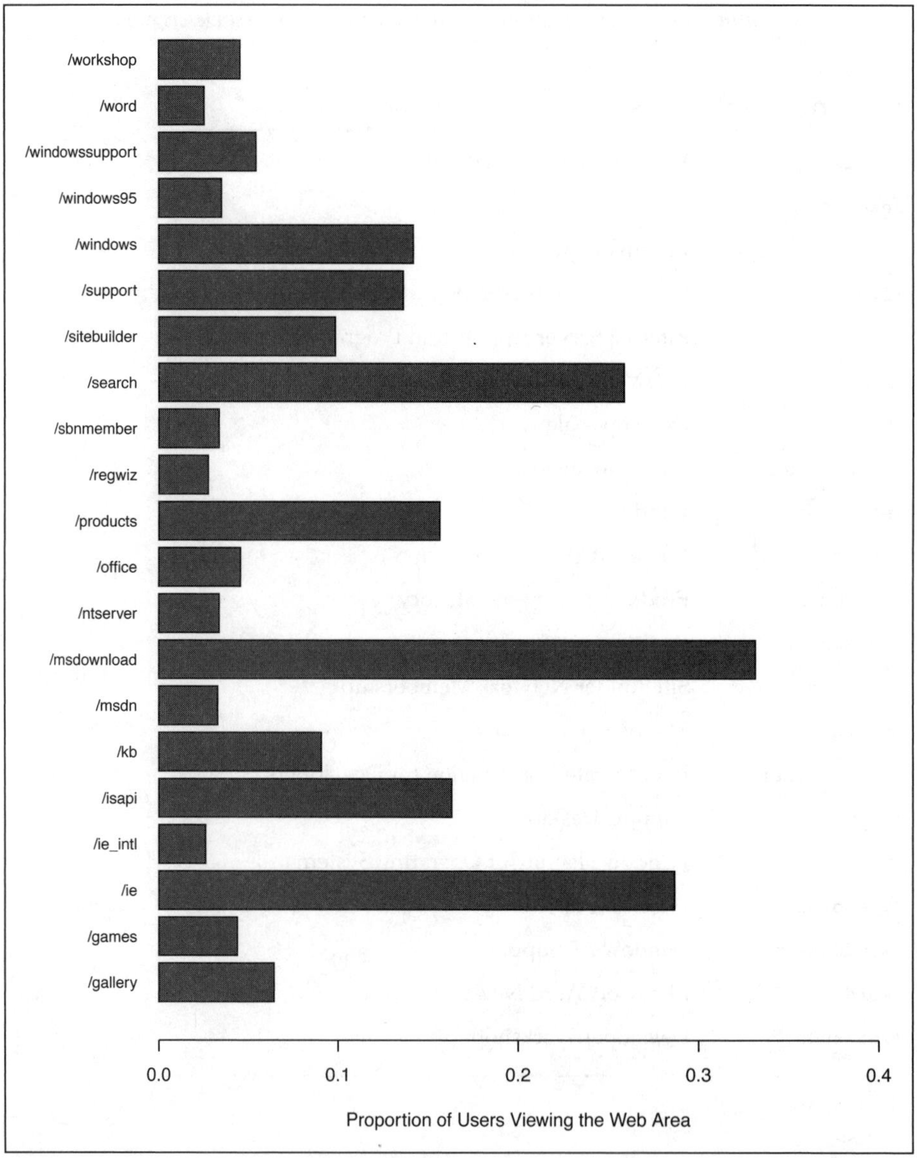

Figure 10.2. *Association Rule Support and Confidence*

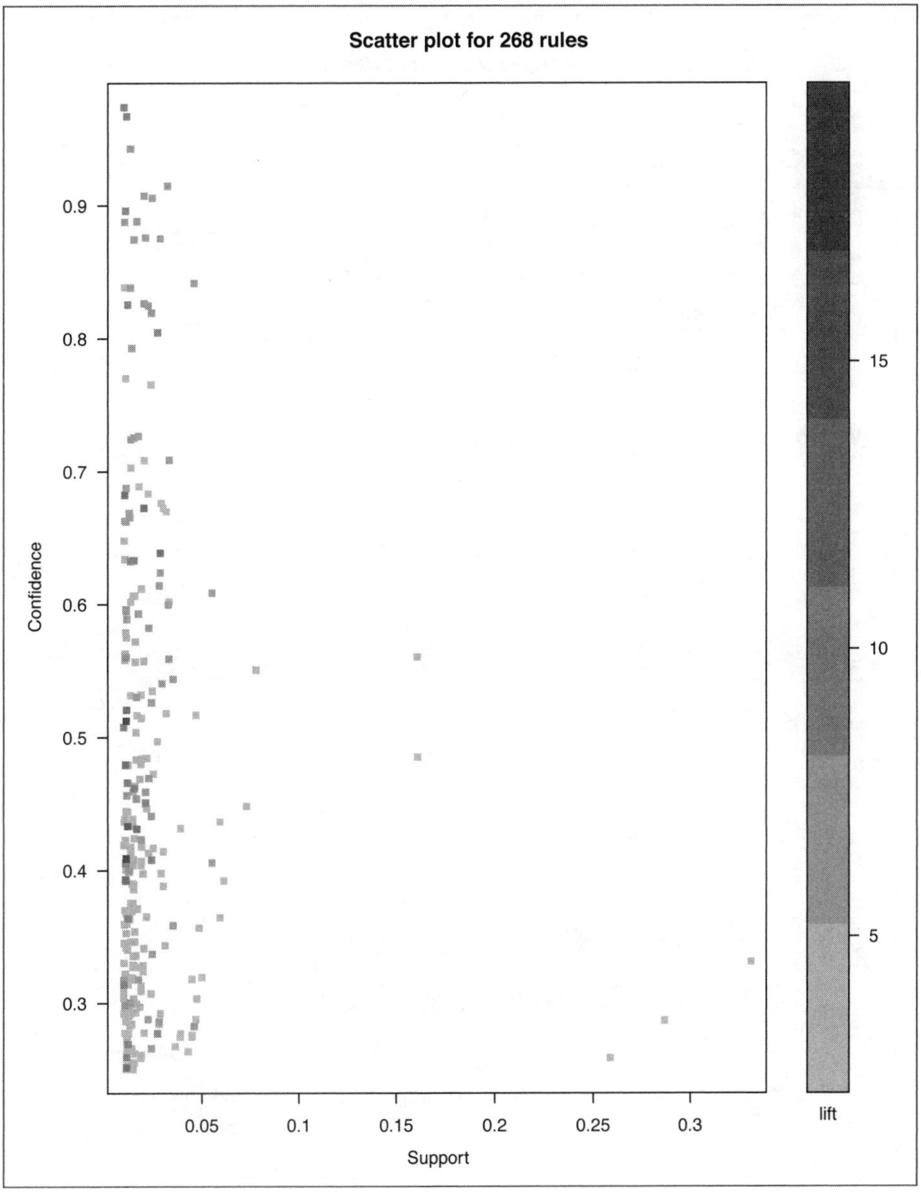

Figure 10.3. Association Rule Antecedents and Consequents

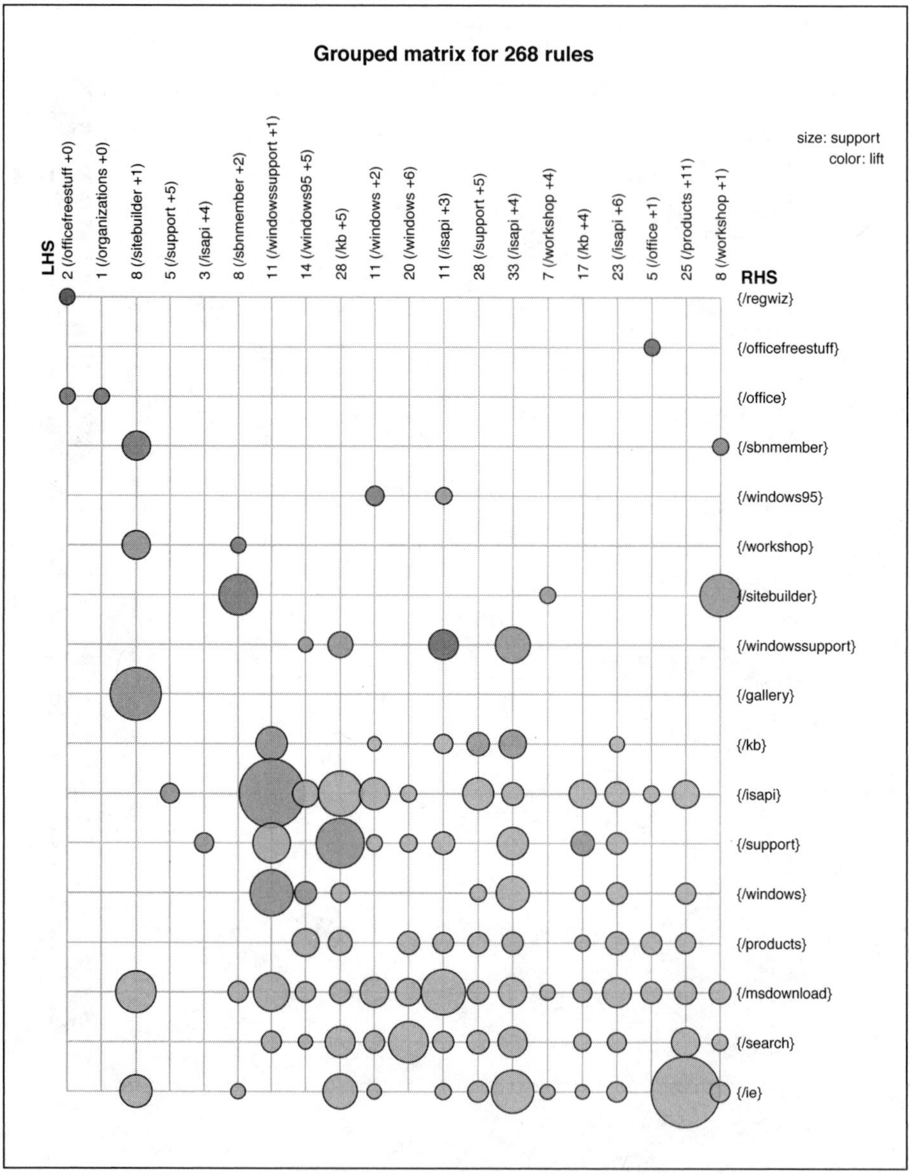

Reviewing the reported measures for association rules, support is prevalence. As a relative frequency or probability estimate it takes values from zero to one. Low values are tolerated, but values that are extremely low relative to other website areas may indicate a lack of importance to website visitors.

Confidence relates to the predictability of the consequent, given the antecedent. As a conditional probability, confidence takes values from zero to one. Higher values are preferred. Lift, being a ratio of nonzero probabilities, takes positive values on the real number line. Lift needs to be above 1.00 to be of use to management. The higher the lift of an association rule, the better.

When working on marketing or customer segmentation problems, it may be useful to employ transaction clustering. That is, we can use cluster analysis prior to association rule analysis to identify groups of website visitors who have similar website area choices. Then we conduct a separate analysis for each group of customers.

Finding the proverbial "needle in the haystack" could be the easy part, given the efficient algorithms at our disposal. But have we learned anything of value in the process? Do we know where to look in the next haystack? How do we turn association rules into recommender systems?

So far, what we have done with the Microsoft case is descriptive rather than predictive. Identified association rules provide a description of website visitor behavior or an analysis of the structure of the website. Some areas are heavily used, others less so. Now we want to see if the identified association rules provide value. Can we use the rules to predict what visitors will do next?

For the Microsoft case, prediction about what a visitor will do next is like a recommendation. The value of such recommendations is more easily understood in the context of web pages, because with web pages we have the potential of preparing for their delivery before visitor requests are made. That is, a recommendation system can use anticipated next requests from users to preload web pages and in this way improve website performance. Preloaded web pages on the server side are ready and waiting to be sent as soon as a request has been received. This was part of the motivation behind the original research study that led to the Microsoft case (Breese, Heckerman, and Kadie 1998).

For the Microsoft case, we have a hold-out test set for testing predictions from models. For the association rules model, we see that more than 36 percent of the predictions are correct. That is, if we know which area a website visitor has viewed, we will be correct in predicting the next area to be viewed 36 percent of the time.

Hastie, Tibshirani, and Friedman (2009) review the theory behind association rules, providing formal definitions for support, confidence (predictability), and lift (departure from independence), as well as a discussion of applications and the Apriori algorithm of Agraval et al. (1996). Additional discussion may be found in the machine learning literature (Tan, Steinbach, and Kumar 2006; Witten, Frank, and Hall 2011; Harrington 2012; Rajaraman and Ullman 2012). Bruzzese and Davino (2008) review data visualization of association rules.

A data scientist may have to look through a lot of rules before finding one that is interesting. It is not surprising, then, that considerable attention has been paid to methods for selecting association rules (Hahsler, Buchta, and Hornik 2008), selecting variables in association rules modeling (Dippold and Hruschka 2013), and combining segmentation and association rules modeling in the same analysis (Boztug and Reutterer 2008).

Discussion of association rules models in R is provided by Hahsler, Grün, and Hornik (2005), Hahsler et al. (2011), and Hahsler et al. (2014a, 2014b), with data visualization support by Hahsler and Chelluboina (2014a, 2014b). Additional association rule algorithms are available through the R interface to Weka (Hornik 2014a, 2014b). Special algorithms are available for working with sparse matrices (Bates and Maechler 2014). Reviews of recommender systems are provided in the edited volumes of Ricci et al. (2011) and Dehuri et al. (2012). R software is provided by Hahsler (2014a, 2014b).

The program for the worked example in this chapter is provided in exhibit 10.1. It draws on code developed by Hahsler et al. (2014a) and Hahsler and Chelluboina (2014a) that uses the Apriori algorithm for association rules. It shows one way to develop website area recommendations for the Microsoft case.

Exhibit 10.1.   *From Rules to Recommendations: The Microsoft Case (R)*

```
# From Rules to Recommendations: The Microsoft Case (R)

# install necessary packages prior to running

library(arules)  # association rules
library(arulesViz)  # data visualization of association rules
library(RColorBrewer)  # color palettes for plots

# set criteria for association rule modeling for this analysis
# set support and confidence settings for input to arules package

low_support_setting <- 0.01
high_support_setting <- 0.025
confidence_setting <- 0.25


# ---------------------------------------------------------
# carmap is modified recode function from car package
# to avoid conflict with recode function from arules
# car package from Fox (2014) and Fox and Weisberg (2011)
# ---------------------------------------------------------

carmap <- function (var, recodes, as.factor.result = FALSE,
    as.numeric.result = FALSE, levels)
{
    lo <- -Inf
    hi <- Inf
    recodes <- gsub("\n|\t", " ", recodes)
    recode.list <- rev(strsplit(recodes, ";")[[1]])
    is.fac <- is.factor(var)
    if (missing(as.factor.result))
        as.factor.result <- is.fac
    if (is.fac)
        var <- as.character(var)
    result <- var

    for (term in recode.list) {
        if (0 < length(grep(":", term))) {
            range <- strsplit(strsplit(term, "=")[[1]][1], ":")
            low <- eval(parse(text = range[[1]][1]))
            high <- eval(parse(text = range[[1]][2]))
            target <- eval(parse(text = strsplit(term, "=")[[1]][2]))
            result[(var >= low) & (var <= high)] <- target
        }
        else if (0 < length(grep("^else=", term))) {
            target <- eval(parse(text = strsplit(term, "=")[[1]][2]))
            result[1:length(var)] <- target
        }
        else {
            set <- eval(parse(text = strsplit(term, "=")[[1]][1]))
            target <- eval(parse(text = strsplit(term, "=")[[1]][2]))
```

```
                for (val in set) {
                    if (is.na(val))
                        result[is.na(var)] <- target
                    else result[var == val] <- target
                }
            }
    }
    if (as.factor.result) {
        result <- if (!missing(levels))
            factor(result, levels = levels)
        else as.factor(result)
    }
    else if (as.numeric.result && (!is.numeric(result))) {
        result.valid <- na.omit(result)
        opt <- options(warn = -1)
        result.valid <- as.numeric(result.valid)
        options(opt)
        if (!any(is.na(result.valid)))
            result <- as.numeric(result)
    }
    result
}


# ---------------------------------
# Data preparation work
# ---------------------------------
#
# note that the initial data file has seven initial records of data
# that are identifiers only ... these were deleted...
# then the records that define the attributes (A records) are placed
# in a file <000_attribute_data.csv>
# the base training data for the study are the web activity records
# (C and V records) which are in the original data files
#
# from the documentation we have the following information
#
#  Case and Vote Lines:
#  For each user, there is a case line followed by zero or more vote lines.
#
#  For example:
#      C,"10164",10164
#      V,1123,1
#      V,1009,1
#      V,1052,1
#          Where:
#              'C' marks this as a case line,
#                '10164' is the case ID number of a user,
#                'V' marks the vote lines for this case,
#                '1123', 1009', 1052' are the attributes ID's
#                      of Vroots that a user visited.
#                    '1' may be ignored.
#    The datasets record which Vroots [web site areas]
#    each user visited in a one-week timeframe in Feburary 1998.
```

```
# note that all C and V data were set as numeric data on input
# for both training and test data files
# this was done in a plain text editor
#
# our editing resulted in files <000_training_data.csv> and
#    <000_test_data.csv>
# the former is used to define association rules
# the latter can be used to test predictions/recommendations
# based upon these association rules
#
# we detected what appear to be two miscodings in the attribute data:
# "/security." was changed to "/security" and "/msdownload." to "/msdownload"

# ----------------------------------
# User-defined functions
# ----------------------------------
# function for replacing web area numeric IDs with meaningful area names
map_microsoft <- function(x) {
carmap(var = x,
recodes = '1287 = "/autoroute";1288 = "/library";1289 = "/masterchef";
1297 = "/centroam";1215 = "/developer";1279 = "/msgolf";1239 = "/msconsult";
1282 = "/home";1251 = "/referencesupport";1121 = "/magazine";
1083 = "/msaccesssupport";1145 = "/vfoxprosupport";1276 = "/vtestsupport";
1200 = "/benelux";1259 = "/controls";1155 = "/sidewalk";1092 = "/vfoxpro";
1004 = "/search";1057 = "/powerpoint";1140 = "/netherlands";
1198 = "/pictureit";1147 = "/msft";1005 = "/norge";1026 = "/sitebuilder";
1119 = "/corpinfo";1216 = "/vrml";1218 = "/publishersupport";
1205 = "/hardwaresupport";1269 = "/business";1031 = "/msoffice";1003 = "/kb";
1238 = "/exceldev";1118 = "/sql";1242 = "/msgarden";1171 = "/merchant";
1175 = "/msprojectsupport";1021 = "/visualc";1222 = "/msofc";
1284 = "/partner";1294 = "/bookshelf";1053 = "/visualj";1293 = "/encarta";
1167 = "/hwtest";1202 = "/advtech";1234 = "/off97cat";1054 = "/exchange";
1262 = "/chile";1074 = "/ntworkstation";1027 = "/intdev";1061 = "/promo";
1236 = "/globaldev";1212 = "/worldwide";1204 = "/msscheduleplus";
1196 = "/ie40";1188 = "/korea";1228 = "/vtest";1078 = "/ntserversupport";
1008 = "/msdownload";1052 = "/word";1091 = "/hwdev";1280 = "/music";
1247 = "/wineguide";1064 = "/activeplatform";1065 = "/java";
1133 = "/frontpagesupport";1102 = "/homeessentials";1132 = "/msmoneysupport";
1240 = "/thailand";1225 = "/piracy";1130 = "/syspro";1157 = "/win32dev";
1058 = "/referral";1076 = "/ntwkssupport";1163 = "/opentype";1187 = "/odbc";
1152 = "/rus";1139 = "/k-12";1223 = "/finland";1001 = "/support";
1043 = "/smallbiz";1165 = "/poland";1194 = "/china";1138 = "/mind";
1158 = "/imedia";1094 = "/mshome";1055 = "/kids";1277 = "/stream";
1143 = "/workshoop";1068 = "/vbscript";1229 = "/uruguay";1177 = "/events";
1014 = "/officefreestuff";1019 = "/mspowerpoint";1122 = "/mindshare";
1041 = "/workshop";1033 = "/logostore";1233 = "/vbscripts";
1211 = "/smsmgmtsupport";1199 = "/feedback";1024 = "/iis";1179 = "/colombia";
1067 = "/frontpage";1181 = "/kidssupport";1174 = "/nz";
1162 = "/infoservsupport";1046 = "/iesupport";1197 = "/sqlsupport";
1231 = "/win32devsupport";1141 = "/europe";1120 = "/switch";1112 = "/canada";
1142 = "/southafrica";1250 = "/middleeast";1214 = "/finserv";
1190 = "/repository";1098 = "/devonly";1263 = "/services";
1049 = "/supportnet";1073 = "/taiwan";1166 = "/mexico";
```

```
1226 = "/msschedplussupport";1184 = "/msexcelsupport";1025 = "/gallery";
1160 = "/visualcsupport";1156 = "/powered";1268 = "/javascript";
1220 = "/macofficesupport";1060 = "/msword";1203 = "/danmark";
1176 = "/jscript";1168 = "/salesinfo";1066 = "/musicproducer";1128 = "/msf";
1275 = "/security";1136 = "/usa";1146 = "/msp";1237 = "/devdays";
1081 = "/accessdev";1016 = "/excel";1069 = "/windowsce";
1148 = "/channel_resources";1161 = "/workssupport";1013 = "/vbasicsupport";
1116 = "/switzerland";1093 = "/vba";1249 = "/fortransupport";
1095 = "/catalog";1023 = "/spain";1192 = "/visualjsupport";1080 = "/brasil";
1050 = "/macoffice";1255 = "/msmq";1273 = "/mdn";1206 = "/select";
1230 = "/mailsupport";1172 = "/belgium";1011 = "/officedev";1009 = "/windows";
1096 = "/mspress";1235 = "/onlineeval";1070 = "/activex";1154 = "/project";
1099 = "/cio";1186 = "/college";1291 = "/news";1256 = "/sia";
1270 = "/developr";1232 = "/standards";1159 = "/transaction";
1035 = "/windowssupport";1164 = "/smsmgmt";
1077 = "/msofficesupport";1295 = "/train_cert";1056 = "/sports";
1006 = "/misc";1272 = "/softlib";1123 = "/germany";
1151 = "/mspowerpointsupport";1103 = "/works";1243 = "/usability";
1244 = "/devwire";1260 = "/trial";1258 = "/peru";1208 = "/israel";
1106 = "/cze";1124 = "/industry";1114 = "/servad";1012 = "/outlookdev";
1045 = "/netmeeting";1082 = "/access";1261 = "/diyguide";1137 = "/mscorp";
1059 = "/sverige";1037 = "/windows95";1227 = "/argentina";
1281 = "/intellimouse";1134 = "/backoffice";1044 = "/mediadev";
1028 = "/oledev";1248 = "/softimage";1085 = "/exchangesupport";
1131 = "/moneyzone";1079 = "/australia";1048 = "/publisher";
1042 = "/vstudio";1075 = "/jobs";1201 = "/hardware";1105 = "/france";
1153 = "/venezuela";1292 = "/northafrica";1015 = "/msexcel";
1290 = "/devmovies";1017 = "/products";1010 = "/vbasic";
1126 = "/mediamanager";1144 = "/devnews";1191 = "/management";
1002 = "/athome";1213 = "/corporate_solutions";1084 = "/uk";
1178 = "/msdownload";1036 = "/organizations";1257 = "/devvideos";
1180 = "/slovenija";1246 = "/gamesdev";1088 = "/outlook";
1117 = "/sidewinder";1097 = "/latam";1266 = "/licenses";
1072 = "/vinterdev";1169 = "/msproject";1107 = "/slovakia";
1089 = "/officereference";1038 = "/sbnmember";1224 = "/atec";
1086 = "/oem";1108 = "/teammanager";1007 = "/ie_intl";1252 = "/giving";
1283 = "/cinemania";1127 = "/netshow";1189 = "/internet";1110 = "/mastering";
1090 = "/gamessupport";1109 = "/technet";1040 = "/office";1150 = "/infoserv";
1195 = "/portugal";1111 = "/ssafe";1274 = "/pdc";1267 = "/caribbean";
1113 = "/security";1245 = "/ofc";1253 = "/worddev";1087 = "/proxy";
1185 = "/sna";1209 = "/turkey";1063 = "/intranet";1101 = "/oledb";
1264 = "/se_partners";1032 = "/games";1173 = "/moli";1051 = "/scheduleplus";
1278 = "/hed";1062 = "/msaccess";1020 = "/msdn";1104 = "/hk";
1071 = "/automap";1000 = "/regwiz";1135 = "/mswordsupport";1207 = "/icp";
1217 = "/ireland";1254 = "/ie3";1022 = "/truetype";1183 = "/italy";
1170 = "/mail";1241 = "/india";1149 = "/adc";1029 = "/clipgallerylive";
1221 = "/mstv";1115 = "/hun";1125 = "/imagecomposer";1039 = "/isp";
1034 = "/ie";1265 = "/ssafesupport";1271 = "/mdsn";1129 = "/ado";
1018 = "/isapi";1193 = "/offdevsupport";1219 = "/ads";1030 = "/ntserver";
1182 = "/fortran";1100 = "/education";1210 = "/snasupport"')
}
```

```
# -----------------------------------------------------------------
# user-defined function to identify antecedent items and
# consequent items lists... as many items as are given in rules

make_lists_for_all_rules <- function(association_rules) {
    # function returns items lists for antecedents and consequents
    # note. default apriori algorithm finds single-item consequents
    #       but we will return lists for consequents as well
    all_antecedent_matrix <- as(lhs(association_rules), "matrix")
    all_consequent_matrix <- as(rhs(association_rules), "matrix")
    # create list of lists for antecedent items in each association rule
    all_antecedent_items <- NULL
    for (irule in 1:nrow(all_antecedent_matrix)) {
        this_rule_antecedent_items <- NULL
        for (jitem in 1:ncol(all_antecedent_matrix)) {
            if (all_antecedent_matrix[irule, jitem] == 1)
            this_rule_antecedent_items <-
                c(this_rule_antecedent_items,
                    colnames(all_antecedent_matrix)[jitem])
        }  # end inner for-loop for gathering up items in this rule
        all_antecedent_items <-
            rbind(all_antecedent_items,list(this_rule_antecedent_items))
    }  # end outter for-loop for rules

    all_consequent_items <- NULL
    for (irule in 1:nrow(all_consequent_matrix)) {
        this_rule_consequent_items <- NULL
        for (jitem in 1:ncol(all_consequent_matrix)) {
            if (all_consequent_matrix[irule, jitem] == 1)
                this_rule_consequent_items <-
                c(this_rule_consequent_items,
                    colnames(all_consequent_matrix)[jitem])
        }  # end inner for-loop for gathering up items in this rule
            all_consequent_items <-
                rbind(all_consequent_items,list(this_rule_consequent_items))
    }  # end outter for-loop for rules

    list(all_antecedent_items,all_consequent_items)

    }  # end of function make.item.lists.for.all rules

# -------------------------------------------------------------------------

make_lists_for_single_item_antecedent_rules <- function(association_rules) {
    # function returns items lists for antecedents and consequents
    # note. default apriori algorithm finds single-item consequents
    #       but we will return lists for consequents as well
    all_antecedent_matrix <- as(lhs(association_rules), "matrix")
    all_consequent_matrix <- as(rhs(association_rules), "matrix")

    # create list of lists for antecedent items in each association rule
    indices_for_single_antecedent_rules <- NULL
    all_antecedent_items <- NULL
```

```
    for (irule in 1:nrow(all_antecedent_matrix)) {
        this_rule_antecedent_items <- NULL
        for (jitem in 1:ncol(all_antecedent_matrix)) {
            if (all_antecedent_matrix[irule, jitem] == 1)
            this_rule_antecedent_items <-
                c(this_rule_antecedent_items,
                    colnames(all_antecedent_matrix)[jitem])
            } # end inner for-loop for gathering up items in this rule
        all_antecedent_items <-
            rbind(all_antecedent_items,list(this_rule_antecedent_items))
        if (length(this_rule_antecedent_items) == 1)
            indices_for_single_antecedent_rules <-
                c(indices_for_single_antecedent_rules, irule)
        } # end outter for-loop for rules

    all_consequent_items <- NULL
    for (irule in 1:nrow(all_consequent_matrix)) {
        this_rule_consequent_items <- NULL
        for (jitem in 1:ncol(all_consequent_matrix)) {
            if (all_consequent_matrix[irule, jitem] == 1)
                this_rule_consequent_items <-
            c(this_rule_consequent_items,
                colnames(all_consequent_matrix)[jitem])
            } # end inner for-loop for gathering up items in this rule
        all_consequent_items <-
            rbind(all_consequent_items,list(this_rule_consequent_items))
        } # end outter for-loop for rules

    antecedent_items <-
        all_antecedent_items[indices_for_single_antecedent_rules,]
    consequent_items <-
        all_consequent_items[indices_for_single_antecedent_rules,]
    list(antecedent_items,consequent_items)

    } # end of function make_lists_for_single_item_antecedent_rules


# -------------------------------------------
# user-defined function to track processing
# used for long-running loop processing code
multiple_of_one_thousand <- function(x) {
    # return true if x is a multiple of 1000
    returnvalue <- FALSE
    if(trunc(x/1000)==(x/1000)) returnvalue <- TRUE
    returnvalue
    } # end of function multiple_of_one_thousand


# ----------------------------------------------------
# Read in training data... create transactions data
# ----------------------------------------------------
# read in the input data for attribute names
# we do not use these data in the jump-start program
# it could be used later to provide an interpretation of association rules
# the rules are identified by numbers only in training and test data
```

```
# the attribute data show what those names mean in terms of the
# subdirectory name (area_name) and its description
# we will use the area_name as our web_area_id because it is
# short enough for listings and more meaningful than numbers
# in order to do this, we will need to define the mapping
# much as we would with dictionaries and the map method in Python
attribute_input_data <-
    read.csv(file = "000_attribute_data.csv", header = FALSE,
        col.names = c("A_type", "numeric_value", "ignore_value",
            "description", "area_name"), stringsAsFactors = FALSE)

# read in the input data for users and web areas visited
training_input_data <- read.csv(file = "000_training_data.csv",
    header = FALSE, col.names = c("CV_type","value","ignore_value"),
    stringsAsFactors = FALSE)

# the following code creates a transaction data frame for the training data
# as needed for association rule analysis with the R package arules
# the transaction data frame has two columns, the first being user_id
# and the second being the web_area_id, a web site area visited by the user

n_records <- nrow(training_input_data )  # stopping rule for while-loop

alphanumeric_transaction_data_frame <- NULL  # initialize object
n_record_count <- 0  # intialize record count

# note that this step can take a while to run...
# we set things up to report on the process, but this reporting
# may not work on all operating systems... be patient with this step
# note also that this step may be skipped on subsequent runs
# because you will have the transactions_file_name in place
cat("\n\nTransaction processing percentage complete: ")
while (n_record_count < n_records) {
  n_record_count <- n_record_count + 1
  # report to screen the proportion of job completed
  if(multiple_of_one_thousand(n_record_count))
      cat(" ",trunc(100 * (n_record_count /  n_records)))
  # read one record from input data
  this_record <- training_input_data[n_record_count,]
  if (this_record$CV_type == "C") user_id <-
      this_record$value  # C for user_id
  if (this_record$CV_type == "V") {  # V record provides web_area_id
    web_area_id <- map_microsoft(this_record$value)  # meaningful area name
    # add one record to the transaction.data.frame
    alphanumeric_transaction_data_frame <-
      rbind(alphanumeric_transaction_data_frame, data.frame(user_id, web_area_id))
    }
  }
# write the transactions to a comma-delimited file
transactions_file_name <- "000_training_transactions.csv"
write.csv(alphanumeric_transaction_data_frame,
    file = "000_training_transactions.csv", row.names = FALSE)
cat("\nTransactions sent to ",transactions_file_name,"\n")
```

```
# --------------------------------------------------
# Use transactions data to find association rules
# --------------------------------------------------
transactions_object <-
    read.transactions(file = transactions_file_name, cols = c(1,2),
        format = "single", sep = ",", rm.duplicates = TRUE)

cat("\n",dim(transactions_object)[1],"unique user_id values")
cat("\n",dim(transactions_object)[2],"unique web_area_id values")

# examine frequency for each item with support
# greater than high_support_setting
pdf(file = "fig_wnds_recommend_web_area_support_bar_plot.pdf",
  width = 8.5, height = 11)
itemFrequencyPlot(transactions_object, support = high_support_setting,
  cex.names=0.8, xlim = c(0,0.4),
  type = "relative", horiz = TRUE, col = "darkblue", las = 1,
  xlab = "Proportion of Users Viewing the Web Area")
 title("Association Rules Analysis (Areas with Highest Support)")
dev.off()

# obtain large set of association rules for web areas
# this is done by setting very low criteria for support and confidence
# set support at low_support_setting and confidence at confidence_setting
# using our best judgment
association_rules <- apriori(transactions_object,
  parameter =
      list(support = low_support_setting, confidence = confidence_setting))
print(summary(association_rules))

# data visualization of association rules in scatter plot
pdf(file = "fig_wnds_recommend_web_area_rules_scatter_plot.pdf",
    width = 8.5,height = 11)
plot(association_rules,
  control=list(jitter=2, col = rev(brewer.pal(9, "Blues")[4:9])),
  shading = "lift")
dev.off()

# grouped matrix of rules
pdf(file = "fig_wnds_recommend_web_area_bubble_chart.pdf",
    width = 8.5, height = 11)
plot(association_rules, method="grouped",
  control=list(col = rev(brewer.pal(9, "Blues")[4:9])))
dev.off()

# route association rule results to text file
sink(file = "fig_wnds_recommend_partial_list_of_rules.txt")
cat("Web Usage Analysis by Association Rules\n")
cat("\n",dim(transactions_object)[1],"unique user_id values")
cat("\n",dim(transactions_object)[2],"unique web_area_id values")
cat("\n\nCriteria set for discovering association rules:")
cat("\nMinimum support set to:   ",low_support_setting)
cat("\nMinimum confidence set to:",confidence_setting,"\n\n")
```

```
print(summary(association_rules))
cat("\n","Resulting Web Area Association Rules\n\n")
inspect(association_rules)
sink()

# ------------------------------------------------
# Select association rules for recommender model
# ------------------------------------------------
# let's focus upon rules with a single-item/area-antecedent rules

selected_rules <-
    make_lists_for_single_item_antecedent_rules(association_rules)
antecedent_areas <- as.numeric(unlist(selected_rules[[1]]))
consequent_areas <- as.numeric(unlist(selected_rules[[2]]))

# ----------------------------------------------------
# Read in test data and create users-by-areas matrix
# ----------------------------------------------------

test_input_data <- read.csv(file = "000_test_data.csv", header = FALSE,
  col.names = c("CV_type","value","ignore_value"), stringsAsFactors = FALSE)

case_test_data <- subset(test_input_data, subset = (CV_type == "C"))
list_of_case_id_names <- as.character(sort(unique(case_test_data$value)))

web_area_test_data <- subset(test_input_data, subset = (CV_type == "V"))
list_of_web_area_id_names <-
    as.character(sort(unique(web_area_test_data$value)))

# the following code creates a test data matrix for the test data
# rows correspond to the user_ids
# columns correspond to the web_area_ids

# initialize test matrix as matrix of zeroes
test_data_matrix <- matrix(0, nrow = length(list_of_case_id_names),
  ncol = length(list_of_web_area_id_names),
  dimnames = list(as.character(list_of_case_id_names),
                  as.character(list_of_web_area_id_names)))

n_records <- nrow(test_input_data )  # used as stopping rule for while-loop

n_record_count <- 0  # intialize record count

# note that this step should go pretty fast...
cat("\n\nTest data processing percentage complete: ")
while (n_record_count < n_records) {
  n_record_count <- n_record_count + 1
  # report to screen the proportion of job completed
  if(multiple_of_one_thousand(n_record_count))
      cat(" ",trunc(100 * (n_record_count / n_records)))
  # read one record from input data
  this_record <- test_input_data[n_record_count,]
```

```
   if (this_record$CV_type == "C")
      name_user_id <- as.character(this_record$value)  # C record for user_id
   if (this_record$CV_type == "V") {  # V record provides web_area_id
     name_web_area_id <- map_microsoft(this_record$value)
     # enter 1 in appropriate cell of test_data_matrix
     test_data_matrix[name_user_id, name_web_area_id] <- 1
     }
   }


# quick check of the test_data_matrix
# there should be as many 1s in test_data_matrix
#    as there are V records in the test data file
if(nrow(web_area_test_data) != sum(test_data_matrix))
    stop("test_data_matrix problem")

# useful rows for testing are those with at least two web areas visited
# sum of rows is number of 1s
user_areas_visited <- apply(test_data_matrix, 1, FUN = sum)

# keep track of the antecedent that is selected for testing
selected_antecedent <- rep(NA, length = nrow(test_data_matrix)) # initialize

# set up binary indicator for a correct prediction
# from any of the association rules
correct_predictions <- rep(0, length = nrow(test_data_matrix)) # initialize

# set up a counter for the number of rules
# that match the antecedent for this user
antecedent_rule_matches <-
    rep(0, length = nrow(test_data_matrix))  # initialize

# set up character string vector for storing consequents correctly predicted
correctly_predicted_consequents <-
    rep("", length = nrow(test_data_matrix))  # initialize

# for each user visiting more than one area
# select an area from his/her visited list at random
# predict from this area other areas likely to have been visited
# using the selected association rules from previous work
# determine if, in fact, these predictions are correct

set.seed(9999)  # for reproducible results
for(iuser in 1:nrow(test_data_matrix)) {
  if(user_areas_visited[iuser] > 1) {
    this_user_row <- test_data_matrix[iuser,]
    this_user_web_areas <-
        as.numeric(names(this_user_row[(this_user_row == 1)]))
    # pick one of the areas at random
    randomized_web_areas <- sample(this_user_web_areas)
    selected_antecedent[iuser] <-
        randomly_selected_antecedent <- randomized_web_areas[1]
    randomly_selected_consequents <-
        randomized_web_areas[2:length(randomized_web_areas)]
```

```
    # apply association rules to obtain predicted consequents
    for (iarea in seq(along = antecedent_areas)) {
      if (randomly_selected_antecedent == antecedent_areas[iarea]) {
        antecedent_rule_matches[iuser] <- antecedent_rule_matches[iuser] + 1
        predicted_consequent <- consequent_areas[iarea]
        if (predicted_consequent %in% randomly_selected_consequents) {
          correct_predictions[iuser] <- correct_predictions[iuser] + 1
          correctly_predicted_consequents[iuser] <-
            paste(correctly_predicted_consequents[iuser],
                  predicted_consequent, " ")
        }
      }
    } # end for-loop for checking association rules

  } # end of if-block for users with more than one site visited
 } # end of testing for-loop

prediction_data_frame <- na.omit(data.frame(user.id = list_of_case_id_names,
                                  user_areas_visited,
                                  selected_antecedent,
                                  antecedent_rule_matches,
                                  correct_predictions,
                                  correctly_predicted_consequents))

cat("\n\nPercentage predicted accurately by recommendation model: ",
  round((100 * sum(correct_predictions) /
      sum(antecedent_rule_matches)),digits = 3),"\n")

# Suggestions for the student: Try alternative values for support
# and confidence and note changes in the set off association rules.
# Develop a formal assessment strategy for evaluating the accuracy
# of recommendations. How much better are we doing with the model
# than we would expect from the levels of support across website areas?
# We see Percentage predicted accurately by recommendation model: 36.461
# How good is that value compared to what we could get without a model?
# What about recommendation system models other than association rules?
# Could any of those models be used given the data we have here?
```

11 Playing Network Games

"I'm also worried about Walt Waldowski—Painless. His poker players got in an argument and asked him for a ruling, and he said what difference did it make, it was just a card game."

—Rene Auberjonois as Father John Patrick Mulcahy in
MASH (1970)

Most days I go for an afternoon swim in the pool at the apartment complex where I live. I usually have the pool to myself except for a few formerly flying insects, failed swimmers all. If one of those creatures invades my space, I decide if it lives or dies. Most often, feeling benevolent, I endeavor to throw every insect out of the pool in a handful of water. Such is my life. I spend a good deal of time debugging things of one type or another.

Agent-based models are another place where I get to play god. If I can define the initial structure of a network, the nature of the links and the characteristics of the nodes, then I can use that information to build a behavioral model of the network. I can simulate the behavior of individual nodes or agents, including their interactions with one another.

With agent-based models, we create a microworld intended to simulate real world phenomena. This presents a rich platform for modeling complexity, most useful in the domain of network data science. We can set the stage for agent-based models and for models of dynamic networks by first working with discrete event simulation and fixed network models.

Figure 11.1. Network Modeling Techniques

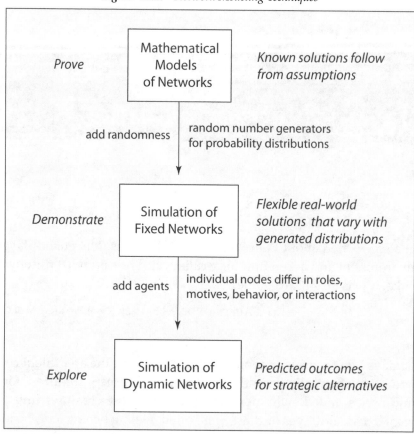

Figure 11.1 provides an overview of network modeling techniques, beginning with mathematical models and ending with agent-based models. With mathematical models, known solutions follow from assumptions—they are proven to be correct. The Erdős-Rényi random graph model (Erdős and Rényi, 1959, 1960, 1961) is an example of a mathematical model for networks. Mathematical models of networks provide the foundation of network science (Lewis 2009; Kolaczyk 2009; Newman 2010).

Standard mathematical models for networks include the random graph, preferential attachement (Barabási and Albert 1999), and small-world (Watts and Strogatz 1998) models reviewed earlier in this book. Discrete event simulation of fixed networks builds on mathematical models and provides a flexible structure for modeling. With discrete event simulation, we demonstrate solutions rather than prove solutions. We define a known, fixed network structure in advance and study performance within that fixed network structure.

Discrete event simulation of fixed networks is most useful in the study of websites and physical communication networks, transportation and production problems. The nodes and links are fixed. Resource limitations are known. And traffic or demand for resources varies across simulated time.

Suppose we were designing a website for an online retailer with the objective of moving customers from landing page to order page as quickly as possible, thus reducing the likelihood of abandoned online shopping carts. We might begin by representing the problem as a directed graph. Nodes are ordering milestones, such as arrival at a landing page, product information page, or payment information page. Links represent activities, each associated with a time to move from one node to the next.

The total time to traverse the network is of special interest in this context. We seek efficient performance so that there are few abandoned shopping carts. The network structure is fixed by the current website. Alternative paths through the network are associated with various landing pages (points of origin) and choices made by users. We use random number generators from various probability distributions to simulate user choices, events, and traffic across the network.

A well-understood component of management science, project management problems may be addressed by mathematical programming (Bradley, Hax, and Magnanti 1977; Williams 2013; Vanderbei 2014) and by discrete event simulation (Burt and Garman 1971). Projects are defined by activities, some of which may be completed in parallel, others sequentially. Alternative representations include task-oriented networks in which nodes represent activities, and event-oriented networks in which links represent activities. To solve these problems by simulation, we set the structure of the project network, generate random variates for the duration of activities, and look for the critical path through the network. The objective may be to identify the shortest path, lowest cost path, maximum flow through the network, or the distribution of project completion times.

Agent-based modeling of networks can build on discrete event simulation of fixed networks. We acknowledge the fact that individual nodes may differ from one another in roles, motives, behavior, or interactions. Agent-based techniques provide a facility for modeling networks that change with time (dynamic networks).

Many questions of interest to management involve agents of various types and networks that change with time. By introducing agents into a simulation model, we can explore alternative agent inputs and management actions, and in the process predict outcomes associated with strategic alternatives (North and Macal 2007).

Suppose a consumer electronics firm plans to offer a new product in the wearable technology space. The firm is interested in testing alternative methods for promoting the product. Mass media offers wide coverage of the population with limited contact with each individual consumer. Direct marketing to target groups entails narrow coverage. But with a narrow, more focused target group, it is possible to tailor a message for that group, resulting in more extensive contact with individual consumers.

The success of mass versus targeted marketing will depend on the coverage of the population, the levels of exposure or contact with individuals, and the responsiveness of individuals to the message. Just as the spread of disease in a population depends on the structure of the network and the susceptibility of individuals to the disease, the adoption of a new product depends upon the willingness of individual consumers to purchase and the structure of the network of consumers.

Some consumers are more willing to accept new ideas, information, and products than others. And we can use agent-based models to explore individual differences among consumers and their effect on the network.

The response of consumers to the initial marketing message begins the process. Some purchase the wearable technology product and wear it. They interact with other consumers, influencing their purchases. The process plays out in simulation time. As the simulation progresses, we note the numbers of consumers exposed to the idea and their reaction to the idea (purchase or not).

An agent-based framework allows us to fine tune consumer characteristics to match our understanding of individual characteristics in the marketplace. We can test alternative marketing actions (mass versus direct), and their effect on consumer adoption of the product.

Discussions of agent-based models and complexity go hand-in-hand. This is because agent-based models are most useful when the entire system is less well understood than its individual components. So rather than build a complex systems model, we build many small models. Agent-based models also have a role to play when underlying processes are coincidental rather than sequential.

Validation of agent-based models presents challenges. Is it sufficient to see the simulation acting like what we see in the real world? And how shall we judge the degree of similarity?

Agent-based models remain in an experimental state, not widely employed in business research. To move from the academic world into business—to be viewed as more than a modeler's playground—agent-based models need to do things that are useful and understood by people outside the modeling community.

Law (2014) discusses the domain of simulation modeling, including agent-based modeling. For discussion of complexity theory and agent-based modeling, see North and Macal (2007), Miller and Page (2007), and Šalamon (2011). Much of the work in this area is biologically inspired, modeling the behavior of individual organisms to see what happens in the community (Resnick 1998; Mitchell 2009).

Agent-based studies, like many simulation studies, are set up as factorial experiments. We look at alternative network configurations and agent characteristics and see how they play out in simulation time. An example would be an information cascade, which often builds on a biological model of disease. We might consider factors such as the number of nodes in the network, network density, and the number of individuals who are initially influenced by the idea or infected by the virus. We could also consider agent characteristics such as risk-aversion or resistance to disease. Factors may be systematically varied across the experiment with each run or trial of the experiment beginning with a distinct random number seed, ensuring that experimental observations are independent.

We can learn more about agent-based simulation as we learn about many things—by doing, by programming. Agent-based simulation requires an environment or framework unto itself, and the programming can take a good deal of time. We will play simulation games later as we add exemplary code to the website for the book.

We analyze the results of agent-based simulations much as we would any other experimental study. Analysis of variance and linear models work for continuous response variables. Analysis of deviance and generalized linear models work for binary response variables, proportions, and counts.

Exhibit 11.1 shows an example of analysis of deviance using Python, and exhibit 11.2 shows the corresponding analysis with R. With factorial designs, interaction plots can provide a useful summary of experimental results. The R program shows how to generate a two-way or two-factor interaction plot using software developed by Wickham and Chang (2014).

Exhibit 11.1. *Analysis of Agent-Based Simulation (Python)*

```
# Analysis of Agent-Based Simulation (Python)

# install necessary packages

# import packages into the workspace for this program
import numpy as np
import pandas as pd
import statsmodels.api as sm


# ----------------------------
# Simulation study background
# ----------------------------
# an agent-based simulation was run with NetLogo, a public-domain
# program available from Northwestern University (Wilensky 1999)

# added one line of code to the Virus on a Network program:

if ticks = 200 [stop]

# this line was added to stop the simulation at exactly 200 ticks
# the line was added to the <to go> code block as shown here:
#
# to go
#   if all? turtles [not infected?]
#     [ stop ]
#   ask turtles
#   [
#     set virus-check-timer virus-check-timer + 1
#     if virus-check-timer >= virus-check-frequency
#       [ set virus-check-timer 0 ]
#   ]
#   if ticks = 200 [stop]
#   spread-virus
#   do-virus-checks
#   tick
# end

# the simulation stops when no nodes/turtles are infected
# or when the simulation reaches 200 ticks

# To see the results of the simulation at 200 ticks, we route the simulation
# world to a file using the GUI File/Export/Export World
# this gives an a comma-delimited text file of the status of the network
# at 200 ticks. Specifically, we enter the following Excel command into
# cell D1 of the results spreadsheet to compute the proportion of nodes
# infected:  = COUNTIF(N14:N163, TRUE)/M10

# NetLogo turtle infected status values were given in cells N14 through N163.
# The detailed results of the simulation runs or trials are shown in the files
# <trial01.csv> through <trial20.csv> under the directory NetLogo_Results
```

```
# this particular experiment, has average connectivity or node degree
# at 3 or 5 and the susceptibility or virus spread chance to 5 or 10 percent.
# we have a completely crossed 2 x 2 design with 5 replications of each cell
# that is, we run each treatment combination 5 times, obtaining 20 independent
# observations or trials. for each trial, we note the percentage of infected
# nodes after 200 ticks---this is the response variable
# results are summarized in the comma-delimited file <virus_results.csv>.

# ----------------------------
# Analysis of Deviance
# ----------------------------

# read in summary results and code the experimental factors
virus = pd.read_csv("virus_results.csv")

# check input DataFrame
print(virus)

Intercept = np.array([1] * 20)

# use dictionary object for mapping to 0/1 binary codes
degree_to_binary = {3 : 0, 5 : 1}
Connectivity = np.array(virus['degree'].map(degree_to_binary))

# use dictionary object for mapping to 0/1 binary codes
spread_to_binary = {5 : 0, 10 : 1}
Susceptibility = np.array(virus['spread'].map(spread_to_binary))

Connectivity_Susceptibility = Connectivity * Susceptibility

Design_Matrix = np.array([Intercept, Connectivity, Susceptibility,\
    Connectivity_Susceptibility]).T

print(Design_Matrix)

Market_Share = np.array(virus['infected'])

# generalized linear model for a response variable that is a proportion
glm_binom = sm.GLM(Market_Share, Design_Matrix, family=sm.families.Binomial())
res = glm_binom.fit()
print res.summary()
```

Exhibit 11.2. *Analysis of Agent-Based Simulation (R)*

```
# Analysis of Agent-Based Simulation (R)

# install necessary packages

library(ggplot2)  # for pretty plotting

# -------------------------------
# Simulation study background
# (same as under Python exhibit)
# -------------------------------
# note. it is possible to run NetLogo simulations within R
# using the RNetLogo package from Thiele (2014)

# ----------------------------
# Analysis of Deviance
# ----------------------------
options(warn = -1)  # drop glm() warnings about non-integer responses

# read in summary results and code the experimental factors
virus <- read.csv("virus_results.csv")

# define factor variables
virus$Connectivity <- factor(virus$degree,
  levels = c(3, 5), labels = c("LOW", "HIGH"))
virus$Susceptibility <- factor(virus$spread,
  levels = c(5, 10), labels = c("LOW", "HIGH"))

virus$Market_Share <- virus$infected

# show the mean proportions by cell in the 2x2 design
with(virus, print(by(Market_Share, Connectivity * Susceptibility, mean)))

# generalized linear model for response variable that is a proportion
virus_fit <- glm(Market_Share ~ Connectivity + Susceptibility +
  Connectivity:Susceptibility,
  data = virus, family = binomial(link = "logit"))
print(summary(virus_fit))

# analysis of deviance
print(anova(virus_fit, test="Chisq"))

# compute market share cell means for use in interaction plot
virus_means <- aggregate(Market_Share ~ Connectivity * Susceptibility,
  data = virus, mean)

# ----------------------------
# Interaction Plotting
# ----------------------------
# generate an interaction plot for market share as a percentage
```

```
interaction_plot <- ggplot(virus_means,
  aes(x = Connectivity, y = 100*Market_Share,
    group = Susceptibility, fill = Susceptibility)) +
  geom_line(linetype = "solid", size = 1, colour = "black") +
  geom_point(size = 4, shape = 21) +
  ylab("Market Share (Percentage)") +
  ggtitle("Network Interaction Effects in Innovation") +
  theme(plot.title = element_text(lineheight=.8, face="bold"))
print(interaction_plot)
```

What's Next for the Web?

[Bar scene. Princeton, New Jersey]

Nash: "Adam Smith needs revision."

Hansen: "What are you talking about?"

Nash: "If we all go for the blonde, we block each other and not a single one of us is going to get her. So then we go for her friends. But they will all give us the cold shoulder because nobody likes to be second choice. But what if no one goes for the blonde? We don't get in each other's way, and we don't insult the other girls. That's the only way we win. That's the only way we all get laid.

"Adam Smith said the best result comes from everyone in the group doing what is best for himself. Right? Incomplete. Incomplete. Because the best result will come from everyone in the group doing what's best for himself and the group."

Hansen: "Nash, if this is some way for you to get the blonde on your own, you can go to hell."

—RUSSELL CROWE AS JOHN NASH AND JOSH LUCAS AS HANSEN
IN *A Beautiful Mind* (2001)

The psychologist Jean Piaget, inspired by biological thinking, would speak of assimilation and accommodation as two ways of learning. Assimilation involves taking in new information and superimposing it on already established cognitive structures. We take in more information, but we keep thinking the same way.

And then there is the learning that forces us to change, what Piaget called accommodation. When new information arrives that does not fit with what we know, we must revamp our cognitive structures.

A teacher's job is not to make learning easy. Transformative learning, the learning that matters, is never easy. A teacher's job is to help students change, to open their eyes to new worlds and new ways of thinking.

When I think of the teachers who made a difference in my life, I realize they were teachers who helped me think in new ways. They were not particularly strict, not taskmasters, although they certainly had standards. What they were most of all was supportive. They made it easier for me to do the hard learning I had to do.

There is learning to be done on the web. It is hard. It will take years to pay dividends, and it will never be finished. But it is learning that will change the way we live and work and think. This learning is machine learning, and it involves the *semantic web*.

We often describe web data as unstructured or semi-structured text. In this book we have worked numerous examples, crawling, scraping, parsing, and analyzing, all the while trying to make sense of things.

The Document Object Model may help us to traverse the text domain. XML and HTML and their tags for nodes provide a structure for storing and displaying disparate pieces of information. These standards for information interchange promote communication and collaboration among research entities, just as they facilitate interprocess communication among computers. But recognize that these are merely text formatting rules. They were never intended to be a map of the knowledge of the web.

We lament the difficulties of text analytics. We endure the idiosyncrasies of natural language, developing specialized utilities and processing schemes. But what if we could take text relating to what we know and store it in a way that could be more easily accessed by computers and people? What if we revamped the cognitive structure of the web, organizing information in a systematic way, rendering it as knowledge?

We might think of the future research world as a collection of web services linking networks of information providers and clients. The World Wide Web itself holds promise as an information store or semantic web, a vast repository of machine-retrievable and machine-understandable data (Fenzel et al. 2003; Daconta, Obrst, and Smith 2003).

Structuring text to represent knowledge is the goal of the semantic web. More than an idea about what might be possible for organizing text, it is a set of technologies that are being used by researchers today. Allemang and Hendler (2007) and Wood, Zaidman, and Ruth (2014) provide an overview, with additional information about data representation and programming techniques available in various sources (Powers 2003; Lacy 2005; Segaran, Evans, and Taylor 2009; DuCharme 2013).

Semantic databases, natural language processing, and machine learning come together in the domain of question-answering. We have a long way to go before computers querying a web can match what we do as thinkers, but there has been great progress with the artificial intelligence for question-answering. Examples include Cyc (Lenat et al. 2010), Halo (Gunning et al. 2010), and IBM's Watson (Ferrucci et al. 2010; Lopez et al. 2013). And extracting linked information from the web is the purpose of open-source projects such as DBpedia (Bizar et al. 2009, Lehmann et al. 2014), a semantic data source for Watson. Imagine a Wikipedia bot that knows how facts fit together and can answer your every question. Making this happen is just a small matter of programming (we used to call it "SMOP").

Software is a collective endeavor. Sharing code in an open-source environment is the key to working efficiently today. We build on a foundation of tools created by others. The growth of the Python and R communities and the story of GitHub (Bourne 2013) show the promise of collaborative software environments.

Evan "Ev" Williams and Jack Dorsey talk about the origin of the microblogging that made their firm famous. Jack says he was the inventor. Ev replies, "No, you didn't invent Twitter. I didn't invent Twitter either. Neither did Biz [Christopher "Biz" Stone]. People don't invent things on the Internet. They just expand on an idea that already exists" (Bilton 2013, p. 203).

This is the way of the web—collaboration, teamwork, sharing programs and ideas. Along the way a few people get rich, and many others are listed as contributors. The data train moves forward.

Only time will tell what new community endeavors emerge. But I suspect we will see products and services that turn information into knowledge and formulaic reasoning into intelligence.

There is a child's circle game, the story game. Boys and girls form a circle. A story starts, one child whispering in the ear of the next. Around the circle the story goes, until it returns to the first child.

"Oh, no," she says, "that isn't the story I told. Not at all."

Some think the story game a lesson in information loss or distortion. But what if the opposite were true? What if the story got better as it traveled the circle—embellished, improved, the product of a learning community? What if the lesson were about collaborative creation?

Looking north from my kitchen window, I see the San Gabriel Mountains. From a distance they are very beautiful. That is how I feel about the future. No fears. Just wonder. I wonder what is next to learn.

A

Data Science Methods

Artim: "Data, ... haven't you ever just played for fun?"

Data: "Androids... don't have... fun."

Artim: "Why not...?"

Data: "No one's ever asked me that before."

—MICHAEL WELCH AS ARTIM AND BRENT SPINER AS DATA
IN *Star Trek: Insurrection* (1998)

Doing data science means implementing flexible, scalable, extensible systems for data preparation, analysis, visualization, and modeling. We are empowered by the growth of open source. Whatever the modeling technique or application, there is likely a relevant package, module, or library that someone has written or is thinking of writing. Doing data science for the web and networks means working in Python and R, and drawing on other languages as needed.

Data scientists, those working in the field of predictive analytics, speak the language of business—accounting, finance, marketing, and management. They know about information technology, including data structures, algorithms, and object-oriented programming. They understand statistical modeling, machine learning, and mathematical programming.

These are the things that data scientists do:

- **Finding out about.** This is the first thing we do—information search, finding what others have done before, learning from the literature. We draw on the work of academics and practitioners in many fields of study, contributors to predictive analytics and data science.

- **Preparing text and data.** Text is unstructured or partially structured. Data are often messy or missing. We extract features from text. We define measures. We prepare text and data for analysis and modeling.

- **Looking at data.** We do exploratory data analysis, data visualization for the purpose of discovery. We look for groups in data. We find outliers. We identify common dimensions, patterns, and trends.

- **Predicting how much.** We are often asked to predict how many units or dollars of product will be sold, the price of financial securities or real estate. Regression techniques are useful for making these predictions. Prediction is distinct from explanation.[1] We may not know why models work, but we need to know when they work and when to show others how they work. We identify the most critical components of models and focus on the things that make a difference.

- **Predicting yes or no.** Many business problems are classification problems. We use classification methods to predict whether or not a person will buy a product, default on a loan, or access a web page.

- **Testing it out.** We examine models with diagnostic graphics. We see how well a model developed on one data set works on other data sets. We employ a training-and-test regimen with data partitioning, cross-validation, or bootstrap methods.

- **Playing what-if.** We manipulate key variables to see what happens to our predictions. We play what-if games in simulated marketplaces. We employ sensitivity or stress testing of mathematical programming models. We see how values of input variables affect outcomes, payoffs, and predictions. We assess uncertainty about forecasts.

- **Explaining it all.** Data and models help us understand the world. We turn what we have learned into an explanation that others can understand. We present project results in a clear and concise manner.

---

[1] Statisticians distinguish between explanatory and predictive models. Explanatory models are designed to test causal theories. Predictive models are designed to predict new or future observations. See Geisser (1993), Breiman (2001b), and Shmueli (2010).

Data scientists are methodological eclectics, drawing from many scientific disciplines and translating the results of empirical research into words and pictures that management can understand. These presentations benefit from well-constructed data visualizations. In communicating with management, data scientists need to go beyond formulas, numbers, definitions of terms, and the magic of algorithms. Data scientists convert the results of predictive models into simple, straightforward language that others can understand.

The data scientists are knowledge workers par excellence. They are communicators playing a critical role in today's data-intensive world. Data scientists turn data into models and models into plans for action.

The role of data science in business has been discussed by many (Davenport and Harris 2007; Laursen and Thorlund 2010; Davenport, Harris, and Morison 2010; Franks 2012; Siegel 2013; Maisel and Cokins 2014; Provost and Fawcett 2014). In-depth reviews of methods include those of Izenman (2008), Hastie, Tibshirani, and Friedman (2009), and Murphy (2012). For data science with open-source tools, additional discussion is available in Conway and White (2012), Putler and Krider (2012), James et al. (2013), Kuhn and Johnson (2013), Lantz (2013), and Ledoiter (2013).

This appendix identifies classes of methods and reviews selected methods in databases and data preparation, statistics, machine learning, data visualization, and text analytics. We provide an overview of these methods and cite relevant sources for further reading.

A.1 Databases and Data Preparation

As noted earlier, there have always been more data than we can use. What is new today is the ease of collecting data and the low cost of storing data. Data come from many sources. There are unstructured text data from online systems. There are pixels from sensors and cameras. There are data from mobile phones, tablets, and computers worldwide, located in space and time. Flexible, scalable, distributed systems are needed to accommodate these data.

Relational databases have a row-and-column table structure, similar to a spreadsheet. We access and manipulate these data using structured query language (SQL). Because they are transaction-oriented with enforced data integrity, relational databases provide the foundation for sales order processing and financial accounting systems.

It is easy to understand why non-relational (NoSQL) databases have received so much attention. Non-relational databases focus on availability and scalability. They may employ key-value, column-oriented, document-oriented, or graph structures. Some are designed for online or real-time applications, where fast response times are key. Others are well suited for massive storage and off-line analysis, with map-reduce providing a key data aggregation tool.

Many firms are moving away from internally owned, centralized computing systems and toward distributed cloud-based services. Distributed hardware and software systems, including database systems, can be expanded more easily as the data management needs of organizations grow.

Doing data science means being able to gather data from the full range of database systems, relational and non-relational, commercial and open source. We employ database query and analysis tools, gathering information across distributed systems, collating information, creating contingency tables, and computing indices of relationship across variables of interest. We use information technology and database systems as far as they can take us, and then we do more, applying what we know about statistical inference and the modeling techniques of predictive analytics.

Regarding analytics, we acknowledge an unwritten code in data science. We do not select only the data we prefer. We do not change data to conform to what we would like to see or expect to see. A two of clubs that destroys the meld is part of the natural variability in the game and must be played with the other cards. We play the hand that is dealt. The hallmarks of science are an appreciation of variability, an understanding of sources of error, and a respect for data. Data science is science.

We are often asked to make a model out of a mess. Management needs answers, and the data are replete with miscoded and missing observations, outliers and values of dubious origin. We use our best judgement in preparing data for analysis, recognizing that many decisions we make are subjective and difficult to justify.

Missing data present problems in applied research because many modeling algorithms require complete data sets. With large numbers of explanatory variables, most cases have missing data on at least one of the variables. Listwise deletion of cases with missing data is not an option. Filling in missing data fields with a single value, such as the mean, median, or mode, would distort the distribution of a variable, as well as its relationship with other variables. Filling in missing data fields with values randomly selected from the data adds noise, making it more difficult to discover relationships with other variables. The method preferred by statisticians is multiple imputation.

Garcia-Molina, Ullman, and Widom (2009) provide an overview of database systems. White (2011), Chodorow (2013), and Robinson, Webber, and Eifrem (2013) review selected non-relational systems. For map-reduce operations, see Dean and Ghemawat (2004) and Rajaraman and Ullman (2012)

Osborne (2013) provides an overview of data preparation issues, and the edited volume by McCallum (2013) provides much needed advice about what to do with messy data. Missing data methods are discussed by Rubin (1987), Little and Rubin (1987), Schafer (1997), Groves et al. (2009), and Lumley (2010) and implemented in R packages from Gelman et al. (2014), Honaker, King, and Blackwell (2014), and Lumley (2014).

A.2 Classical and Bayesian Statistics

How shall we draw inferences from data? Formal scientific method suggests that we construct theories and test those theories with sample data. The process involves drawing statistical inferences as point estimates, interval estimates, or tests of hypotheses about the population. Whatever the form of inference, we need sample data relating to questions of interest. For valid use of statistical methods we desire a random sample from the population.

Which statistics do we trust? Statistics are functions of sample data, and we have more faith in statistics when samples are representative of the population. Large random samples, small standard errors, and narrow confidence intervals are preferred.

Classical and Bayesian statistics represent alternative approaches to inference, alternative ways of measuring uncertainty about the world. Classical hypothesis testing involves making null hypotheses about population parameters and then rejecting or not rejecting those hypotheses based on sample data. Typical null hypotheses (as the word *null* would imply) state that there is no difference between proportions or group means, or no relationship between variables. Null hypotheses may also refer to parameters in models involving many variables.

To test a null hypothesis, we compute a special statistic called a test statistic along with its associated p-value. Assuming that the null hypothesis is true, we can derive the theoretical distribution of the test statistic. We obtain a p-value by referring the sample test statistic to this theoretical distribution. The p-value, itself a sample statistic, gives the probability of rejecting the null hypothesis under the assumption that it is true.

Let us assume that the conditions for valid inference have been satisfied. Then, when we observe a very low p-value (0.05, 0.01, or 0.001, for instance), we know that one of two things must be true: either (1) an event of very low probability has occurred under the assumption that the null hypothesis is true or (2) the null hypothesis is false. A low p-value leads us to reject the null hypothesis, and we say the research results are statistically significant. Some results are statistically significant and meaningful. Others are statistically significant and picayune.

For applied research in the classical tradition, we look for statistics with low *p*-values. We define null hypotheses as straw men with the intention of rejecting them. When looking for differences between groups, we set up a null hypothesis that there are no differences between groups. In studying relationships between variables, we create null hypotheses of independence between variables and then collect data to reject those hypotheses. When we collect sufficient data, testing procedures have statistical power.

Variability is both our enemy and our friend. It is our enemy when it arises from unexplained sources or from sampling variability—the values of statistics vary from one sample to the next. But variability is also our friend because, without variability, we would be unable to see relationships between variables.[2]

While the classical approach treats parameters as fixed, unknown quantities to be estimated, the Bayesian approach treats parameters as random variables. In other words, we can think of parameters as having probability distributions representing of our uncertainty about the world.

The Bayesian approach takes its name from Bayes' theorem, a famous theorem in statistics. In addition to making assumptions about population distributions, random samples, and sampling distributions, we can make assumptions about population parameters. In taking a Bayesian approach, our job is first to express our degree of uncertainty about the world in the form of a probability distribution and then to reduce that uncertainty by collecting relevant sample data.

How do we express our uncertainty about parameters? We specify prior probability distributions for those parameters. Then we use sample data and Bayes' theorem to derive posterior probability distributions for those same parameters. The Bayesian obtains conditional probability estimates from posterior distributions.

Many argue that Bayesian statistics provides a logically consistent approach to empirical research. Forget the null hypothesis and focus on the research question of interest—the scientific hypothesis. There is no need to talk

---

[2] To see the importance of variability in the discovery of relationships, we can begin with a scatter plot of two variables with a high correlation. Then we restrict the range of one of the variables. More often than not, the resulting scatter plot within the window of the restricted range will exhibit a lower correlation.

about confidence intervals when we can describe uncertainty with a probability interval. There is no need to make decisions about null hypotheses when we can view all scientific and business problems from a decision-theoretic point of view (Robert 2007). A Bayesian probabilistic perspective can be applied to machine learning as well as traditional statistical models (Murphy 2012).

It may be a challenge to derive mathematical formulas for posterior probability distributions. Indeed, for many research problems, it is impossible to derive formulas for posterior distributions. This does not stop us from using Bayesian methods, however, because computer programs can generate or estimate posterior distributions. Markov chain Monte Carlo simulation is at the heart of Bayesian practice (Tanner 1996; Albert 2009; Robert and Casella 2009; Suess and Trumbo 2010).

Bayesian statistics is alive and well today because it helps us solve real-world problems (McGrayne 2011; Flam 2014). As Efron (1986) pointed out, however, there are reasons everyone is not a Bayesian. There are many works from which to learn about classical inference (Fisher 1970; Fisher 1971; Snedecor and Cochran 1989; Hinkley, Reid, and Snell 1991; Stuart, Ord, and Arnold 2010; O'Hagan 2010; Wasserman 2010). There are also many good sources for learning about Bayesian methods (Geisser 1993; Gelman, Carlin, Stern, and Rubin 1995; Carlin and Louis 1996; Robert 2007).

When asked if the difference between two groups could have arisen by chance, we might prefer a classical approach. We estimate a p-value as a conditional probability, given a null hypothesis of no difference between the groups. But when asked to estimate the probability that the share price of Apple stock will be above \$100 at the beginning of the next calendar year, we may prefer a Bayesian approach. Which is better, classical or Bayesian? It does not matter. We need both. Which is better, Python or R? It does not matter. We need both.

A.3 Regression and Classification

Much of the work of data science involves a search for meaningful relationships between variables. We look for relationships between pairs of continuous variables using scatter plots and correlation coefficients. We look for relationships between categorical variables using contingency tables and the methods of categorical data analysis. We use multivariate methods and multi-way contingency tables to examine relationships among many variables. And we build predictive models.

There are two main types of predictive models: *regression* and *classification*. Regression is prediction of a response of meaningful magnitude. Classification involves prediction of a class or category. In the language of machine learning, these are methods of supervised learning.

The most common form of regression is *least-squares regression*, also called ordinary least-squares regression, linear regression, or multiple regression. When we use ordinary least-squares regression, we estimate regression coefficients so that they minimize the sum of the squared residuals, where residuals are differences between the observed and predicted response values. For regression problems, we think of the response as taking any value along the real number line, although in practice the response may take a limited number of distinct values. The important thing for regression is that the response values have meaningful magnitude.

Poisson regression is useful for counts. The response has meaningful magnitude but takes discrete (whole number) values with a minimum value of zero. Log-linear models for frequencies, grouped frequencies, and contingency tables for cross-classified observations fall within this domain.

For models of events, duration, and survival, as in *survival analysis*, we must often accommodate censoring, in which some observations are measured precisely and others are not. With left censoring, all we know about imprecisely measured observations is that they are less than some value. With right censoring, all we know about imprecisely measured observations is that they are greater than some value.

Most traditional modeling techniques involve *linear models* or linear equations. The response or transformed response is on the left-hand side of the linear model. The *linear predictor* is on the right-hand side. The linear predictor involves explanatory variables and is linear in its parameters. That is, it involves the addition of coefficients or the multiplication of coefficients by the explanatory variables. The coefficients we fit to linear models represent estimates of population parameters. Least-squares regression, Poisson regression, logistic regression, and survival models fall within the class of *generalized linear models* which lie at the core of traditional statistical inference, both classical and Bayesian.

For regression examples in this book, we use R-squared or the coefficient of determination as an index of goodness of fit. This is a quantity that is easy to explain to management as the proportion of response variance accounted for by the model. An alternative index that many statisticians prefer is the *root mean-squared error of prediction* (RMSE), which is an index of badness or lack of fit. Other indices of badness of fit, such as the percentage error in prediction, are sometimes preferred by managers.

The method of *logistic regression,* although called "regression," is actually a classification method. It involves the prediction of a binary response. Ordinal and multinomial logit models extend logistic regression to problems involving more than two classes. Linear discriminant analysis is another classification method from the domain of traditional statistics. The benchmark study of text classification in the chapter on sentiment analysis employed logistic regression and a number of machine learning algorithms for classification.

Evaluating classifier performance presents a challenge because many problems are low base rate problems. Fewer than five percent of customers may respond to a direct mail campaign. Disease rates, loan default, and fraud are often low base rate events. When evaluating classifiers in the context of low base rates, we must look beyond the percentage of events correctly predicted. Based on the four-fold table known as the *confusion matrix*, figure A.1 provides an overview of various indices available for evaluating binary classifiers.

Figure A.1. Evaluating the Predictive Accuracy of a Binary Classifier

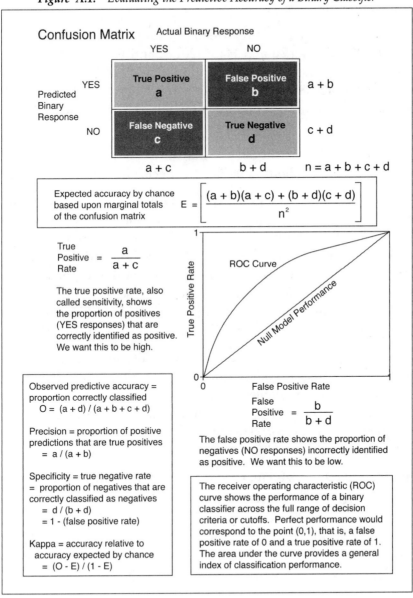

Summary statistics such as Kappa (Cohen 1960) and the area under the receiver operating characteristic (ROC) curve are sometimes used to evaluate classifiers. Kappa depends on the probability cut-off used in classification. The area under the ROC curve does not.[3]

Useful references for linear regression include Draper and Smith (1998), Harrell (2001), Chatterjee and Hadi (2012), and Fox and Weisberg (2011). Data-adaptive regression methods and machine learning algorithms are reviewed in Berk (2008), Izenman (2008), and Hastie, Tibshirani, and Friedman (2009). For traditional nonlinear models, see Bates and Watts (2007).

Of special concern to data scientists is the structure of the regression model. Under what conditions should we transform the response or selected explanatory variables? Should interaction effects be included in the model?

Regression diagnostics are data visualizations and indices we use to check on the adequacy of regression models. Discussion may be found in Belsley, Kuh, and Welsch (1980) and Cook (1998). The base R system provides many diagnostics, and Fox and Weisberg (2011) provide additional diagnostics. Diagnostics may suggest that transformations of the response or explanatory variables are needed in order to meet model assumptions or improve predictive performance. A theory of power transformations is provided in Box and Cox (1964) and reviewed by Fox and Weisberg (2011).

When defining parametric models, we would like to include the right set of explanatory variables in the right form. Having too few variables or omitting key explanatory variables can result in biased predictions. Having too many variables, on the other hand, may lead to over-fitting and high out-of-sample prediction error. This bias-variance tradeoff, as it is sometimes called, is statistical fact of life.

Shrinkage and regularized regression methods provide mechanisms for tuning, smoothing, or adjusting model complexity (Tibshirani 1996; Hoerl and

---

[3] The area under the ROC curve is a preferred index of classification performance for low base rate problems. The ROC curve is a plot of the true positive rate against the false positive rate. It shows the tradeoff between sensitivity and specificity and measures how well the model separates positive from negative cases. The area under the curve provides an index of predictive accuracy independent of the probability cut-off that is being used to classify cases. Perfect prediction corresponds to an area of 1.0 (curve that touches the top-left corner). An area of 0.5 depicts random (null-model) predictive accuracy, with the curve being the diagonal line from bottom left to top right and with the area associated with lower triangle.

Kennard 2000). Alternatively, we can select subsets of explanatory variables to go into predictive models. Special methods are called into play when the number of parameters being estimated is large, perhaps exceeding the number of observations (Bühlmann and van de Geer 2011). For additional discussion the bias-variance tradeoff, regularized regression, and subset selection, see Izenman (2008) and Hastie, Tibshirani, and Friedman (2009).

Graybill (1961, 2000) and Rencher and Schaalje (2008) review of linear models. Generalized linear models are discussed in McCullagh and Nelder (1989) and Firth (1991). Kutner, Nachtsheim, Neter, and Li (2004) provide a comprehensive review of linear and generalized linear models, including discussion of their application in experimental design. R methods for the estimation of linear and generalized linear models are reviewed in Chambers and Hastie (1992) and Venables and Ripley (2002).

See Christensen (1997) and Hosmer, Lemeshow, and Sturdivant (2013) for discussion of logistic regression. See Fawcett (2003) and Sing et al. (2005) for further discussion of the ROC curve. Discussion of alternative methods for evaluating classifiers is provided in Hand (1997) and Kuhn and Johnson (2013).

For Poisson regression and the analysis of multi-way contingency tables, useful references include Bishop, Fienberg, and Holland (1975), Fienberg (2007), Tang, He, and Tu (2012), and Agresti (2013). Reviews of survival data analysis have been provided by Andersen, Borgan, Gill, and Keiding (1993), Le (1997), Therneau and Grambsch (2000), Harrell (2001), Nelson (2003), Hosmer, Lemeshow, and May (2013), and Allison (2010), with programming solutions provided by Therneau (2014) and Therneau and Crowson (2014).

We sometimes consider robust regression methods when there are influential outliers or extreme observations. Robust methods represent an active area of research using statistical simulation tools (Fox 2002; Koller and Stahel 2011; Maronna, Martin, and Yohai 2006; Maechler 2014b; Koller 2014). Huet et al. (2004) and Bates and Watts (2007) review nonlinear regression, and Harrell (2001) discusses spline functions for regression problems.

A.4 Machine Learning

Recommender systems, collaborative filtering, association rules, optimization methods based on heuristics, as well as a myriad of methods for regression, classification, and clustering fall under the rubric of machine learning. These are data-adaptive methods, also called data mining.

The benchmark study of text classification in the chapter on sentiment analysis employed a number of machine learning algorithms, including support vector machines, classification trees, and random forests. Other machine learning tools for classification include Naïve Bayes classifiers and neural networks.

Machine learning methods often perform better than traditional linear or logistic regression methods, but explaining why they work is not easy. Machine learning models are sometimes called *black box models* for a reason. The underlying algorithms can yield thousands of formulas or nodal splits fit to the training data.

Extensive discussion of machine learning algorithms may be found in Duda, Hart, and Stork (2001), Izenman (2008), Hastie, Tibshirani, and Friedman (2009), Kuhn and Johnson (2013), Tan, Steinbach, and Kumar (2006), and Murphy (2012). In addition to the wide variety of machine learning tools available in R, there are many in Python (Pedregosa et al. 2011, Demšar and Zupan 2013) and Java (Witten, Frank, and Hall 2011). Graphical user interfaces to machine learning algorithms in Python and R may be found in open-source solutions KNIME (Berthold et al. 2007) and Orange (Demšar and Zupan 2013) and in commercial solutions from Alteryx, IBM, and SAS.

Hothorn et al. (2005) review principles of benchmark study design, and Schauerhuber et al. (2008) show a benchmark study of classification methods. Alfons (2014a) provides cross-validation tools for benchmark studies. Benchmark studies, also known as statistical simulations or statistical experiments, may be conducted with programming packages designed for this type of research (Alfons 2014b; Alfons, Templ, and Filzmoser 2014).

Duda, Hart, and Stork (2001), Tan, Steinbach, and Kumar (2006), Hastie, Tibshirani, and Friedman (2009), and Rajaraman and Ullman (2012) introduce clustering from a machine learning perspective. Everitt, Landau, Leese, and Stahl (2011), Kaufman and Rousseeuw (1990) review traditional clustering methods. Izenman (2008) provides a review of traditional clustering, self-organizing maps, fuzzy clustering, model-based clustering, and biclustering (block clustering). Within the machine learning literature, cluster analysis is referred to as *unsupervised learning* to distinguish it from classification, which is *supervised learning*, guided by known, coded values of a response variable or class.

Leisch and Gruen (2014) describe programming packages for various clustering algorithms. Methods developed by Kaufman and Rousseeuw (1990) have been implemented in programs by Maechler (2014a), including silhouette modeling and visualization techniques for determining the number of clusters. Silhouettes were introduced by Rousseeuw (1987), with additional documentation and examples provided in Kaufman and Rousseeuw (1990) and Izenman (2008).

Another unsupervised technique, principal component analysis, draws on linear algebra and gives us a way to reduce the number of measures or quantitative features we use to describe domains of interest. Long a staple of measurement experts and a prerequisite of factor analysis, principal component analysis has seen recent applications in latent semantic analysis, a technology for identifying important topics across a document corpus (Blei, Ng, and Jordan 2003; Murphy 2012; Ingersoll, Morton, and Farris 2013).

When some observations in the training set have coded responses and others do not, we employ a *semi-supervised learning* approach. The set of coded observations for the supervised component can be small relative to the set of uncoded observations for the unsupervised component (Liu 2011).

Thinking more broadly about machine learning, we see it as a subfield of artificial intelligence (Luger 2008; Russell and Norvig 2009). Machine learning encompasses biologically-inspired methods, genetic algorithms, and heuristics, which may be used to address complex optimization, scheduling, and systems design problems. (Mitchell 1996; Engelbrecht 2007; Michalawicz and Fogel 2004; Brownlee 2011).

A.5 Data Visualization

Data visualization is critical to the work of data science. Examples in this book demonstrate the importance of data visualization in discovery, diagnostics, and design. We employ tools of exploratory data analysis (discovery) and statistical modeling (diagnostics). In communicating results to management, we use presentation graphics (design).

Statistical summaries fail to tell the story of data. To understand data, we must look beyond data tables, regression coefficients, and the results of statistical tests. Visualization tools help us learn from data. We explore data, discover patterns in data, identify groups of observations that go together and unusual observations or outliers. We note relationships among variables, sometimes detecting underlying dimensions in the data.

Graphics for exploratory data analysis are reviewed in classic references by Tukey (1977) and Tukey and Mosteller (1977). Regression graphics are covered by Cook (1998), Cook and Weisberg (1999), and Fox and Weisberg (2011). Statistical graphics and data visualization are illustrated in the works of Tufte (1990, 1997, 2004, 2006), Few (2009), and Yau (2011, 2013). Wilkinson (2005) presents a review of human perception and graphics, as well as a conceptual structure for understanding statistical graphics. Cairo (2013) provides a general review of information graphics. Heer, Bostock, and Ogievetsky (2010) demonstrate contemporary visualization techniques for web distribution. When working with very large data sets, special methods may be needed, such as partial transparency and hexbin plots (Unwin, Theus, and Hofmann 2006; Carr, Lewin-Koh, and Maechler 2014; Lewin-Koh 2014).

R is particularly strong in data visualization. An R graphics overview is provided by Murrell (2011). R lattice graphics, discussed by Sarkar (2008, 2014), build on the conceptual structure of an earlier system called S-Plus Trellis[TM] (Cleveland 1993; Becker and Cleveland 1996). Wilkinson's (2005) "grammar of graphics" approach has been implemented in the Python ggplot package (Lamp 2014) and in the R ggplot2 package (Wickham and Chang 2014), with R programming examples provided by Chang (2013). Cairo (2013) and Zeileis, Hornik, and Murrell (2009, 2014) provide advice about colors for statistical graphics. Ihaka et al. (2014) show how to specify colors in R by hue, chroma, and luminance.

A.6 Text Analytics

Text analytics draws from a variety of disciplines, including linguistics, communication and language arts, experimental psychology, political discourse analysis, journalism, computer science, and statistics. And, given the amount of text being gathered and stored by organizations, text analytics is an important and growing area of predictive analytics.

We have discussed web crawling, scraping, and parsing. The output from these processes is a document collection or text corpus. This document collection or corpus is in the natural language. The two primary ways of analyzing a text corpus are the *bag of words* approach and *natural language processing*. We parse the corpus further, creating commonly formatted expressions, indices, keys, and matrices that are more easily analyzed by computer. This additional parsing is sometimes referred to as text annotation. We extract features from the text and then use those features in subsequent analyses.

Natural language is what we speak and write every day. Natural language processing is more than a matter of collecting individual words. Natural language conveys meaning. Natural language documents contain paragraphs, paragraphs contain sentences, and sentences contain words. There are grammatical rules, with many ways to convey the same idea, along with exceptions to rules and rules about exceptions. Words used in combination and the rules of grammar comprise the linguistic foundations of text analytics as shown in figure A.2.

Linguists study natural language, the words and the rules that we use to form meaningful utterances. "Generative grammar" is a general term for the rules; "morphology," "syntax," and "semantics" are more specific terms. Computer programs for natural language processing use linguistic rules to mimic human communication and convert natural language into structured text for further analysis.

Natural language processing is a broad area of academic study itself, and an important area of computational linguistics. The location of words in sentences is a key to understanding text. Words follow a sequence, with earlier words often more important than later words, and with early sentences and paragraphs often more important than later sentences and paragraphs.

Figure A.2. *Linguistic Foundations of Text Analytics*

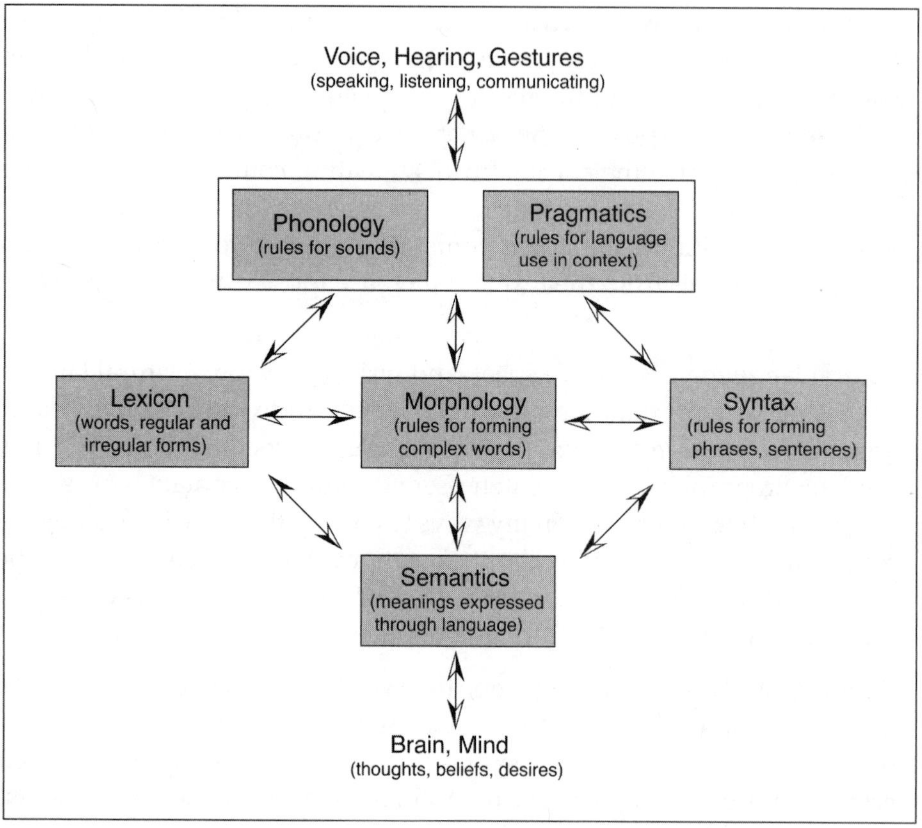

Source: Adapted from Pinker (1999).

Words in the title of a document are especially important to understanding the meaning of a document. Some words occur with high frequency and help to define the meaning of a document. Other words, such as the definite article "the" and the indefinite articles "a" and "an," as well as many prepositions and pronouns, occur with high frequency but have little to do with the meaning of a document. These *stop words* are dropped from the analysis.

The features or attributes of text are often associated with terms—collections of words that mean something special. There are collections of words relating to the same concept or word stem. The words "marketer," "marketeer," and "marketing" build on the common word stem "market." There are syntactic structures to consider, such as adjectives followed by nouns and nouns followed by nouns. Most important to text analytics are sequences of words that form terms. The words "New" and "York" have special meaning when combined to form the term "New York." The words "financial" and "analysis" have special meaning when combined to form the term "financial analysis." We often employ *stemming,* which is the identification of word stems, dropping suffixes (and sometimes prefixes) from words. More generally, we are parsing natural language text to arrive at structured text.

In English, it is customary to place the subject before the verb and the object after the verb. In English verb tense is important. The sentence "Daniel carries the Apple computer," can have the same meaning as the sentence "The Apple computer is carried by Daniel." "Apple computer," the object of the active verb "carry" is the subject of the passive verb "is carried." Understanding that the two sentences mean the same thing is an important part of building intelligent text applications.

A key step in text analysis is the creation of a terms-by-documents matrix (sometimes called a lexical table). The rows of this data matrix correspond to words or word stems from the document collection, and the columns correspond to documents in the collection. The entry in each cell of a terms-by-documents matrix could be a binary indicator for the presence or absence of a term in a document, a frequency count of the number of times a term is used in a document, or a weighted frequency indicating the importance of a term in a document.

Figure A.3 illustrates the process of creating a terms-by-documents matrix. The first document comes from Steven Pinker's *Words and Rules* (1999, p. 4), the second from Richard K. Belew's *Finding Out About* (2000, p. 73). Terms correspond to words or word stems that appear in the documents. In this example, each matrix entry represents the number of times a term appears in a document. We treat nouns, verbs, and adjectives similarly in the definition of stems. The stem "combine" represents both the verb "combine" and the noun "combination." Likewise, "function" represents the verb, noun, and adjective form "functional." An alternative system might distinguish among parts of speech, permitting more sophisticated syntactic searches across documents. After being created, the terms-by-documents matrix is like an index, a mapping of document identifiers to terms (keywords or stems) and vice versa. For information retrieval systems or search engines we might also retain information regarding the specific location of terms within documents.

Typical text analytics applications have many more terms than documents, resulting in sparse rectangular terms-by-documents matrices. To obtain meaningful results for text analytics applications, analysts examine the distribution of terms across the document collection. Very low frequency terms, those used in few documents, are dropped from the terms-by-documents matrix, reducing the number of rows in the matrix.

Unsupervised text analytics problems are those for which there is no response or class to be predicted. Rather, as we showed with the movie taglines, the task is to identify common patterns or trends in the data. As part of the task, we may define text measures describing the documents in the corpus.

For supervised text analytics problems there is a response or class of documents to be predicted. We build a model on a training set and test it on a test set. Text classification problems are common. Spam filtering has long been a subject of interest as a classification problem, and many e-mail users have benefitted from the efficient algorithms that have evolved in this area. In the context of information retrieval, search engines classify documents as being relevant to the search or not. Useful modeling techniques for text classification include logistic regression, linear discriminant function anal-

Figure A.3. *Creating a Terms-by-Documents Matrix*

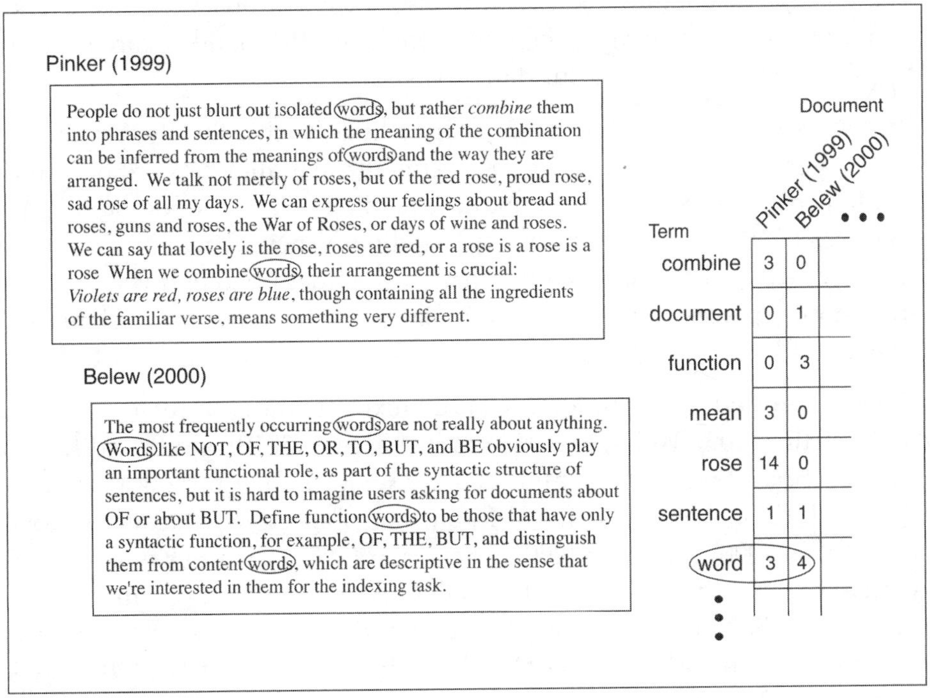

Pinker (1999)

People do not just blurt out isolated words, but rather *combine* them into phrases and sentences, in which the meaning of the combination can be inferred from the meanings of words and the way they are arranged. We talk not merely of roses, but of the red rose, proud rose, sad rose of all my days. We can express our feelings about bread and roses, guns and roses, the War of Roses, or days of wine and roses. We can say that lovely is the rose, roses are red, or a rose is a rose is a rose When we combine words, their arrangement is crucial: *Violets are red, roses are blue*, though containing all the ingredients of the familiar verse, means something very different.

Belew (2000)

The most frequently occurring words are not really about anything. Words like NOT, OF, THE, OR, TO, BUT, and BE obviously play an important functional role, as part of the syntactic structure of sentences, but it is hard to imagine users asking for documents about OF or about BUT. Define function words to be those that have only a syntactic function, for example, OF, THE, BUT, and distinguish them from content words, which are descriptive in the sense that we're interested in them for the indexing task.

Term	Pinker (1999)	Belew (2000)	···
combine	3	0	
document	0	1	
function	0	3	
mean	3	0	
rose	14	0	
sentence	1	1	
word	3	4	

Source: Adapted from Miller (2005).

ysis, classification trees, and support vector machines. Various ensemble or committee methods may be employed.

Automatic text summarization is an area of research and development that can help with information management. Imagine a text processing program with the ability to read each document in a collection and summarize it in a sentence or two, perhaps quoting from the document itself. Today's search engines are providing partial analysis of documents prior to their being displayed. They create automated summaries for fast information retrieval. They recognize common text strings associated with user requests. These applications of text analysis comprise tools of information search that we take for granted as part of our daily lives.

Programs with syntactic processing capabilities, such as IBM's Watson, provide a glimpse of what intelligent agents for text analytics are becoming. These programs perform grammatical parsing with an understanding of the roles of subject, verb, object, and modifier. They know parts of speech (nouns, verbs, adjective, adverbs). And, using identified entities representing people, places, things, and organizations, they perform relationship searches.

Those interested in learning more about text analytics can refer to Jurafsky and Martin (2009), Weiss, Indurkhya, and Zhang (2010) and the edited volume by Srivastava and Sahami (2009). Reviews may be found in Miller (2005), Trybula (1999), Witten, Moffat, and Bell (1999), Meadow, Boyce, and Kraft (2000), Sullivan (2001), Feldman (2002), and Sebastiani (2002). Hausser (2001) gives an account of generative grammar and computational linguistics. Statistical language learning and natural language processing are discussed by Charniak (1993), Manning and Schütze (1999), and Indurkhya and Damerau (2010).

The writings of Steven Pinker (1994, 1997, 1999) provide insight into grammar and psycholinguistics. Maybury (1997) reviews data preparation for text analytics and the related tasks of source detection, translation and conversion, information extraction, and information exploitation. Detection relates to identifying relevant sources of information; conversion and translation involve converting from one medium or coding form to another.

Belew (2000), Meadow, Boyce, and Kraft (2000) and the edited volume by Baeza-Yates and Ribeiro-Neto (1999) provide reviews of computer technologies for information retrieval, which depend on text classification, among other technologies and algorithms.

Authorship identification, a problem addressed a number of years ago in the statistical literature by Mosteller and Wallace (1984), continues to be an active area of research (Joula 2008). Merkl (2002) provides discussion of clustering techniques, which explore similarities between documents and the grouping of documents into classes. Dumais (2004) reviews latent semantic analysis and statistical approaches to extracting relationships among terms in a document collection.

Special topics from computational linguistics provide additional insights into working with text. Text tiling (Hearst 1997) involves the automatic division of text documents into blocks or units for further analysis. Adjacent blocks of text are more likely to have words in common with one another and thus be topically related. Text tiling can be used in text summarization and information retrieval, as well as stylistic analysis of literature and discourse (Youmans 1990; Youmans 1991).

For an overview of text analytics in R, refer to Feinerer, Hornik, and Meyer (2008), Feinerer and Hornik (2014a), and Feinerer (2014). Of special interest are books that deal with the overlap between linguistics and statistics with R (Baayen 2008; Johnson 2008; Gries 2013). The work of Gries (2009) is of special note as it shows how to work with document corpuses. Text analytics utilities are provided by Grothendieck (2014a, 2014b) and Wickham (2010, 2014b). Understanding regular expression coding in R can be of special value to anyone interested in doing text analytics (Friedl 2006; Wickham 2014b). Dictionary capabilities are provided through an interface to WordNet (Miller 1995; Fellbaum 1998; Feinerer 2012; Feinerer and Hornik 2014b). For fun, we have word clouds with utilities for plotting non-overlapping text in scatter plots (Fellows 2014a; Fellows 2014b).

B

Primary Research Online

Primary research online builds on the web. Applications are web-browser based. In the early years of the web, research applications would start by using a web browser for access and then switch to another application or applet (often built on the Java programming language). Today's applications are largely browser- and JavaScript-based on the client side, with server interaction for data requests. Hypertext transfer protocol (HTTP) provides formatting codes for data presentation in browsers. The data model is the Document Object Model (DOM). In addition to JavaScript, scripting languages such as Perl, PHP, and Python, working in conjunction with web servers and database systems, serve up the dynamic web pages of online research applications, including online focus groups, focused conversations, chats, blogs, bulletin boards, in-depth interviews, and online surveys.

Methods of survey research, sometimes called primary quantitative research, are well understood and well documented. Groves et al. (2009) provides a comprehensive review of survey research, including sampling, data collection modalities, data quality, nonresponse, and analysis issues. Miller (2015a) provides discussion of measurement, item formats, and alternative scales of measurement for business research.

This appendix draws from the book *Qualitative Research Online* (Miller and Walkowski 2004) with permission from Research Publishers LLC. Thanks to Jeff Walkowski of QualCore.com Inc. for introducing me to online qualitative research more than ten years ago.

We focus on online qualitative research in this appendix, placing it within the broader context of qualitative research in the social sciences and business. Qualitative research online is a growth area. More than just a new technology or modality of research, qualitative research online represents a cultural change in the way we do qualitative research and the way we think about qualitative research and research in general.

Things happen faster when we go online. We have access to networked resources. We have the computer to collect and analyze data. We take words from e-mail messages and online discussions and automatically convert them into transcripts for text analysis. Raw data are at our fingertips, a few mouse clicks or touch-screen taps away.

While many researchers recognize the time and cost advantages of online research, some decry its claimed methodological flaws. Some are concerned that online methods may replace what they think of as more appropriate traditional methods.

Traditional qualitative methods—face-to-face interviews, focus groups, case studies, field observation, and ethnography—remain important methods for learning about people and cultures. But qualitative research online has a role to play in the social sciences and business. What role can it play, and how can it play that role most convincingly or effectively? Miller and Walkowski (2004) addressed these questions by interviewing many qualitative research practitioners. Dale and Abbott (2014) provide additional discussion of qualitative research online.

Like traditional qualitative research, qualitative research online should be purposeful research. We watch what people do. We listen to what they say through their typing and clicking. We collect data and interpret those data. We try to be objective, knowing that what we learn from data is often guided by the subjective categories we impose on those data. Qualitative research online, like all research, requires care in design, analysis, and interpretation.

Traditional qualitative research comes in many varieties. Ethnographers are the prototypical observers, watching what people do in their natural and social environments: "What do people do at work and play? How do they live their lives? How do they organize themselves into groups? What is the meaning of culture?" Using in-depth interviews, phenomenologists probe the intricacies of human cognition: "How do people make sense of what they see and hear? What do they believe? How do they view the world? What are the meanings of signs and symbols?" Research practitioners lead focus group discussions about topics of interest to their clients: "What does the voter think about the current candidates for governor? What would prompt the consumer to use these products and services? How does the viewer feel about the new TV commercial?"

What makes research qualitative? Qualitative data come to us unstructured, disorganized, free-form. We start with events as they occur, words as spoken, and we tell a story about them. It is not unusual, of course, for different researchers to tell different stories from the same data. But if we do our jobs right, if we are good researchers, we will stay close to the data, and our stories will ring true.

Qualitative research online can involve the analysis of secondary data, such as computer logs of user group discussions or e-mail communication. The researcher can conduct observational studies about human-computer interaction and online communication. Much online qualitative research is primary research, involving customized studies designed for specific purposes, such as in-depth interviews and focus groups conducted online. We can set up usability studies in which users are given explicit tasks to perform on a website and observers record their behavior.

E-mail, a common store-and-forward communications tool, provides a convenient method for interpersonal communications. E-mail is asynchronous, meaning sender and receiver do not need to be online at the same time, and users do not need to be online to receive messages. To send or receive an e-mail message, the user establishes a connection to the network and executes the e-mail application. To send an e-mail message, the user must have the address of the intended recipient. To receive a message, the user runs the e-mail application, locates the message in an in-tray, and opens the message—like going to a mailbox, reaching in to get a letter, and opening the letter. Many e-mail facilities have utilities for notifying users that e-mail messages have been received.

Listservs are programs for creating automated multi-user e-mail groups. User and special interest group messages sent to a listserv are automatically distributed to the e-mail addresses of list members.

Text messaging, like e-mail, is an interpersonal communications tool. Those who send and receive messages must have unique names or addresses. But text messaging provides synchronous or real-time communications.

A chat room is a synchronous or real-time facility for communication within a group; it is an application running on a server on the Internet. To gain access to the application, chat room members type the URL for the chat room. After they enter the chat room, they are identified by their user names or by pseudonyms provided by the chat room administrator. Public chat rooms are open to all individuals with access to the web.

Bulletin boards, also called message boards or web boards, are asynchronous facilities for communication among groups. Like chat rooms, these are application programs running on a server on the Internet. Access is via URLs, with public bulletin boards being open to anyone who has the URL. As with chat rooms, organizations can develop customized bulletin board applications using public domain or commercial software tools.

Unlike public chat rooms and bulletin boards, which are open to anyone with access to the Internet, most qualitative research online is designed for qualified respondents, members of target populations recruited for research purposes. Recruitment may be conducted by telephone, mail, e-mail, web-based surveys, or a combination of modalities. There are also automated recruitment services such as Amazon's Mechanical Turk.

Much qualitative research online is professionally moderated. Working from a discussion guide that has been prepared in advance and from an understanding of the information needs of research clients, a moderator keeps participants active, interested, and focused on selected research topics.

A weblog or blog is a website with online information in a dated log format. Weblogs are useful for individuals and organizations wanting to provide current information relating to a particular content domain. Blog postings may be viewed as personal statements or monologues, rather than online conversations. As facilities for communities of users focused on a common theme, posting information relevant to that theme, weblogs represent rich sources of data for ethnographic analysis and qualitative research online.

Lindahl and Blount (2003) provided an overview of the history and technology of weblogs, noting that they had grown in number from a handful in 1998 to more than a million in 2003. There are many more today.

Traditional ethnography involves field observation, the study of social and consumer behavior in real life settings, as introduced by Wolcott (1994, 1999, 2001). Cell phones, including those with integrated cameras and personal data assistants, are changing the face of traditional ethnography. Traditional ethnography is becoming "digital ethnography," observation assisted by digital technology that may be easily transported into the field.

Online ethnography, virtual ethnography, or "netnography," as it has been called by Sherry and Kozinets (2001), concerns the study of online behavior itself. This is ethnographic research that draws on computer- or network-based data and artifacts.

A burgeoning literature deals with online communication, culture, and social theory (Walther 1996; Davis and Brewer 1997; Jones 1997; Johnson 1997; Lévy 1997; Holeton 1998; Jones 1999; Jonscher 1999; Smith and Kollock 1999; Herman and Swiss 2000; Mann and Stewart 2000; Miller and Slater 2000; Huberman 2001; Lévy 2001). There are numerous studies of virtual communities or aggregations of individuals that emerge through communication over the Internet (Porter 1997; Markham 1998; Cherny 1999; Smith and Kollock 1999; Rheingold 2000). Ellis, Oldridge, and Vasconcelos (2004) provide a useful review of this literature.

Generally speaking, an online community is a group of individuals that interacts online over an extended period of time. Large online communities, composed of thousands of participants, are possible. Werry and Mowbray (2001) provide numerous examples of large corporate-sponsored and nonprofit online communities. When used for primary research, online communities often consist of groups of fifty to two hundred people who have been recruited to participate in research about related topics over a period of three to twelve months.

Research with online communities may incorporate a variety of methods, including online surveys, focus groups, and bulletin boards. Preece (2000), Powazak (2001), and McArthur and Bruza (2001) review critical success factors in developing online communities, including shared purpose, knowledge, and beliefs; participant commitment to repeated access over time;

and technology for communication, shared resources, and data storage and retrieval.

Taking their name from traditional focus groups, online focus groups are group interviews guided by a moderator. We identify two major types: synchronous (real-time) and asynchronous focus groups. An online focus group is a moderated group interview conducted online either synchronously or asynchronously.

There are parallels between online and traditional face-to-face focus groups. In an online focus group, like a face-to-face group, there is a moderator, a group of participants, and observers. The observers are not visible to the participants. Participants are pre-qualified for the group to ensure that they have relevant demographics and experience for the group. The moderator is a trained discussion leader. The moderator begins the discussion by introducing himself or herself, explaining the purpose of the group, disclosing the fact that there are observers, and covering "ground rules" for all to follow. The discussion covers particular topic areas, agreed to in advance by the moderator and the research client. At the end of the online focus group, the moderator thanks participants for their time and opinions, and guides them out of the focus group.

There are differences between online and traditional focus groups. Face-to-face moderators often work from abbreviated discussion guides. Many moderator comments are extemporaneous and personal. When working online, the discussion guide is typically more explicit and complete. Things that might be said informally or extemporaneously in face-to-face communication should be formalized and typed out for an online group. Because the moderator isn't meeting with participants in person, the moderator may try to put his or her personality into the guide. The moderator's initial monologue, usually unwritten for face-to-face focus groups, needs to be part of the online discussion guide. As a result, online discussion guides are usually longer than face-to-face guides.

Face-to-face participants see name cards and faces around a table, whereas online participants see a computer screen with group members identified by their first names or pseudonyms. Face-to-face participants may be rewarded with good food, but they must travel to a common location to get the food. Online respondents often participate from their homes or workplaces. For face-to-face groups, moderators often show photographs, concept boards, videotapes, and other physical stimuli as part of the focus group. Online focus group applications provide virtual white boards for the display of digital stimulus materials, such as computer graphics and images.

Consider a generic information systems architecture for qualitative research online. Users, including moderators, participants, and observers on the client side, connect to a network through web browsers. Conversational content is formatted, bundled in HTML tags, and communicated over the Internet using the HTTP protocol. Apache and Microsoft web servers manage the interchange of dynamic conversational content across the Internet channel. Other information systems components include data collection applications, database services, and analytical services for text and data analysis.

To implement qualitative research online, systems programmers and administrators are responsible for system, network, and database management. The qualitative researcher (often the same person as the moderator) analyzes transcripts and data from online bulletin boards, synchronous focus groups, and surveys. Technical specialists sometimes work alongside moderators, providing software and typing assistance with research applications. And there is the usual bevy of administrative support staff engaged in recruiting and paying focus group participants.

We can provide a general classification of methods for qualitative research online by noting (1) whether or not a researcher, interviewer, or moderator is actively involved with participants and (2) the synchronous or asynchronous time frame for the research.

An interview involves an interviewer (moderator) and one or more interviewees (participants). Online focus groups, reviewed in previous sections, are group online interviews. It is also possible to conduct online individual interviews or what are often called "one-on-ones." The interviewer leads an interviewee in a conversation related to research topics of interest. Online individual interviews can make use of specialized software for online focus groups or general-purpose software for e-mail, instant messaging, listservs, chat rooms, and bulletin boards.

Interviews put the researcher (interviewer, moderator) in direct contact with research subjects (respondents, participants). The researcher is active, asking questions, listening, and probing. Synchronous interviews and conversations take place in real time with fully overlapping time intervals for moderators and participants. Asynchronous interviews and conversations take place across the same period of time, but do not require overlapping time intervals for moderators or participants. Mann and Stewart (2002) provide a review of online interviewing.

Traditional field observation has its online analogue in online observation. Online observation is conducted by observing the behavior of computer and network users as it happens. More common is observation of online behavior after it happens, which is a form of database research. Databases include computer records of past organizational behavior and text documents relevant to the purposes of qualitative research. These data, collected as part of the day-to-day communication activities of organizations, arise from synchronous applications like instant messaging and chat, and asynchronous applications like e-mail, listservs, and bulletin boards.

Focused conversations and observational studies do not put the researcher in direct contact with research subjects. Open-ended text responses from online surveys provide an additional data source for qualitative research online.

The real-time or synchronous focus group is a widely used online method. The real-time focus group is a guided group interview with all participants logged on at the same time. It is common to have from six to twelve participants and a moderator involved in a real-time focus group.

Real-time focus groups are usually conducted with specialized software. Like chat room applications, real-time focus group applications permit users to connect to a synchronous online discussion using a standard web browser. In addition to providing a facility for an online real-time discussion over the Internet, real-time focus group applications include special facilities for moderators and observers.

Figure B.1 shows time lines for a real-time focus group, illustrating system availability, time online (connected), and time active (reading or typing messages) for three participants and a moderator. Although participants are online for most of the one- to two-hour focus group, they are not

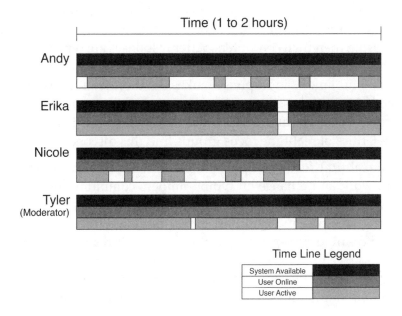

Figure B.1. *Participant and Moderator Time Lines for a Real-Time (Synchronous) Focus Group*

equally active participants. Andy appears to be paying attention to something other than the focus group. Erika was online and active for most of the session, although she had a system failure preventing her from participating for a short period of time. Nicole had frequent interruptions and logged off early. Tyler, as we might hope for a moderator, was online and active for the entire period except for brief periods of inactivity or inattention.

System failures with real-time focus groups may result from computer or network problems. One participant having a system failure for a short period of time, as shown in the time lines for Erika, may be tolerated. Given the one- to two-hour duration of real-time focus groups, system failures of more than ten or fifteen minutes and system failures that affect all participants, such as server or network failures, can be very disruptive, affecting the flow of conversation. It is sometimes necessary to ignore data from groups run while experiencing system problems.

The experience of a text-based real-time focus group is qualitatively different from a face-to-face group. For a face-to-face focus group, the moderator is likely to encourage participants to take turns, to speak one at a time. Not so with a real-time focus group. Online participants type responses simultaneously using their own computers. After responses are typed, participants send them to the discussion window, where they can be viewed by everyone. This simultaneity in response makes the discussion transcript appear disjointed or less like a sequential conversation. Because participants aren't taking turns in conversation, many discussion entries will appear to be out of order. Critics of real-time focus groups find this disconcerting.

Critics of real-time focus groups say the Internet is an alien environment in which to work and reliance on typing hinders communication. The environment is impersonal, which keeps moderators from digging deeper into the real motivations and attitudes of respondents. Researchers may get short, top-of-mind responses rather than the rich responses they can get with face-to-face groups. With research participants working online, moderators can miss subtle cues in body language and tone of voice, which can be important in drawing valid inferences from the discussion.

Real-time focus group usage has grown despite the criticism. It has grown because well-designed software applications for online research have many advantages. In addition to utilizing the Internet with its worldwide reach, online focus groups offer opportunities for stimulus display, experimentation, and control not available with traditional focus groups. Because participants type their messages, a printed transcript of the discussion is easy to obtain.

From a moderator's point of view, the real-time focus group involves two concurrent discussions: (1) the focus group discussion among research participants and moderator, viewable by participants, moderator, and clients, and (2) the discussion among clients and moderator, viewable only by clients and moderator. Online focus group applications empower clients. Rather than having selected times for clients to send questions or suggestions to the moderator, as with a traditional face-to-face group, clients have continuous access to the moderator. They can ask questions, make suggestions, or send requests at any time without respondents knowing about these communications. If the moderator is willing to accommodate their requests, clients can have a substantial impact on the direction of a focus group discussion.

Well-designed focus group applications also provide facilities for sending private messages to participants or clients. If a participant is dominating the discussion or making inappropriate remarks, the moderator can send a private message to encourage more cooperative and appropriate behavior. If the participant doesn't "shape up," the moderator can discretely drop the participant from the group. Private messages may also be sent to clients.

The bulletin board or asynchronous focus group has similarities to the real-time focus group. A moderator leads the discussion, using a guide prepared in advance. There is interaction among participants and probing from the moderator. Participants type their responses, which are visible to participants, clients, and the moderator. Because responses are typed, electronic transcripts are generated automatically by the software.

Asynchronous focus groups are usually conducted with specialized software. Like general-purpose bulletin board or message board applications, asynchronous focus group applications permit users to connect to an asynchronous discussion over the Internet using a standard web browser. But, in addition to providing a facility for an online discussion, asynchronous focus group applications include special utilities for moderators and observers. Well-designed applications include facilities for private moderator-client and moderator-participant communications, as well as for the white board display of computer graphics and images.

The moderator encourages asynchronous focus group participants to log on once or twice a day for the duration of the discussion. With each log-on, participants might be expected to spend fifteen to thirty minutes reading and responding to messages. Accordingly, the total time participants devote to the asynchronous focus group is likely to be greater than the time devoted to a real-time focus group.

Figure B.2 shows an asynchronous focus group with system availability, time online (connected), and time active (reading or typing messages) for three participants and a moderator. Except for a short period of unavailability (perhaps due to a network or server failure), the focus group facility is available for days or weeks at a time. Participants log on for short periods of time and are active (reading and posting messages) for part of the

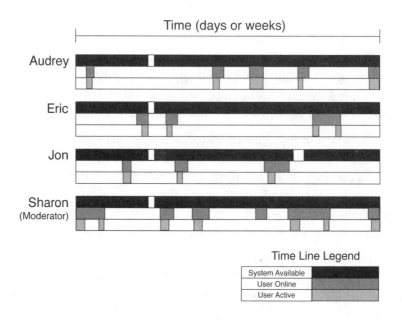

Figure B.2. *Participant and Moderator Time Lines for a Bulletin Board (Asynchronous Focus Group)*

time they are logged on. Individual system failures sometimes prevent a participant from being online, as shown in the time lines for Jon.

The moderator of an asynchronous focus group is likely to log in two or more times a day, responding to participant messages and promoting involvement and interaction. The moderator also sends e-mail reminders to inactive participants, encouraging their involvement in the discussion. Accordingly, the total time a moderator devotes to an asynchronous focus group is likely to be greater than the time devoted to a real-time focus group.

Well-designed applications for asynchronous focus groups afford moderators an opportunity to post messages at predetermined times. We use the word "post" rather than "send" when talking about an asynchronous focus group because participants do not actually see moderator messages until they log on to the focus group and request to see them. To accompany the discussion guide, software for asynchronous focus groups provides a schedule for the posting of moderator messages and white board stimuli.

Asynchronous focus groups have an epistolary quality. The moderator sends a message, beginning a sequence of related statements or discussion thread. Time passes, and participants post their replies. Time passes, more replies are posted. Time passes, and the moderator sends another message, beginning a second discussion thread. Like letters between distant friends or lovers, accumulated over time and saved for future reading, the transcript for an asynchronous focus group becomes the record of a virtual relationship, a meeting of minds or sharing of opinions over time.

Because participants can join in or continue with the discussion at any place or time, a mechanism is needed for providing order to the discussion. What provides order in an asynchronous focus group discussion is its organization along discussion threads. Each new message from the moderator begins a new discussion thread. Discussion threads remain active for the duration of the asynchronous focus group. Relying, as we often do, on the intelligence and goodwill of participants, we hope that messages falling under a discussion thread are indeed relevant to that thread.

Asynchronous focus groups share many of the advantages and disadvantages of real-time focus groups and online research in general. In addition, asynchronous focus groups have special value for working with professional or hard-to-reach participant groups. Busy people find it hard to attend real-time focus groups, which require all participants to be logged on concurrently. An asynchronous focus group gives participants the opportunity to log on at various times during the day or night, weekday or weekend, for the duration of the study.

Asynchronous focus groups are also good for discussions of difficult topics. Because they occur over a period of days or weeks, participants don't feel rushed and don't need to wait for others to type their responses. Asynchronous focus groups give participants time to describe not just what they think, but also how they came to their conclusions. Lengthy, in-depth responses are common as people discuss their reasoning.

Moderators and research clients want to hear about discussion topics in the participants' words. The objective is to get respondents to talk freely without making assumptions about their understanding of topics and without introducing concepts or words that reflect the biases of the moderator or client.

Focus group moderators strive to intervene as little as possible in focus group conversations. They try to ask the one "magic" question that will allow a group of participants (a dyad or triad perhaps) to continue with its own lengthy conversation, hitting important topics of discussion. Moderators in training are encouraged to strive for "rolls," times in the discussion when a group is, without knowing it, self-moderating. These are the times when entire pages of a transcript show no moderator interruptions.

A moderator might say, "Imagine that I'm a fly on the wall, just listening in. Talk with each other, not to me. But, like a fly, I might occasionally 'land' on the table, ask a question, or steer your group in a direction it needs to go. Remember, you are the center of attention, not me."

Despite the best of intentions, good moderators will never be "flies on the wall." They are not invisible, not so long as they are in the room or identified by name on the computer screen. Experimenter effects are alive and well in qualitative research, with participants trying to please moderators, interviewers, and observers. Is there a way to have a more natural conversation among participants, an unmoderated conversation?

Online conversations without moderators are commonplace. They may be observed in e-mail and instant messages between friends and acquaintances. They are seen in public chat rooms, listservs, and bulletin boards. People talk with people, sharing opinions and information about many topics of interest to social and business researchers. What these free-flowing conversations lack is structure and focus, qualities that can be provided by a conversation guide.

Although the term "focused conversation" could easily be used to describe many qualitative research methods, including in-depth interviews and moderated focus groups, we use the term to describe online conversations without an interviewer or moderator interacting with participants. A focused conversation is an open-ended conversation among research participants with the focus provided by a conversation guide. Because there is no interviewer or moderator guiding the conversation, the focused conversation is participant self-directed or self-moderated with a conversation guide to provide focus.

A focused conversation is first and foremost a conversation. The objective is to learn about people in their own words. In his book on interviewing, Kvale (1996) argued for the primacy of conversation in qualitative research. In his words, "Conversation is a basic mode of human interaction. Human beings talk with each other—they interact, pose questions, and answer questions. Through conversations we get to know people, get to learn about their experiences, feelings, and hopes" (Kvale 1996, p. 5).

Imagine a focused conversation between two people, much like a series of letters or messages being exchanged over time. The time frame could be short (twenty to sixty minutes for a real-time focus group) or long (days or weeks for an asynchronous focus group). What matters is not the time, but the depth of conversation, the extent to which participants challenge one another to explain their points of view.

A focused conversation between two people is like an in-depth, open-ended interview without the interviewer. We can think of it as concurrent individual interviews with peers taking turns as interviewer and interviewee. What we hope to achieve from focused conversation is a medium in which participants freely pose and respond to questions in their own words. There will be interaction among participants, but that interaction is not directly influenced by a moderator.

To have a focused conversation, the researcher prepares a conversation guide, consisting of stimulus events and a schedule for their presentation. Stimulus events include participant instructions, questions for discussion, and materials for review. Stimulus events may be set to occur at pre-specified times or triggered by participant requests or conversation conditions, such as the number of words typed in response to a prior question, the number of messages sent by participants, or observed lulls in real-time conversation.

Online technology helps to implement the focused conversation. E-mail, instant messaging, chat, or online focus group software could be used. Asynchronous focus group applications may be used, given their capability for organizing discussion along threads and posting text messages and white board stimulus materials at designated times. Still better is specialized software designed for focused conversation.

In a focused conversation, research participants read from the conversation guide and type their responses. They have conversations with one another about the topics listed in the guide. As with online focus groups, participants' typed responses are collected and stored by the software application, producing an automatic transcript of the conversation. But unlike online focus groups, there is no moderator for the conversation itself.

A focused conversation between two people is likely to show explicit turn-taking. There is the natural reciprocity of a two-person dialogue. With each new discussion topic, there is an initial opportunity for participants to respond simultaneously. But as the discussion continues, turn-taking emerges. Suppose Jennifer and Shawn respond to a question from a conversation guide. Jennifer reads Shawn's response and sends her reply. Shawn replies to Jennifer's message. Jennifer replies to Shawn's message, and so on, until Jennifer and Shawn address the next topic of conversation from the guide. Turn-taking in focused conversation results in a transcript that reads like a natural conversation. There is sequence; there is flow.

Figure B.3 shows time lines for a synchronous focused conversation, such as one that might be conducted using a real-time focus group application, chat room, or instant messaging software. With a synchronous focused conversation of twenty to sixty minutes, we should expect participants to be online and active most of the time. Periods of inactivity are sometimes coordinated, with participants agreeing to take short breaks at the same time, as shown for Jennifer and Shawn. Figure B.4 shows time lines for an asynchronous focused conversation, such as one that might be conducted using asynchronous focus group, e-mail, or bulletin board software.

Participants in asynchronous focused conversations can coordinate schedules, sharing their plans for online activity over the days or weeks of a study. There may be times when participants are online at the same time, experiencing the immediacy of a synchronous conversation.

Tools for qualitative research online may be understood as general-purpose tools for collaboration. They have the potential of fostering productive communication across a wide range of application areas.

What we learn from qualitative research online can contribute to qualitative research in general. Successful experiments with online focused conversation, for example, suggest new ways of doing offline research. Consider the possibilities—two participants in a room opening a series of envelopes with questions for discussion, self-moderating an offline focused conversation.

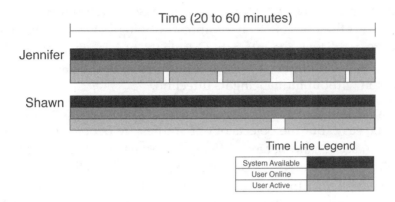

Figure B.3. *Participant Time Lines for a Synchronous Focused Conversation*

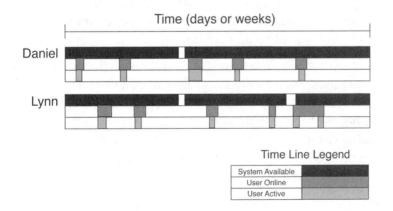

Figure B.4. *Participant Time Lines for an Asynchronous Focused Conversation*

What do we mean by qualitative research? We start with observation, watching behavior in natural and social contexts, seeing communities at work or play, learning from what we see. Standing nearby the objects of our study, we strive to be unobtrusive, involved in observation but not directly involved with the objects of observation. Moving beyond observation, we interview, we listen, we interact. Our goal—to hear the voices of people in their own words—is consistent with the role of facilitator, encouraging discussion without influencing its direction.

If we were biologists studying ants, being unobtrusive and nondirective would be easy. But the objects of our studies are people. Research participants can see us and behave differently because they see us. Much qualitative research involves collaboration between researcher and participant. Interviewees know they're being interviewed. Focus group volunteers see the one-way mirrors. They expect to be observed, audio-taped and videotaped for research purposes. Accepting this truism of social research, we take the roles of participatory observers, interviewers, and moderators. We interact, we react to what we see and hear, and, like it or not, we influence what people do and say. It is a fine line between passive observation and active participation, between objective and subjective, between science and art. It is a line often crossed in qualitative research.

We have the participant and the researcher, and now we have the computer and the network. To say that the qualitative researcher's life will not change with technology is to ignore most of what has been happening for the last fifty years with computers and communications.

There is a need for research on research and research comparing online and offline modalities, such as the study by Esipova et al. (2004). A number of issues merit study. Online qualitative proponents claim that respondents are more honest online than in traditional settings. Critics speculate that in the online environment respondents may actually take on different personas and be less honest than in face-to-face settings. We need to find out if there is evidence to back up either view.

More than twenty-five years since its first application, qualitative research online is still in its infancy. Exciting work awaits those willing to embrace technological innovation. The research world, along with the world of communication, is changing quickly.

Panels of experienced respondents have replaced samples of recruited respondents in many research settings. Building on self-motivated online communities and public chat rooms, some research providers have proposed the development of online communities of research respondents, groups of concerned consumers who are paid to discuss selected topics, such as their feelings about a brand or product category, over a long period of time.

The data of qualitative research online are words—text. Words of consumers and research subjects, user group discussions, logs from online chats and e-mail, real-time focus group and bulletin board transcripts—text from many sources is being stored electronically, available for future research and analysis. Methods of text analytics reviewed in this book are generally applicable to the analysis of qualitative research data.

Useful reviews of methods for qualitative data analysis and interpretation have been presented by others (Silverman 1993; Georgakopoulou and Goutsos 1997; Roberts 1997; Ryan and Bernard 2000; Silverman 2000a). Especially relevant to social scientists and marketing researchers is grounded theory construction (Glaser and Strauss 1967; Strauss and Corbin 1998; Dey 1999; Charmaz 2002).

Word counts, automated text summarization, and text analytics procedures search for regularities in word and phrase usage, turning words into numerical summaries. The distinction between qualitative and quantitative research becomes blurred when we work with various numerical approaches to qualitative data analysis. Reviews of content analysis methods have been presented by Schrott and Lanoue (1994), Silverman (2000b, 2001), and Neuendorf (2002). Computer-aided methods of text analysis have been discussed by others (Kelle 1995; Weitzman and Miles 1995; Roberts 1997; Stone 1997; Fielding and Lee 1998; Popping 2000; Weitzman 2000; West 2001; Seale 2002). And our review of text analytics in the data science appendix A (page 253) applies to primary as well as secondary research.

Standard research issues remain, with sampling being one of those issues. When conducting primary research, we often begin by defining a target population and a sampling method. We try to obtain samples that are representative of the population. An unfortunate fact about secondary research on the web is that sampling questions are rarely asked. Much secondary research is carried out with the data that just happen to be available—convenience samples. It is hard to generalize from a convenience sample because we do not know to what population we are generalizing.

We are at the forefront of a new era in research. Call it ultramodern rather than postmodern. Rather than decrying the limitations of science, we see the emergence of a technology-facilitated science with online methods permeating many forms of research, quantitative and qualitative.

Some online research systems rely on reading and typing of text. Others combine text, streaming audio and video, and teleconferencing utilities. The web as a transmission medium continues to expand through higher bandwidth and wider distribution. Computer system performance continues to improve, and many online systems are being built around highly responsive client-side applications.

What shall we do about change in research modalities and methods? Adapt and accommodate. How shall we respond to new media? Use it in creative ways to enhance communication and research. Accept new technologies for what they can do. Be open to a variety of methods, both offline and online. Recognize that, despite the many ways we have of communicating and collaborating, we must continue to search for better ways of understanding one another.

C

Case Studies

This appendix reviews cases in web and network data science and modeling techniques in predictive analytics, as we have in the chapters of the book. Data for the cases are available in the public domain or are provided on the book's website: http://www.ftpress.com/miller/.

C.1   E-Mail or Spam?

This case concerns the automatic detection of junk e-mail or spam. Spam is unsolicited and unwanted commercial e-mail, mass mailings such as advertisements for products, get-rich schemes, chain letters, and adult erotic literature. Spam is usually sent to people on mailing lists and newsgroups. This type of e-mail activity is considered unethical because the full cost of sending the messages is not borne by the senders and because recipients have not agreed to receive the messages.

Adapted from Miller (2005). According to Hormel Foods Corporation (http://www.spam.com), the use of the word "spam" to describe unsolicited commercial e-mail resulted from a Monty Python skit featuring Hormel's meat product SPAM (spelled with all capital letters). In the skit a group of Vikings sang a chorus of "spam, spam, spam," drowning out other conversation. By analogy, unsolicited commercial e-mail can drown out good e-mail. For additional background on e-mail spam, see Cranor and LaMacchia (1998) and the *InformationWeek* online site (http://information.week.com/spam). Ethical issues regarding spam, including research uses of spam, are discussed on Web sites for the Interactive Marketing Research Organization (http://www.imro.org) and the Council of American Survey Research Organizations (http://www.casro.org). The original data for this case were generated by Mark Hopkins, Erik Reeber, George Forman, and Jaap Suermondtat at Hewlett-Packard Laboratories. George Forman donated the data to the Machine Learning Repository at the University of California–Irvine, thus placing them in the public domain. Exemplary analyses of these data may be found in Izenman (2008) and Hastie, Tibshirani, and Friedman (2009).

E-mail text categorization is important to business. As reported by *InformationWeek* in August 2003, levels of spam reached epidemic proportions as spam-promoting viruses and worms attacked e-mail applications and servers on the Internet. Overrun by bogus messages, many organizations were forced to shut down their e-mail systems. Interest grew in software products designed to filter or block spam, as well as software patches to server systems to prevent further security violations.

We can use the text characteristics of e-mail messages to identify spam versus normal e-mail. Spam indicators could be words relating to money, a preponderance of capital letters, or special characters that are used to garner the reader's attention. Indicators of normal e-mail include personal names, work-related terms, and work-related numbers, such as telephone area codes for business contacts.

The data for this case consist of descriptors of 4,601 e-mail messages collected at Hewlett-Packard Laboratories in 1999. Of these messages, 1,813 (39.4 percent) were classified as spam. There are fifty-seven potential explanatory variables, all continuous, described in Exhibit C.1. The response variable is the binary classification of the message as spam or normal e-mail.

The fundamental problem in the case concerns the development of a model for classifying e-mail messages. Many methods may be explored, both traditional and data adaptive. What type of method works best? How shall we select an appropriate model, and how shall we evaluate its performance?

Table C.1. *Data Coding: E-mail or Spam?*

Variable Name	Description
make	Percentage of words or character strings in the message that correspond to make
address	Percentage of words or character strings in the message that correspond to address
all	Percentage of words or character strings in the message that correspond to all
...	Forty-five additional percentage-of variables follow the structure above, with variable names associated with words and character strings (variable names are set to begin with the letter x in front of numbers)

```
3rd our over remove internet

report addresses free business email

you credit your font 000

money hp hpl george 650

lab labs telnet 857 data

415 85 technology 1999 parts

pm direct cs meeting original

project re edu table conference
```

Variable Name	Description
semicolon	Percentage of characters in message that match the semicolon character ;
openparen	Percentage of characters in message that match the open parenthesis character (
openbracket	Percentage of characters in message that match the open bracket character [
exclamation	Percentage of characters in message that match the exclamation point !
dollarsign	Percentage of characters in message that match the dollar sign character $
pound	Percentage of characters in message that match the pound or number character #
avecaprun	Average length of uninterrupted sequences of capital letters
maxcaprun	Maximum length of uninterrupted sequences of capital letters
totcap	Total number of capital letters in the message
class	Classification of the message ("email" or "spam")

C.2 ToutBay Begins

In October 2014, ToutBay LLC, a publisher and distributor of data science applications, remains in start-up mode awaiting the release of its first products. In what is becoming an increasingly data-driven world, ToutBay co-founders Greg Blence and Tom Miller see opportunities for *data science as a service (DSaaS)*, a term they use to describe ToutBay's business. ToutBay's goal is to be a market maker in the data science space, publishing and distributing time-sensitive information and competitive intelligence.

The ToutBay website www.toutbay.com tells the story of a company founded in December 2013 to provide access to applications developed by analysts, researchers, and data scientists across the world. These subject matter experts or touts work with data and develop models that are of use to many people. As a two-minute video on the website claims, *ToutBay gets people together—people who have answers and people who need answers.* The video introduces the firm and explains why information from ToutBay can be more valuable than information freely obtained from search engines.

There are four application areas: sports, finance, marketing, and general information. Sports touts go beyond raw data about players and teams to build models that predict future performance. ToutBay works with sports touts to make their predictive models available to players, owners, managers, and sports enthusiasts.

Finance touts help individuals and firms make informed decisions about when and where to make investments. These touts have expertise in econometrics and time series analysis. They understand markets and predictive models. They detect trends in the past and make forecasts about the future.

One of ToutBay's first financial products, the Stock Portfolio Constructor, is the work of Dr. Ernest P. Chan, a recognized expert in the area of quantitative finance and author of two books on the subject (Chan 2009, 2013). The idea behind this product is to allow a stock investor to specify his/her investment objectives and time horizon, as well as the domain of stocks being considered and the number of stocks desired in a portfolio. Then,

This case draws from information at the ToutBay website http://www.toutbay.com and from Google Analytics reports, including reports summarizing Scroll Depth plug-in data.

using current information about stock prices and performance, as well as selected economic factors, the Stock Portfolio Constructor creates a customized stock portfolio for the investor. It lists the selected stocks and shows their expected future return over the investor's time horizon, assuming an equal level of investment in each stock. The Stock Portfolio Constructor also shows what would have been the historical performance of that portfolio in recent years.

Marketing touts play a similar expert role, going beyond raw sales data to provide consumer and marketplace insights. They have formal training in measurement, statistics, or machine learning, as well as extensive business consulting experience. The results of their models for site selection, product positioning, segmentation, or target marketing are of special interest to business managers.

General science, a fourth product area, involves scientists of all stripes, building models of human interest. ToutBay intends to give scientific thinking and models wider distribution than they are likely to receive when published in academic journals.

ToutBay's major public event to date has been the R User Conference, also known as UseR!, June 30 through July 3, 2014. The conference was held on the UCLA campus in Los Angeles, California, and attracted around 700 scientists and software engineers, people who write programs (scripts) in the open-source language R (a widely used language in statistics and data science). ToutBay was one of the sponsors of UseR!, along with major software developers and publishers.

ToutBay's goal at UseR! was to introduce itself to potential touts. The company's message was simple: *You do the research and modeling, and we do the rest. We turn scripts into products.* The idea is that, by working with Tout-Bay, data scientists can focus on data science and ToutBay will take care of marketing, communications, sales, order processing, distribution, and customer support. The ToutBay website has a *For Touts* page that provides the details.

Because ToutBay operates entirely online, its business depends on having a website that conveys a clear message to visitors or guests. Success means converting website guests into ToutBay account holders. And after information products become available, success will mean converting account holders into subscribers to information products.

Revenues will come from customer subscriptions, with touts setting prices for their information products and ToutBay charging a fee for online sales and distribution of those products. In recruiting future touts, ToutBay has a simple message: *If you were the author of a book, you would look for a publisher, and you would hope that the publisher would work with bookstores to sell your book. But what if you are the author of a predictive model? Where do you go to publish your model? Where do you go to sell the results of your model? ToutBay— that's where.*

Since opening its website in April 2014, ToutBay has been tracking user traffic with Google Analytics. Recently, the firm has been reviewing data relating to visits, page views, and time on the site. There may have been a slight increase in traffic around the time of the UseR! conference. Otherwise, traffic has been limited, which is a source of concern for the company.

The ToutBay website employs a single-page design, with extensive information on the home page, including the two-minute video introduction to the company. A single-page approach to website design provides better overall performance than a multi-page approach because a single-page approach requires fewer data transmissions between the client browser and the website server.

One difficulty in employing a single-page approach, however, is that standard page-view statistics provide an incomplete picture of website usage. Recognizing this, ToutBay website developers employed JavaScript code to detect how far down users were scrolling on the home page. These scrolling data are included in user traffic information for the site. Table C.2 shows variables and variable definitions for website data under review.

The ToutBay co-founders hope that a detailed analysis of website content and structure, as well as data about website usage, will provide guidance in developing a new version of the website, coinciding with the introduction of the company's first products.

Table C.2. *ToutBay Begins: Website Data*

Variable	Description
date	Date coded as mm/dd/yy
sessions	Number of sessions (a session is a defined period of time, such as 30 minutes, that a user is actively engaged with the website)
users	Number of users who have at least one session
new_sessions	Estimate of the percentage of first-time visits
pageviews	Number of pages viewed (repeated views of the same page are counted)
pages_per_session	Average number of pages viewed during a session (repeated views of the same page are counted)
ave_session_duration	Average length of a session in hours, minutes, and seconds (hh:mm:ss)
bounce_rate	Percentage of single-page visits (user enters and leaves from the same page, usually the home page)
scroll_videopromo	Number of users who scroll down as far as the video introducing ToutBay on the home page
scroll_whatstoutbay	Number of users who scroll down as far as the *What's ToutBay* section of the home page
scroll_howitworks	Number of users who scroll down as far as the *How It Works* section of the home page
scroll_faq	Number of users who scroll down as far as the *FAQ* section of the home page
scroll_latestfeeds	Number of users who scroll down as far as the *Press Releases* section of the home page
chrome	Number of sessions from the Google Chrome browser
safari	Number of sessions from the Apple Safari browser
firefox	Number of sessions from the Mozilla Firefox browser
internet_explorer	Number of sessions from the Microsoft Internet Explorer browser
windows	Number of sessions from Microsoft Windows operating systems
macintosh	Number of sessions from Apple Macintosh operating systems
ios	Number of sessions from Apple iOS operating systems
android	Number of sessions from Android operating systems

In preparing these data, we first created an external traffic reporting segment by filtering out traffic coming from website developers and ToutBay principals. The variables in the data set include data gathered from Google Analytics reports for www.toutbay.com for the period from April 12, 2014 through September 19, 2014. Also included are counts from Scroll Depth, a Google Analytics plug-in that tracks how far users scroll down a page. Scroll Depth is especially useful for a website that puts a lot of information on individual pages such as the home page (a single-page approach). Documentation for Scroll Depth is available at http://scrolldepth.parsnip.io/. When using Google Analytics, we do not have access to the original data that have been collected. Rather, we use the variables and reporting aggregates that Google Analytics defines. Documentation for Google Analytics measures (dimensions and metrics) is available at https://developers.google.com/analytics/devguides/reporting/core/dimsmets.

C.3 Keyword Games: Dodgers and Angels

Google provides search traffic estimates for website owners wishing to use paid advertising on Google. Traffic estimates are based on past user query behavior. Popular keywords have higher estimated traffic values and higher bid prices. These are associated with higher observed search frequencies. Online advertisers pay more to have their ads displayed in response to search queries using popular keywords. A keyword is one word or a group of words used in search queries and indexed by search engine providers.

Paid search results appear on the same screens as organic search results, usually at the top of the screen or to the right of organic search results. Google AdWords employs a cost-per-click (CPC) pricing model. The price an advertiser pays for an online advertisement depends on the cost per click that is set through the online bidding process and the actual number of clicks received by the ad (click-through rate). Data from the Google AdWords Keyword Planner helps advertisers in placing bids for keywords and in picking the spots where their ads will be displayed.

Keyword traffic estimates reflect what users do in search. They show the relative standing of words and word combinations employed in user queries. While these data are certainly useful in planning for online advertising, they are also useful for designing website content and improving organic search performance. Keywords in the data set are coded keywords as used in the search engine provider's index.

We should expect keywords with higher traffic estimates to be associated with more competition, where competition is a scaled value between zero and one based on the number of advertisers bidding for a keyword. We should see the suggested higher cost-per-click bidding prices for keywords with higher competition values. That is, the relationship between traffic, competition, and suggested bidding price should be positive.

Data for this case were obtained from the Google AdWords Keyword Planner on November 1, 2014. The planning tool is located at `https://adwords.google.com/KeywordPlanner`. Google provides online documentation and training for the planning tool, which helps marketing professionals to design online advertising campaigns with Google AdWords. The data reported here were obtained from the public version of the Google AdWords Keyword Planner. These data and other data like them are free and available to anyone with a Google e-mail account.

There is great variability in keyword data, so that keywords with the same traffic estimates can have very different values for competition and very different suggested bidding prices. To get the most from their advertising budgets, online advertisers look for undervalued keywords, those with high estimated traffic and low suggested bidding prices, recognizing that data provided by Google are not observed data from search. Rather, they are estimates that flow from Google pricing models.

To collect the data for this case we conducted twelve keyword studies. For each study we completed three fields in an online form: (1) your product or service, (2) your landing page, and (3) your product category. For each study, the product or service being sold was specified as "tickets." Half the studies used the Los Angeles Dodgers landing page: `www.dodgers.com`. The other half of the studies used the Los Angeles Angels landing page: `losangeles.angels.mlb.com`.

Product or service category varied from one search to the next across six values: "Arts and Entertainment," "Sports Entertainment," "Sports Event Tickets and Ticketing Services," "Sports and Fitness," "Sports," and "Baseball." Product and service categories are arranged in a hierarchy. "Sports Event Tickets and Ticketing Services" is a subset of "Sports Entertainment," which is a subset of "Arts and Entertainment." "Baseball" is a subset of "Sports," which is a subset of "Sports and Fitness." All other things being equal, broader product categories are associated with higher keyword traffic estimates and higher price-per-click values.

The Major League Baseball season extends from April through September, but tickets are sold throughout the year. Google provides monthly traffic estimates, as well as average monthly traffic estimates. Table C.3 shows the structure of the aggregate data across the twelve studies.

Table C.3. *Data Dictionary for Keyword Games: Dodgers versus Angels*

Variable Name	Description
group	Name of advertising keyword group
keyword	Keyword (coded word or set of words for query and index)
traffic	Estimated average monthly traffic (clicks)
october	October traffic estimate
november	November traffic estimate
december	December traffic estimate
january	January traffic estimate
february	February traffic estimate
march	March traffic estimate
april	April traffic estimate
may	May traffic estimate
june	June traffic estimate
july	July traffic estimate
august	August traffic estimate
september	September traffic estimate
competition	Scaled value for amount of competition (range 0–1)
cpcbid	Suggested cost-per-click bid price for keyword (dollars)
study	Name of the product or service category used in study
	Arts and Entertainment
	Sports Entertainment
	Sports Events Ticketing
	Sports and Fitness
	Sports
	Baseball
team	Team name corresponding to the landing page URL
	Dodgers: `www.dodgers.com`
	Angels: `losangeles.angels.mlb.com`

Data source: `https://adwords.google.com/KeywordPlanner`

C.4   Enron E-Mail Corpus and Network

Enron Corporation was an energy and commodity services company based in Houston, Texas. It began business in 1985 and grew to be one of the ten largest companies in the United States, achieving a stock price of $90.75 in mid 2000. By November 2001, however, its stock price fell to less than $1 a share. Enron filed for bankruptcy on December 2, 2001, prompting investigations by the U.S. Securities and Exchange Commission and the Federal Energy Regulatory Commission.

Enron's fraudulent practices became a matter of public record and brought corporate accounting and auditing practices into question. The Enron scandal provides a lesson in business ethics, well documented in popular books and articles.

Most of the executive employees of Enron and had nothing to do with the scandal, but their e-mail records became a matter of public record when the Federal Energy Regulatory Commission released them as part of its investigation. The Enron case data are Enron-centric—for non-Enron actors, we see only their communications with Enron executives.

As one of the few sources of real e-mail data in the public domain, the Enron e-mail archive and network represent a substantial opportunity for research in text analytics, online communications, and social networks. The current Enron e-mail corpus, occupying more than two gigabytes of storage and 500 thousand files, contains folders for 158 executives and over 200 thousand e-mail messages. The e-mail network consists of more than 36 thousand nodes and 183 thousand links.

McLean and Elkind (2003) and Eichenwald (2005) provided popular business books about the Enron scandal. Tim Grieve's *Salon* article quoted selected messages from the e-mail archive (2003), focusing on Ken Lay (1942–2006), CEO of Enron for most of the period from 1985 through 2002. Original source materials relating to the Enron case are available from the Federal Energy Regulatory Commission at http://www.ferc.gov/industries/electric/indus-act/wec/enron/info-release.asp. The Enron e-mail corpus, maintained by William W. Cohen of Carnegie Mellon University, is available at http://www.cs.cmu.edu/~enron/. Data showing the from-node and to-node structure of the e-mail network data, drawn from Leskovec et al. (2009), are available as part of the Stanford Large Network Dataset Collection at http://snap.stanford.edu/data/email-Enron.html. An overview of the Enron data is provided by Klimmt and Yang (2004). The Enron data have been the source of many studies over the past decade (Leber 2013), with a special issue of *Computational and Mathematical Organization Theory* devoted to their analysis (Carley and Skillicorn 2005).

C.5 Wikipedia Votes

The Wikipedia online encyclopedia is a collaborative writing project open to all. Jimmy Wales and Larry Sanger started Wikipedia January 15, 2001 using wiki software from Ward Cunningham.

The Wikipedia website grew slowly during its first four years. Between its fifth and sixth years of operation, however, the website doubled in size, growing from 500 thousand articles to more than one million articles. By September 2014, Wikipedia consisted of more than 33 million articles in 287 languages, with more than 48 million contributors.

Wikipedia is maintained by a set of elected administrators. Votes are cast by existing administrators and by non-administrator users. A set of votes over any selected period of time may be used to define a social network. The act of voting defines a link in a directed network, with a user/voter linked to another user/candidate.

Wikipedia Votes represents a network data set of 7,115 nodes (users, voters, candidates) and 103,689 links (votes). The data span the first seven years of Wikipedia, January 2001 through January 2008, documenting the early growth of the site and user collaboration in building the site.

Table C.4 lists the top ten websites worldwide in September 2014 according to Alexa Internet, Inc., a subsidiary of Amazon.com. The ranking is based on page view and daily visitor counts. Wikipedia ranks sixth on the list and is the only member not maintained by a corporation.

Data showing the from-node and to-node structure of this social network are drawn from Lestovec, Huttenlocher, and Kleinbert (2010a, 2010b) and are available as part of the Stanford Large Network Dataset Collection at `https://snap.stanford.edu/data/wiki-Vote.html`. Background information about Wikipedia was obtained from Wikipedia (2014c).

Table C.4. *Top Sites on the Web, September 2014*

Rank	Name	Description
1	Google.com	Search engine
2	Facebook.com	Social network, photo sharing
3	Youtube.com	Online videos
4	Yahoo.com	Internet portal
5	Baidu.com	Chinese search engine
6	Wikipedia.org	Online encyclopedia
7	Twitter.com	Social network, microblogging
8	Amazon.com	Online retailer
9	Qq.com	Chinese Internet portal
10	Linkedin.com	Social network for professionals

Adapted from Alexa Internet (2014).

C.6 Quake Talk

On August 23, 2011, at 1:51p.m. Eastern Daylight Time, there was a magnitude 5.8 earthquake centered in Louisa County, Virginia. It was the largest magnitude earthquake east of the Rocky Mountains since a 1944 5.8 magnitude earthquake on the New York-Ontario boarder. Geological research of the Virginia quake indicated that landslides occurred 150 miles from the epicenter (latitude 37.936111, longitude -77.933056).

Due to its magnitude, location and time of occurrence, more people felt the Virginia earthquake than any other earthquake in the history of the United States. There was extensive social media activity relating to the event, with Twitter reporting that users were sending more than five thousand messages a second immediately following the quake.

A public repository of Twitter tweets allows us to analyze a sample of social media data on the day of the quake. These data are provided in JavaScript Object Notation (JSON), readable by people and machines. Many of the tweets permit analysis in terms of time and location. As we might expect of social media text data, these tweets are replete with the street talk, idioms, and profanities of English as it is spoken in the United States.

The Quake Talk data are available at `https://github.com/maksim2042/earthquake`. An overview of these data is provided by Tsvetovat and Kouznetsov (2011). Additional information about the August 23, 2011 earthquake may be found at Wikipedia (2014a).

C.7 POTUS Speeches

The American Presidency Project resides at the University of California, Santa Barbara. Researchers at the project maintain an archive of more than 100 thousand public documents relating to the United States Presidency. The archive represents an important resource for any student of government or history, as well as for ordinary citizens.

For this case, we selected State of the Union addresses delivered by the President of the United States (POTUS). Taken as a group these addresses represent a rich corpus for text analytics.

We choose the State of the Union addresses for a number of reasons. They relate to our interest in learning more about the political process and the United States government's role in both domestic and foreign affairs. They extend over a number of years, allowing for the study of trends. They are public documents, available to all with no copyright restrictions. And, with the help of The American Presidency Project, text from these documents is easily accessed online.

We chose to store the State of the Union addresses as plain text files with each file corresponding to a separate speech. File names were coded to provide easy access to document metadata. The first character of the file name was used for the political party of the President ("D" for Democrat or "R" for Republican). Underline separators were used between metadata items, with the method for delivering the address ("O" for oral, "W" for written), a unique name coding for the President, and the year of the address. For example, the file R_O_BushGW_2008.txt is the file for Republican President George W. Bush, who delivered an oral address in 2008.

The Constitution of the United States requires that the President provide a message to Congress every year. Accordingly, the State of the Union addresses represent a consistent political record from the formation of the Union in 1790 to date. The intent of this case is to provide the complete corpus in a form that is easy to input into programs for text analytics.

The data for this case come from the American Presidency Project at the University of California, Santa Barbara. Access to these data is provided at http://www.presidency.ucsb.edu/. For additional information about the papers see reports prepared by Gerhard Peters and John T. Woolley (Peters and Woolley 2014).

C.8 Anonymous Microsoft Web Data

For one week in February 1998, researchers sampled and parsed data from the server logs of the Microsoft website `www.microsoft.com`. There were 37,711 users in the sample, and the case is called "anonymous" because the data files contain no personally identifiable information for these users.

While we often think of website nodes as being pages, in this study it is website areas that are the nodes. Page-view requests of users were categorized to reflect areas of the Microsoft website visited. There were 294 distinct areas, identified by name and number.

To provide an honest evaluation of predictive models, Microsoft user data are partitioned into training and test sets, with 32,711 users in the training file and 5,000 users in the test file. Data rows in these files shows a user identification number along with a website area visited during the one-week time frame of the study. A separate file shows area identification numbers, directory names for website areas, and area descriptions.

Analysis of the Microsoft data can begin by looking at areas visited and characterizing user behavior. Network area structure may be gleaned from user behavior by construing joint area usage as a link between area nodes.

A more ambitious goal, one consistent with published studies drawing on these data, would be to predict which areas of the website a user will visit based upon other areas visited. For predictive models of this type, we could utilize methods of association rule analysis and recommender systems.

The data for this case come from the University of California–Irvine Machine Learning Repository of the Center for Machine Learning and Intelligent Systems (Bache and Lichman 2013). The data sets and documentation are available at `http://archive.ics.uci.edu/ml/datasets/Anonymous+Microsoft+Web+Data`. The original data were created by Jack S. Breese, David Heckerman, and Carl M. Kadie of Microsoft Research in Redmond, Washington, and these data were used in testing models for predicting areas of the website a user would visit based on data about other areas that user had visited (Breese, Heckerman, and Kadie 1998).

D

Code and Utilities

"May the Force be with you."

—HARRISON FORD AS HAN SOLO IN *Star Wars* (1977)

This appendix provides code and utilities associated with this book and modeling techniques in predictive analytics. The code and accompanying data sets are open source, downloadable from the publisher's website for the book at `http://www.ftpress.com/miller`.

For the practicing data scientist, there are considerable advantages to being multilingual. As a general-purpose language Python provides a rich selection of tools for networking and text processing. And R is especially strong in modeling and statistical graphics. And by knowing both Python and R, we can test scripts developed in one language against scripts developed in the other.

Doing data science with Python means gathering programs and documentation from GitHub and staying in touch with organizations like PyCon, SciPy and PyData. At the time of this writing, the Python programming environment consists of more than 15,000 packages. There are large communities of open-source developers working on scientific programming packages like NumPy, SciPy, and SciKit-Learn. There is the Python Software Foundation, which supports code development and education. Useful general references for learning Python include Chun (2007), Beazley (2009), and Beazley and Jones (2013).

Of the top five programming languages according to IEEE Specutrum (Cass 2014), four require a compile cycle, two are controlled by major corporations, and three begin with the letter "C." That leaves Python the clear winner. Python gets an "A" in my book. As a general-purpose scripting language, I can do more things with Python than I can with special-purpose languages, and I can do them faster. Python offers a rich set of data structures to work with. It is object-oriented, self-documenting, open-source, and free—the most popular scripting language in the world. JavaScript is sixth in the IEEE list, and R thirteenth—two additional open-source, scripting languages.

Doing data science with R means looking for task views posted with the Comprehensive R Archive Network (CRAN). We go to RForge and GitHub. We read package vignettes and papers in *The R Journal* and the *Journal of Statistical Software*. At the time of this writing, the R programming environment consists of more than 5,000 packages, many of them focused upon modeling methods. Useful general references for learning R include Matloff (2011) and Lander (2014). Venables and Ripley (2002), although written with S/SPlus in mind, remains a critical reference in the statistical programming community.

Researchers specializing in web and network data science have every reason to learn three languages, and they should. Master Python first, then R, and, as time permits, take in a little JavaScript. If there is one thing that we have shown in this book, it is that worlds of information are available to those who can program. So keep coding.

A Python utility for evaluating binary classifiers is provided in exhibit D.1 on page 300. Utilities used in Python programs for sentiment analysis are show in exhibit D.2 beginning on page 301 and in exhibit D.3 on page 303. The corresponding R utilities are provided in exhibit D.7, beginning on page D.7.

R utilities for grid and ggplot2 graphics are provided in exhibits D.4 and **??**. The split-plotting utilities in exhibit D.4 (pages 304 through 306) are used in a number of chapters to render R ggplot2 graphics objects with proper margins and multiple-plot layout.

R utilities for computing distances between points defined by longitude and latitude are useful in spatial data analysis and in displays relating to the origin of online traffic, with IP addresses providing information about geographical location. These utilities are provided in exhibit D.6 on page 308.

Graphical user interfaces and integrated development environments can be of great assistance to programmers when building end-user applications. Tools for building solutions with Python are provided with open-source systems like IPython, KNIME (Berthold et al. 2007), and Orange (Demšar and Zupan 2013) and in commercial systems from Continuum Analytics, Enthought, Microsoft, and IBM. Tools for building solutions with R are provided with open-source systems like KNIME and RStudio and in commercial systems from Alteryx, IBM, Microsoft, and SAS.

Exhibit D.1. *Evaluating Predictive Accuracy of a Binary Classifier (Python)*

```python
# Evaluating Predictive Accuracy of a Binary Classifier (Python)

def evaluate_classifier(predicted, observed):
    import pandas as pd
    if(len(predicted) != len(observed)):
        print('\nevaluate_classifier error:',\
              ' predicted and observed must be the same length\n')
        return(None)
    if(len(set(predicted)) != 2):
        print('\nevaluate_classifier error:',\
              ' predicted must be binary\n')
        return(None)
    if(len(set(observed)) != 2):
        print('\nevaluate_classifier error:',\
              ' observed must be binary\n')
        return(None)

    predicted_data = predicted
    observed_data = observed
    input_data = {'predicted': predicted_data,'observed':observed_data}
    input_data_frame = pd.DataFrame(input_data)

    cmat = pd.crosstab(input_data_frame['predicted'],\
        input_data_frame['observed'])
    a = float(cmat.ix[0,0])
    b = float(cmat.ix[0,1])
    c = float(cmat.ix[1,0])
    d = float(cmat.ix[1,1])
    n = a + b + c + d
    predictive_accuracy = (a + d)/n
    true_positive_rate = a / (a + c)
    false_positive_rate = b / (b + d)
    precision = a / (a + b)
    specificity = 1 - false_positive_rate
    expected_accuracy = (((a + b)*(a + c)) + ((b + d)*(c + d)))/(n * n)
    kappa = (predictive_accuracy - expected_accuracy)\
        /(1 - expected_accuracy)
    return(a, b, c, d, predictive_accuracy, true_positive_rate, specificity,\
        false_positive_rate, precision, expected_accuracy, kappa)
```

Exhibit D.2.   *Text Measures for Sentiment Analysis (Python)*

```python
# Text Measures for Sentiment Analysis (Python)

def get_text_measures(corpus):
    # individually score each of the twenty-five selected positive words
    # for each document in the working corpus... providing new text measures
    # initialize the list structures for each positive word
    beautiful = []; best = []; better = []; classic = [];
    enjoy = []; enough = []; entertaining = []; excellent = [];
    fans = []; fun = []; good = []; great = []; interesting = [];
    like = []; love = []; nice = []; perfect = []; pretty = [];
    right = []; top = []; well = [];
    won = []; wonderful = []; work = []; worth = []
    # initialize the list structures for each negative word
    bad = []; boring = []; creepy = []; dark = [];
    dead = []; death = []; evil = []; fear = [];
    funny = []; hard = []; kill = []; killed = [];
    lack = []; lost = []; mystery = []; plot = [];
    poor = []; problem = []; sad = []; scary = [];
    slow = []; terrible = []; waste = []; worst = []; wrong = []

    for text in corpus:
        beautiful.append(len([w for w in text.split() if w == 'beautiful']))
        best.append(len([w for w in text.split() if w == 'best']))
        better.append(len([w for w in text.split() if w == 'better']))
        classic.append(len([w for w in text.split() if w == 'classic']))

        enjoy.append(len([w for w in text.split() if w == 'enjoy']))
        enough.append(len([w for w in text.split() if w == 'enough']))
        entertaining.append(len([w for w in text.split() if w == 'entertaining']))
        excellent.append(len([w for w in text.split() if w == 'excellent']))

        fans.append(len([w for w in text.split() if w == 'fans']))
        fun.append(len([w for w in text.split() if w == 'fun']))
        good.append(len([w for w in text.split() if w == 'good']))
        great.append(len([w for w in text.split() if w == 'great']))

        interesting.append(len([w for w in text.split() if w == 'interesting']))
        like.append(len([w for w in text.split() if w == 'like']))
        love.append(len([w for w in text.split() if w == 'love']))
        nice.append(len([w for w in text.split() if w == 'nice']))

        perfect.append(len([w for w in text.split() if w == 'perfect']))
        pretty.append(len([w for w in text.split() if w == 'pretty']))
        right.append(len([w for w in text.split() if w == 'right']))
        top.append(len([w for w in text.split() if w == 'top']))

        well.append(len([w for w in text.split() if w == 'well']))
        won.append(len([w for w in text.split() if w == 'won']))
        wonderful.append(len([w for w in text.split() if w == 'wonderful']))
        work.append(len([w for w in text.split() if w == 'work']))
        worth.append(len([w for w in text.split() if w == 'worth']))
```

Weband Network Data Science

```
    # individually score each of the twenty-five selected negative words
    # for each document in the working corpus... poviding new text measures

        bad.append(len([w for w in text.split() if w == 'bad']))
        boring.append(len([w for w in text.split() if w == 'boring']))
        creepy.append(len([w for w in text.split() if w == 'creepy']))
        dark.append(len([w for w in text.split() if w == 'dark']))

        dead.append(len([w for w in text.split() if w == 'dead']))
        death.append(len([w for w in text.split() if w == 'death']))
        evil.append(len([w for w in text.split() if w == 'evil']))
        fear.append(len([w for w in text.split() if w == 'fear']))

        funny.append(len([w for w in text.split() if w == 'funny']))
        hard.append(len([w for w in text.split() if w == 'hard']))
        kill.append(len([w for w in text.split() if w == 'kill']))
        killed.append(len([w for w in text.split() if w == 'killed']))

        lack.append(len([w for w in text.split() if w == 'lack']))
        lost.append(len([w for w in text.split() if w == 'lost']))
        mystery.append(len([w for w in text.split() if w == 'mystery']))
        plot.append(len([w for w in text.split() if w == 'plot']))

        poor.append(len([w for w in text.split() if w == 'poor']))
        problem.append(len([w for w in text.split() if w == 'problem']))
        sad.append(len([w for w in text.split() if w == 'sad']))
        scary.append(len([w for w in text.split() if w == 'scary']))

        slow.append(len([w for w in text.split() if w == 'slow']))
        terrible.append(len([w for w in text.split() if w == 'terrible']))
        waste.append(len([w for w in text.split() if w == 'waste']))
        worst.append(len([w for w in text.split() if w == 'worst']))
        wrong.append(len([w for w in text.split() if w == 'wrong']))

    # creat dictionary data structure as a preliminary
    # to creating the data frame for the fifty text measures
    add_corpus_data = {'beautiful':beautiful,'best':best,'better':better,\
        'classic':classic, 'enjoy':enjoy, 'enough':enough,\
        'entertaining':entertaining, 'excellent':excellent,\
        'fans':fans, 'fun':fun, 'good':good, 'great':great,\
        'interesting':interesting, 'like':like, 'love':love, 'nice':nice,\
        'perfect':perfect, 'pretty':pretty, 'right':right, 'top':top,\
        'well':well, 'won':won, 'wonderful':wonderful, 'work':work,\
        'worth':worth, 'bad':bad, 'boring':boring, 'creepy':creepy,\
        'dark':dark, 'dead':dead, 'death':death, 'evil':evil, 'fear':fear,\
        'funny':funny,'hard':hard, 'kill':kill, 'killed':killed, 'lack':lack,\
        'lost':lost, 'mystery':mystery, 'plot':plot,'poor':poor,\
        'problem':problem, 'sad':sad, 'scary':scary, 'slow':slow,\
        'terrible':terrible, 'waste':waste, 'worst':worst, 'wrong':wrong}

return(add_corpus_data)
```

Exhibit D.3. *Summative Scoring of Sentiment (Python)*

```python
# Summative Scoring of Sentiment (Python)

def get_summative_scores(corpus):
    # individually score each of the positive and negative words/items
    # for each document in the working corpus... providing a summative score
    summative_score = []  # intialize list for summative scores
    for text in corpus:
        score = 0  # initialize for individual document
        # for each document in the working corpus...
        # individually score each of the eight selected positive words
        if (len([w for w in text.split() if w == 'beautiful']) > 0):
            score = score +1
        if (len([w for w in text.split() if w == 'best']) > 0):
            score = score +1
        if (len([w for w in text.split() if w == 'classic']) > 0):
            score = score +1
        if (len([w for w in text.split() if w == 'excellent']) > 0):
            score = score +1
        if (len([w for w in text.split() if w == 'great']) > 0):
            score = score +1
        if (len([w for w in text.split() if w == 'perfect']) > 0):
            score = score +1
        if (len([w for w in text.split() if w == 'well']) > 0):
            score = score +1
        if (len([w for w in text.split() if w == 'wonderful']) > 0):
            score = score +1
    # individually score each of the ten selected negative words
        if (len([w for w in text.split() if w == 'bad']) > 0):
            score = score -1
        if (len([w for w in text.split() if w == 'boring']) > 0):
            score = score -1
        if (len([w for w in text.split() if w == 'funny']) > 0):
            score = score -1
        if (len([w for w in text.split() if w == 'lack']) > 0):
            score = score -1
        if (len([w for w in text.split() if w == 'plot']) > 0):
            score = score -1
        if (len([w for w in text.split() if w == 'poor']) > 0):
            score = score -1
        if (len([w for w in text.split() if w == 'problem']) > 0):
            score = score -1
        if (len([w for w in text.split() if w == 'terrible']) > 0):
            score = score -1
        if (len([w for w in text.split() if w == 'waste']) > 0):
            score = score -1
        if (len([w for w in text.split() if w == 'worst']) > 0):
            score = score -1
        summative_score.append(score)
    summative_score_data = {'summative_score': summative_score}
    return(summative_score_data)
```

Exhibit D.4. *Split-plotting Utilities (R)*

```
# Split-Plotting Utilities with grid Graphics (R)

library(grid)  # grid graphics foundation of split-plotting utilities

# functions used with ggplot2 graphics to split the plotting region
# to set margins and to plot more than one ggplot object on one page/screen

vplayout <- function(x, y)
viewport(layout.pos.row=x, layout.pos.col=y)


# grid graphics utility plots one plot with margins
ggplot.print.with.margins <- function(ggplot.object.name,left.margin.pct=10,
  right.margin.pct=10,top.margin.pct=10,bottom.margin.pct=10)
{ # begin function for printing ggplot objects with margins
  # margins expressed as percentages of total... use integers
 grid.newpage()
pushViewport(viewport(layout=grid.layout(100,100)))
print(ggplot.object.name,
  vp=vplayout((0 + top.margin.pct):(100 - bottom.margin.pct),
  (0 + left.margin.pct):(100 - right.margin.pct)))
} # end function for printing ggplot objects with margins


# grid graphics utility plots two ggplot plotting objects in one column
special.top.bottom.ggplot.print.with.margins <-
  function(ggplot.object.name,ggplot.text.tagging.object.name,
  left.margin.pct=5,right.margin.pct=5,top.margin.pct=5,
  bottom.margin.pct=5,plot.pct=80,text.tagging.pct=10) {
# begin function for printing ggplot objects with margins
# and text tagging at bottom of plot
# margins expressed as percentages of total... use integers
  if((top.margin.pct + bottom.margin.pct + plot.pct + text.tagging.pct) != 100)
    stop(paste("function special.top.bottom.ggplot.print.with.margins()",
    "execution terminated:\n   top.margin.pct + bottom.margin.pct + ",
    "plot.pct + text.tagging.pct not equal to 100 percent",sep=""))
  grid.newpage()
  pushViewport(viewport(layout=grid.layout(100,100)))
  print(ggplot.object.name,
  vp=vplayout((0 + top.margin.pct):
    (100 - (bottom.margin.pct + text.tagging.pct)),
  (0 + left.margin.pct):(100 - right.margin.pct)))

  print(ggplot.text.tagging.object.name,
    vp=vplayout((0 + (top.margin.pct + plot.pct)):(100 - bottom.margin.pct),
    (0 + left.margin.pct):(100 - right.margin.pct)))
} # end function for printing ggplot objects with margins and text tagging
```

```
# grid graphics utility plots three ggplot plotting objects in one column
three.part.ggplot.print.with.margins <- function(ggfirstplot.object.name,
ggsecondplot.object.name,
ggthirdplot.object.name,
left.margin.pct=5,right.margin.pct=5,
top.margin.pct=10,bottom.margin.pct=10,
first.plot.pct=25,second.plot.pct=25,
third.plot.pct=30) {
# function for printing ggplot objects with margins and top and bottom plots
# margins expressed as percentages of total... use integers
if((top.margin.pct + bottom.margin.pct + first.plot.pct +
  second.plot.pct  + third.plot.pct) != 100)
    stop(paste("function special.top.bottom.ggplot.print.with.margins()",
        "execution terminated:\n   top.margin.pct + bottom.margin.pct",
        "+ first.plot.pct + second.plot.pct  + third.plot.pct not equal",
        "to 100 percent",sep=""))
grid.newpage()
pushViewport(viewport(layout=grid.layout(100,100)))

print(ggfirstplot.object.name, vp=vplayout((0 + top.margin.pct):
  (100 - (second.plot.pct  + third.plot.pct + bottom.margin.pct)),
  (0 + left.margin.pct):(100 - right.margin.pct)))

print(ggsecondplot.object.name,
  vp=vplayout((0 + top.margin.pct + first.plot.pct):
  (100 - (third.plot.pct + bottom.margin.pct)),
  (0 + left.margin.pct):(100 - right.margin.pct)))

print(ggthirdplot.object.name,
  vp=vplayout((0 + top.margin.pct + first.plot.pct + second.plot.pct):
  (100 - (bottom.margin.pct)),(0 + left.margin.pct):
  (100 - right.margin.pct)))
}

# grid graphics utility plots two ggplot plotting objects in one row
# primary plot graph at left... legend at right
special.left.right.ggplot.print.with.margins <-
  function(ggplot.object.name, ggplot.text.legend.object.name,
  left.margin.pct=5, right.margin.pct=5, top.margin.pct=5,
  bottom.margin.pct=5, plot.pct=85, text.legend.pct=5) {
# begin function for printing ggplot objects with margins
# and text legend at bottom of plot
# margins expressed as percentages of total... use integers
  if((left.margin.pct + right.margin.pct + plot.pct + text.legend.pct) != 100)
    stop(paste("function special.left.right.ggplot.print.with.margins()",
    "execution terminated:\n   left.margin.pct + right.margin.pct + ",
    "plot.pct + text.legend.pct not equal to 100 percent",sep=""))
  grid.newpage()
  pushViewport(viewport(layout=grid.layout(100,100)))
  print(ggplot.object.name,
  vp=vplayout((0 + top.margin.pct):(100 - (bottom.margin.pct)),
  (0 + left.margin.pct + text.legend.pct):(100 - right.margin.pct)))
```

```
    print(ggplot.text.legend.object.name,
      vp=vplayout((0 + (top.margin.pct)):(100 - bottom.margin.pct),
      (0 + left.margin.pct + plot.pct):(100 - right.margin.pct)))
} # end function for printing ggplot objects with margins and text legend

# save split-plotting utilities for future work
save(vplayout,
  ggplot.print.with.margins,
  special.top.bottom.ggplot.print.with.margins,
  three.part.ggplot.print.with.margins,
  special.left.right.ggplot.print.with.margins,
  file="mtpa_split_plotting_utilities.Rdata")
```

Exhibit D.5. *Correlation Heat Map Utility (R)*

```r
# Correlation Heat Map Utility (R)
#
# Input correlation matrix. Output heat map of correlation matrix.
# Requires R lattice package.

correlation_heat_map <- function(cormat, order_variable = NULL) {
    if (is.null(order_variable)) order_variable = rownames(cormat)[1]
    cormat_line <- cormat[order_variable, ]
    ordered_cormat <-
        cormat[names(sort(cormat_line, decreasing=TRUE)),
            names(sort(cormat_line, decreasing=FALSE))]
    x <- rep(1:nrow(ordered_cormat), times=ncol(ordered_cormat))
    y <- NULL
    for (i in 1:ncol(ordered_cormat))
        y <- c(y,rep(i,times=nrow(ordered_cormat)))
    # use fixed format 0.XXX in cells of correlation matrix
    cortext <- sprintf("%0.3f", as.numeric(ordered_cormat))
    text.data.frame <- data.frame(x, y, cortext)
    text.data.frame$cortext <- as.character(text.data.frame$cortext)
    text.data.frame$cortext <- ifelse((text.data.frame$cortext == "1.000"),
    NA,text.data.frame$cortext)  # define diagonal cells as missing
    text.data.frame <- na.omit(text.data.frame)  # diagonal cells have no text
    # determine range of correlations all positive or positive and negative
    if (min(cormat) > 0)
        setcolor_palette <- colorRampPalette(c("white", "#00BFC4"))
    if (min(cormat) < 0)
        setcolor_palette <- colorRampPalette(c("#F8766D", "white", "#00BFC4"))
    # use larger sized type for small matrices
    set_cex = 1.0
    if (nrow(ordered_cormat) <= 4) set_cex = 1.5
    print(levelplot(ordered_cormat, cuts = 25, tick.number = 9,
        col.regions = setcolor_palette,
        scales=list(tck = 0, x = list(rot=45), cex = set_cex),
        xlab = "",
        ylab = "",
        panel = function(...) {
            panel.levelplot(...)
            panel.text(text.data.frame$x, text.data.frame$y,
            labels = text.data.frame$cortext, cex = set_cex)
            }))
    }

save(correlation_heat_map, file = "correlation_heat_map.RData")
```

Exhibit D.6. *Utilities for Spatial Data Analysis (R)*

```
# Utilities for Spatial Data Analysis (R)

# user-defined function to convert degrees to radians
# needed for lat.long.distance function
degrees.to.radians <- function(x) {
  (pi/180)*x
  } # end degrees.to.radians function

# user-defined function to convert distance between two points in miles
# when the two points (a and b) are defined by longitude and latitude
lat.long.distance <- function(longitude.a,latitude.a,longitude.b,latitude.b) {
  radius.of.earth <- 24872/(2*pi)
  c <- sin((degrees.to.radians(latitude.a) -
    degrees.to.radians(latitude.b))/2)^2 +
    cos(degrees.to.radians(latitude.a)) *
    cos(degrees.to.radians(latitude.b)) *
    sin((degrees.to.radians(longitude.a) -
    degrees.to.radians(longitude.b))/2)^2
  2 * radius.of.earth * (asin(sqrt(c)))
  } # end lat.long.distance function

save(degrees.to.radians,
  lat.long.distance,
  file = "mtpa_spatial_distance_utilities.R")
```

Exhibit D.7. *Word Scoring Code for Sentiment Analysis (R)*

```
# Text Scoring Script for Sentiment Analysis (R)
# -------------------------------------
# Word/item analysis method
# -------------------------------------
# develop simple counts for working.corpus
# for each of the words in the sentiment list
# these new variables will be given the names of the words
# compute the number of words that match each word
amazing <- integer(length(names(working.corpus)))
beautiful <- integer(length(names(working.corpus)))
classic <- integer(length(names(working.corpus)))
enjoy <- integer(length(names(working.corpus)))
enjoyed <- integer(length(names(working.corpus)))
entertaining <- integer(length(names(working.corpus)))
excellent <- integer(length(names(working.corpus)))
fans <- integer(length(names(working.corpus)))
favorite <- integer(length(names(working.corpus)))
fine <- integer(length(names(working.corpus)))
fun <- integer(length(names(working.corpus)))
humor <- integer(length(names(working.corpus)))
lead <- integer(length(names(working.corpus)))
liked <- integer(length(names(working.corpus)))
love <- integer(length(names(working.corpus)))
loved <- integer(length(names(working.corpus)))
modern <- integer(length(names(working.corpus)))
nice <- integer(length(names(working.corpus)))
perfect <- integer(length(names(working.corpus)))
pretty <- integer(length(names(working.corpus)))
recommend <- integer(length(names(working.corpus)))
strong <- integer(length(names(working.corpus)))
top <- integer(length(names(working.corpus)))
wonderful <- integer(length(names(working.corpus)))
worth <- integer(length(names(working.corpus)))
bad <- integer(length(names(working.corpus)))
boring <- integer(length(names(working.corpus)))
cheap <- integer(length(names(working.corpus)))
creepy <- integer(length(names(working.corpus)))
dark <- integer(length(names(working.corpus)))
dead <- integer(length(names(working.corpus)))
death <- integer(length(names(working.corpus)))
evil <- integer(length(names(working.corpus)))
hard <- integer(length(names(working.corpus)))
kill <- integer(length(names(working.corpus)))
killed <- integer(length(names(working.corpus)))
lack <- integer(length(names(working.corpus)))
lost <- integer(length(names(working.corpus)))
miss <- integer(length(names(working.corpus)))
murder <- integer(length(names(working.corpus)))
mystery <- integer(length(names(working.corpus)))
plot <- integer(length(names(working.corpus)))
poor <- integer(length(names(working.corpus)))
sad <- integer(length(names(working.corpus)))
```

```
scary <- integer(length(names(working.corpus)))
slow <- integer(length(names(working.corpus)))
terrible <- integer(length(names(working.corpus)))
waste <- integer(length(names(working.corpus)))
worst <- integer(length(names(working.corpus)))
wrong <- integer(length(names(working.corpus)))

reviews.tdm <- TermDocumentMatrix(working.corpus)

for(index.for.document in seq(along=names(working.corpus))) {
  amazing[index.for.document] <-
    sum(termFreq(working.corpus[[index.for.document]],
    control = list(dictionary = "amazing")))
  beautiful[index.for.document] <-
    sum(termFreq(working.corpus[[index.for.document]],
    control = list(dictionary = "beautiful")))
  classic[index.for.document] <-
    sum(termFreq(working.corpus[[index.for.document]],
    control = list(dictionary = "classic")))
  enjoy[index.for.document] <-
    sum(termFreq(working.corpus[[index.for.document]],
    control = list(dictionary = "enjoy")))
  enjoyed[index.for.document] <-
    sum(termFreq(working.corpus[[index.for.document]],
    control = list(dictionary = "enjoyed")))
  entertaining[index.for.document] <-
    sum(termFreq(working.corpus[[index.for.document]],
    control = list(dictionary = "entertaining")))
  excellent[index.for.document] <-
    sum(termFreq(working.corpus[[index.for.document]],
    control = list(dictionary = "excellent")))
  fans[index.for.document] <-
    sum(termFreq(working.corpus[[index.for.document]],
    control = list(dictionary = "fans")))
  favorite[index.for.document] <-
    sum(termFreq(working.corpus[[index.for.document]],
    control = list(dictionary = "favorite")))
  fine[index.for.document] <-
    sum(termFreq(working.corpus[[index.for.document]],
    control = list(dictionary = "fine")))
  fun[index.for.document] <-
    sum(termFreq(working.corpus[[index.for.document]],
    control = list(dictionary = "fun")))
  humor[index.for.document] <-
    sum(termFreq(working.corpus[[index.for.document]],
    control = list(dictionary = "humor")))
  lead[index.for.document] <-
    sum(termFreq(working.corpus[[index.for.document]],
    control = list(dictionary = "lead")))
  liked[index.for.document] <-
    sum(termFreq(working.corpus[[index.for.document]],
    control = list(dictionary = "liked")))
  love[index.for.document] <-
    sum(termFreq(working.corpus[[index.for.document]],
```

```
      control = list(dictionary = "love")))
loved[index.for.document] <-
  sum(termFreq(working.corpus[[index.for.document]],
  control = list(dictionary = "loved")))
modern[index.for.document] <-
  sum(termFreq(working.corpus[[index.for.document]],
  control = list(dictionary = "modern")))
nice[index.for.document] <-
  sum(termFreq(working.corpus[[index.for.document]],
  control = list(dictionary = "nice")))
perfect[index.for.document] <-
  sum(termFreq(working.corpus[[index.for.document]],
  control = list(dictionary = "perfect")))
pretty[index.for.document] <-
  sum(termFreq(working.corpus[[index.for.document]],
  control = list(dictionary = "pretty")))
recommend[index.for.document] <-
  sum(termFreq(working.corpus[[index.for.document]],
  control = list(dictionary = "recommend")))
strong[index.for.document] <-
  sum(termFreq(working.corpus[[index.for.document]],
  control = list(dictionary = "strong")))
top[index.for.document] <-
  sum(termFreq(working.corpus[[index.for.document]],
  control = list(dictionary = "top")))
wonderful[index.for.document] <-
  sum(termFreq(working.corpus[[index.for.document]],
  control = list(dictionary = "wonderful")))
worth[index.for.document] <-
  sum(termFreq(working.corpus[[index.for.document]],
  control = list(dictionary = "worth")))
bad[index.for.document] <-
  sum(termFreq(working.corpus[[index.for.document]],
  control = list(dictionary = "bad")))
boring[index.for.document] <-
  sum(termFreq(working.corpus[[index.for.document]],
  control = list(dictionary = "boring")))
cheap[index.for.document] <-
  sum(termFreq(working.corpus[[index.for.document]],
  control = list(dictionary = "cheap")))
creepy[index.for.document] <-
  sum(termFreq(working.corpus[[index.for.document]],
  control = list(dictionary = "creepy")))
dark[index.for.document] <-
  sum(termFreq(working.corpus[[index.for.document]],
  control = list(dictionary = "dark")))
dead[index.for.document] <-
  sum(termFreq(working.corpus[[index.for.document]],
  control = list(dictionary = "dead")))
death[index.for.document] <-
  sum(termFreq(working.corpus[[index.for.document]],
  control = list(dictionary = "death")))
evil[index.for.document] <-
  sum(termFreq(working.corpus[[index.for.document]],
```

```
          control = list(dictionary = "evil")))
hard[index.for.document] <-
   sum(termFreq(working.corpus[[index.for.document]],
   control = list(dictionary = "hard")))
kill[index.for.document] <-
   sum(termFreq(working.corpus[[index.for.document]],
   control = list(dictionary = "kill")))
killed[index.for.document] <-
   sum(termFreq(working.corpus[[index.for.document]],
   control = list(dictionary = "killed")))
lack[index.for.document] <-
   sum(termFreq(working.corpus[[index.for.document]],
   control = list(dictionary = "lack")))
lost[index.for.document] <-
   sum(termFreq(working.corpus[[index.for.document]],
   control = list(dictionary = "lost")))
miss[index.for.document] <-
   sum(termFreq(working.corpus[[index.for.document]],
   control = list(dictionary = "miss")))
murder[index.for.document] <-
   sum(termFreq(working.corpus[[index.for.document]],
   control = list(dictionary = "murder")))
mystery[index.for.document] <-
   sum(termFreq(working.corpus[[index.for.document]],
   control = list(dictionary = "mystery")))
plot[index.for.document] <-
   sum(termFreq(working.corpus[[index.for.document]],
   control = list(dictionary = "plot")))
poor[index.for.document] <-
   sum(termFreq(working.corpus[[index.for.document]],
   control = list(dictionary = "poor")))
sad[index.for.document] <-
   sum(termFreq(working.corpus[[index.for.document]],
   control = list(dictionary = "sad")))
scary[index.for.document] <-
   sum(termFreq(working.corpus[[index.for.document]],
   control = list(dictionary = "scary")))
slow[index.for.document] <-
   sum(termFreq(working.corpus[[index.for.document]],
   control = list(dictionary = "slow")))
terrible[index.for.document] <-
   sum(termFreq(working.corpus[[index.for.document]],
   control = list(dictionary = "terrible")))
waste[index.for.document] <-
   sum(termFreq(working.corpus[[index.for.document]],
   control = list(dictionary = "waste")))
worst[index.for.document] <-
   sum(termFreq(working.corpus[[index.for.document]],
   control = list(dictionary = "worst")))
wrong[index.for.document] <-
   sum(termFreq(working.corpus[[index.for.document]],
   control = list(dictionary = "wrong")))
}
```

Glossary

To understand web and network data science, including methods of on-line research, it is helpful to know terms from data science, social science and business research methods, computing, programming languages, social network analysis (network science), communication networks, and the Internet. This glossary defines technical terms and abbreviations used in the book. Readers desiring more complete dictionaries of computer and Internet terms can refer to Hale and Scanlon (1999) and Downing, Covington, Covington, Barrett, and Covington (2012). Moran and Hunt (2009) review terms relevant to website performance in search. Hyslop (2010), Casciano (2011), and Gasston (2011; 2013) review current HTML and CSS concepts. Robbins (2003) provides an introductory treatment for web designers. Amor (2002) and Tannenbaum and Wetherall (2010) review data communications, the Internet, and World Wide Web. As a group, these were useful sources for constructing this glossary.

The World Wide Web arose from the marriage of hypertext and networking technologies. Its network foundation, the Internet, grew out of earlier wide-area networks, such as ARPANET and Usenet, serving the defense establishment, universities, and technical research centers. ARPANET was initially established for defense purposes, but became a wide-area network for many university researchers as well. Usenet linked many computers

---

Part of this glossary appeared in the book *Qualitative Research Online* (Miller and Walkowski 2004). Used with permission from Research Publishers LLC.

together in a simple telephone dial-up mode, permitting e-mail and file transfers. Standards-based networking protocols made Internet communications possible. Lower-level protocols defined how data could be reliably transferred over communication links. Higher-level protocols were developed for communication applications, such as telnet for terminal-to-computer communication and ftp for file transfers.

The idea for the World Wide Web was presented in 1989, while ideas behind hypertext date back to the 1960s. Hypertext—text with links to other text—represented a technology for developing dynamic documents, with the user determining the path between sections of text within documents. The hypertext markup language (HTML) that we use today was designed as a text-only system. Developers added markup codes to ASCII text files to provide formatting for presenting text content on a computer monitor. Of course, software developers have extended these systems to present graphics as well as text. On the World Wide Web, hypertext links include uniform resource locators (URLs), which can refer to remote files and web pages on the network or to local files and web pages on the user's computer. Extensible markup language (XML) holds promise as a standard for information interchange, providing a foundation for interprocess communication and web services. For the history of the World Wide Web, we can refer to its inventor, Tim Berners-Lee (2000). Further information about today's web may be obtained from the World Wide Web Consortium at `www.w3.org`.

adjacency matrix Consider a network with nodes and links between nodes. We let $x_{ij} = 1$ if node i is linked to node j and let $x_{ij} = 0$ otherwise. Undirected networks yield symmetric matrices. Directed networks yield asymmetric matrices.

agent Also called an "artificial agent." A program that automatically performs duties for a computer user or a program that behaves like a person within a limited domain of operation.

ARPANET Military-sponsored wide-area network, precursor of the Internet.

ASP Abbreviation for an Application Services Provider, a firm that provides outsourced application services. Also, in Microsoft systems, Active Server Pages.

asynchronous focus group See bulletin board.

bandwidth An expression that relates to the speed at which information travels over a network. The words "low bandwidth" mean slower speed and "high bandwidth" mean faster speed.

betweenness centrality Usually computed for nodes, this is an index that reflects the degree to which a node lies on the shortest path between other pairs of nodes. An index of node importance. For a network link, betweenness is similarly computed and represents the the degree to which a link lies on the shortest path between pairs of nodes.

blog   Shortened form of "weblog."

boundary (of a network)   The extent of the network with its included nodes and links. Defines where the network begins and ends—the population of interest.

bps   Bits per second. Measure of speed across a communications link. Sometimes referred to as the "baud rate."

browser launch   Some virtual facilities permit the moderator to open up (launch) a new browser window.

bulletin board   Also called a message board or asynchronous focus group. An online application and designated website where participants (moderator, respondents, and observers) read and post messages. Communication can take place over a period of days. As an asynchronous group, it does not require participants to be online at the same time. The discussion is organized along discussion threads.

chat room   A virtual meeting room on the Internet where participants gather and speak to each other. May be public (open to anybody, space permitting) or private (requiring passwords to get in).

client   A computer through which users interact with the network or Internet. Web clients communicate with web servers. "Client" is also used to refer to the person or organization for which research is conducted, the user of research and information services.

client-server application   A computer application implemented with client and server computers communicating over the Internet.

closeness centrality   A measure of node importance that characterizes how close a node is to all other nodes in the network, where closeness is the number of hops or links between nodes.

collage   A projective technique sometimes used in qualitative research. Respondents are given a theme and asked to construct a collage of pictures, words, drawings, or other artifacts, that represents their response to the theme. The collage is then used as a tool to help the respondents verbalize their perceptions, opinions, beliefs, and attitudes about various topics, brands, and issues.

cookie   Data stored on a research user's workstation that may be used to identify the participant and to keep information about past research transactions.

cost per click (CPC)   Pricing model employed in Google AdWords. The online advertiser is charged only when a user clicks on the advertisement, with the price per click determined in a bidding process.

content analysis   The analysis of the meaning of text from qualitative research. Sometimes called thematic, semantic, and network text analysis, content analysis often starts with simple word counts. Also see text measures.

corpus   A document collection (often a collection of transcripts) used in text analysis and text mining. The plural is "corpora."

crawler (web crawler)   A computer program for automated data acquisition from the web. Also called a spider, bot, or robot.

degree   Each node in an undirected network has a degree or number of links to other nodes. An index of node importance.

degree distribution   The distribution or set of degree values for all nodes in a network.

degree centrality For a node, another term for degree. The average degree across all nodes in a network. A summary statistic for networks.

density (of a network) The the proportion of actual links out of the set of all possible links. A completely unconnected or disconnected set of nodes has zero links, while a completely connected network of n nodes or clique has $\frac{n(n-1)}{2}$ links. If l is the number of links in a network, then network density is given by the formula $\frac{l}{n(n-1)/2}$.

discussion thread A thread of discussion or conversation. See thread.

Document Object Model (DOM) Rules that define the proper arrangement or hierarchy of nodes (tree structure) for HTML on web pages.

DSL Digital Subscriber Line. A form of high-speed Internet access offered by phone companies and resellers.

dyad A two-person group. Also used to describe a focus group with one moderator and two participants.

eigenvector centrality For a node, we consider a node as important if it is close to other nodes of importance, which are in turn close to other nodes of importance, and so on. Much as the first principle component characterizes common variability in a set of variables, eigenvector centrality characterizes the degree to which a node is central to the set of nodes comprising the network.

ethnography Observational or field study of social and consumer behavior in real life settings.

e-mail Store-and-forward method for sending messages over the Internet. Sometimes used for interviewing and for surveys.

emoticons Use of characters on a keyboard to indicate emotions in online communication. Capital letters can be used for emphasis or to indicate a raised voice. Some of the more popular examples with special characters include :-) for a smile or happy response, ;-) for a wink or "just kidding," and :-(for an unhappy or disapproving response.

focused conversation An open-ended conversation among research participants with the focus provided by a conversation guide. Because there is no interviewer or moderator guiding the conversation, the focused conversation is participant self-directed.

frame An element of web design. Some web pages are constructed using frames, blocks of a page that a user can navigate within while all other frames remain static. For example, some online facilities utilize frames, with one frame for the chat stream, one for the text entry area, and another for stimuli to be displayed.

ftp File transfer protocol, one of a number of networking standards used over the Internet.

game theory Specialization within mathematics and economics dealing with players, their knowledge, objectives, and strategies for achieving objectives. There are both competitive and cooperative games.

generative grammar Linguistic term referring to the rules that we use to form meaningful utterances. A general term for morphology, syntax, and semantics.

grounded theory A methodology for analyzing and interpreting qualitative data. Theory is derived from the data. The analysis involves identifying categories and coding data by categories. Common themes and concepts emerge from observed relationships among categories.

HTML Hypertext markup language. Standard codes used to format web pages.

HTTP Hypertext transfer protocol. The dominant method for client-server communication over the web.

ICQ I Seek You, a form of chat room software.

IMHO Shorthand jargon developed by online communicators, meaning "in my humble opinion."

Internet A public data communications network consisting of many interconnected networks worldwide.

intranet A data communications network internal to an organization.

IRC Internet Relay Chat, a form of chat room software.

ISP Internet Services Provider. An organization that provides connection services to the Internet. These organizations usually provide software applications that run over the Internet, such as browsers and search engines.

IT Abbreviation for "information technology."

Java A general-purpose programming language introduced into the public domain by Sun Microsystems, with updates and support from Oracle. Especially useful in the development of online applications.

JavaScript An object-oriented scripting language for the web. Preeminent language for client-side event handling. Embedded in web browsers. Implemented as a standalone language in Node.js, which can be used on the server side.

JavaScript Object Notation (JSON) JavaScript data structure with unordered name/value pairs. Provides storage for arrays, strings, numbers, logical (boolean true/false), and the special value *null*. May be utilized directly in JavaScript code used in web pages.

JPEG Graphical bit-mapped file format for images.

kbps Kilobits per second. Measure of speed across a communications link.

keyword One word or a group of words used in search queries and indexed by search engine providers.

LAMP A public-domain suite for the development of online applications, this refers to the Linux operating system, Apache web server, MySQL database, and Perl, PHP, or Python programming languages.

listserv An automated e-mail group. E-mail addresses of all participants (respondents and moderator) are programmed into a listserv. Moderator initiates the thread of discussion. Any response to that message automatically goes to anybody on the listserv.

LOL Shorthand jargon developed by online communicators, meaning "laughing out loud" or "lots of laughs."

Luddite Person opposed to technological change. In nineteenth century England, a worker named Ned Ludd was supposed to be one of a group of workers who destroyed labor-saving machinery in protest for losing their jobs.

MEG Moderated e-mail group. Another name for e-mail group or listserv group.

message board Another term for an asynchronous focus group.

modem Modulator demodulator. Communications equipment that encodes and decodes data for transmission over a communications network.

morphology A branch of linguistics. Rules for forming complex words. Also see generative grammar.

natural language processing Using grammatical rules to mimic human communication and convert natural language into structured text for further analysis. Natural language refers to the words and the rules that we use to form meaningful utterances.

netiquette Social behavior appropriate for working on the Internet. For example, typing USING ALL CAPITAL LETTERS is considered poor netiquette, as it is considered to be the online equivalent of shouting.

network Interconnected objects—nodes and the links connecting them. A data communications network contains computer system nodes and data communication links. The links are electronic circuits, wired or wireless.

offline In terms of qualitative research, refers to any methodology that does not make use of the Internet, such as in-person focus groups or telephone focus groups.

online In terms of qualitative research, refers to any methodology that makes use of the Internet, such as real-time chats, bulletin boards (message boards), and listservs (MEGs).

online community An aggregation of individuals that emerges through communication over the Internet. That is, a group of individuals that interacts online over an extended period of time. Also called virtual community. When used for primary research, an online community may consist of a group of fifty to two hundred people who have been recruited to participate in research about related topics over a period of three to twelve months. The research may incorporate a variety of methods, including online surveys, real-time focus groups, and bulletin boards.

organic search Natural search in which the search engine finds the most relevant matches to the user's query. By far the most important component of relevance is the links a web page receives from other web pages. Organic search is distinct from paid search, in which search results are advertisements and links paid for by advertisers.

page view As the words would imply, a single viewing of a web page by a visitor. The count of page views for a period of time represents a common measure of web traffic.

PageRank Algorithm for computing the importance, credibility, or relevance of links in a web search. Analogous to eigenvector centrality.

paid search Search results that are advertisements and links paid for by advertisers. Distinct from organic or natural search results.

panel A group of respondents that has been recruited to participate in many studies over a period of time.

parser (text parser) A computer program that prepares text for subsequent analysis, replacing one character string sequence with another. Often implemented with regular expressions.

Perl "Practical Extraction and Report Language." Text processing and scripting language used extensively in web applications.

PHP Personal home page. A scripting language alternative to Perl, used extensively in web applications.

post (noun) Posting made by a participant (moderator, respondent, or observer) in a message board discussion. A statement or question made by a participant in a real-time online group. (verb) To send a message.

provider Person or organization providing research and information services on behalf of a client or user.

psychographics Demographics relating to psychological factors.

Python General-purpose programming language. For web applications, a scripting language alternative to Perl and PHP.

real-time focus group An online focus group with the moderator and participants present at the same time. Sometimes called a "real-time chat."

random network (random graph) A set of nodes connected by links in a purely random fashion.

regular expressions Specialized syntax for identifying character strings, as needed for efficient search and text processing. Implemented in operating system shell search tools and various host languages, including Perl, Java, JavaScript, Python, and R.

scraper (web scraper) A computer program for identifying specific portions of data from the web. Utilized in the context of automated data acquisition from the web (web crawling) and often implemented with XPath syntax.

semantics A branch of linguistics. The study of meaning expressed through language. Also see generative grammar.

semantic web An idea for having web-based data defined and linked in a way that can be used by computers as well as by people. This requires wide acceptance of information coding standards such as XML.

stemming (word stemming) In text analytics, stripping affixes and reducing a word to its base or stem.

synchronous focus group See real-time focus group.

syntax A branch of linguistics. Rules for forming phrases and sentences. Also see generative grammar.

TCP/IP Transmission Control Protocol/Internet Protocol. Basic protocol for computer-to-computer communications over the Internet.

telnet Terminal-to-computer communication protocol used over the Internet.

text analysis General term for methods of analysis and interpretation of text, including content analysis, grounded theory construction, and text mining.

text measures Scores on attributes that describe text as in sentiment analysis. Text measures can be used to assess personality, consumer preferences, and political opinions, just as survey instruments can. Text measures begin with unstructured text (documents, transcripts) as their input data, rather than forced-choice questionnaire responses.

text mining The automated or partially automated processing of text. It involves imposing structure upon text and extracting relevant information from text. Like data mining, it is often associated with the analysis of large databases or document collections (corpora).

thread A thread of discussion or conversation. The sequence of online messages relating to a particular topic. Questions and statements that lead to other questions and statements. A term more prevalent in asynchronous focus groups (bulletin boards or message boards) than in real-time focus groups.

transcript The text log of a discussion, such as developed in an online focus group or focused conversation.

transitivity A measure of connectedness of a network. Also known as the average clustering coefficient of a network. Usually computed from the adjacency matrix as the proportion of fully connected triples (triads). A triad or triple is a set of three nodes, and a closed triad is a set of three nodes with links between each pair of nodes.

triad A set of three nodes, a three-person group or triple. In qualitative research, used to describe a focus group with one moderator and three participants.

Usenet Wide-area network precursor of the Internet.

URL Uniform resource locator, address for Internet information resources.

virtual community Another term for online community. Aggregation of individuals that emerges through communication over the Internet

virtual facility Adjective used to describe online the analogue of a real, physical research facility. For example, an online focus group software provides a virtual facility, analogous to a real in-person focus group.

web Refers to the World Wide Web, available over the Internet.

web board Another term for an asynchronous focus group.

weblog A web site with online information in a dated log format. Useful for individuals and organizations wanting to post current information about themselves.

web presence The degree to which an organization, brand, website, or web landing page is getting recognized on the web. What is often called "search engine optimization (SEO)" is one part of web presence. One way to assess web presence is to see where a site falls on the list of organic search results.

web server Networked computer that serves as the host for web applications, such as electronic mail, listservs, and online focus groups. Contrasted with a client or computer where users interact with the network.

web services With web services, processes on one computer interact with processes on a second computer, providing interprocess networked communications. XML standards are used for information interchange.

white board A frame (window of a virtual facility) reserved for displaying images (visual stimuli, concepts, web pages). Virtual analogue of a white board used in a meeting room for writing with dry-erase markers.

Wiki An exercise in collective expression on the web. Software that facilitates the process of generating a website that is publically editable. Wiki web pages may be edited by anyone at any time, from any device with web browser access.

WWW Abbreviation for the World Wide Web.

XML Extensible markup language. Similar in structure and appearance to HTML, XML has a distinct purpose. It is a way of marking up data that describes the nature of the data.

XPath A specialized syntax for navigating across the nodes and attributes of the Document Object Model (DOM) and extracting relevant data.

Bibliography

Ackland, R. 2013. *Web Social Science: Concepts, Data and Tools for Social Scientists in the Digital Age*. Los Angeles: Sage.

Adler, J. 2010. *R in a Nutshell: A Desktop Quick Reference*. Sebastopol, Calif.: O'Reilly.

Agrawal, R., H. Mannila, R. Srikant, H. Toivonen, and A. I. Verkamo 1996. Fast discovery of association rules. In U. M. Fayyad, G. Piatetsky-Shapiro, P. Smyth, and R. Uthurusamy (eds.), *Handbook of Data Mining and Knowledge Discovery*, Chapter 12, pp. 307–328. Menlo Park, Calif. and Cambridge, Mass.: American Association for Artificial Intelligence and MIT Press.

Agresti, A. 2013. *Categorical Data Analysis* (third ed.). New York: Wiley.

Airoldi, E. M., D. Blei, E. A. Erosheva, and S. E. Fienberg (eds.) 2014. *Handbook of Mixed Membership Models and Their Applications*. Boca Raton, Fla.: CRC Press.

Albert, J. 2009. *Bayesian Computation with R*. New York: Springer. 244

Albert, R. and A.-L. Barabási 2002, January. Statistical mechanics of complex networks. *Reviews of Modern Physics* 74(1):47–97. 99

Alexa Internet, I. 2014. The top 500 sites on the web. Retrieved from the World Wide Web, October 21, 2014, at http://www.alexa.com/topsites.

Alfons, A. 2014a. *cvTools: Cross-Validation Tools for Regression Models*. Comprehensive R Archive Network. 2014. http://cran.r-project.org/web/packages/cvTools/cvTools.pdf.

Alfons, A. 2014b. *simFrame: Simulation Framework*. Comprehensive R Archive Network. 2014. http://cran.r-project.org/web/packages/simFrame/simFrame.pdf. 250

Alfons, A., M. Templ, and P. Filzmoser 2014. *An Object-Oriented Framework for Statistical Simulation: The R Package simFrame*. Comprehensive R Archive Network. 2014. http://cran.r-project.org/web/packages/simFrame/vignettes/simFrame-intro.pdf. 250

Allemang, D. and J. Hendler 2007. *Semantic Web for the Working Ontologist: Effective Modeling in RDFS and OWL* (second ed.). Boston: Morgan Kaufmann.

Allison, P. D. 2010. *Survival Analysis Using SAS: A Practical Guide* (second ed.). Cary, N.C.: SAS Institute Inc.

Amor, D. 2002. *The E-business (R)evolution: Living and Working in an Interconnected World* (second ed.). Upper Saddle River, N.J.: Prentice Hall.

Andersen, P. K., Ø. Borgan, R. D. Gill, and N. Keiding 1993. *Statistical Models Based on Counting Processes*. New York: Springer.

Appleton, D. R. 1995. May the best man win? *The Statistician* 44(4):529–538. 103

Asur, S. and B. A. Huberman 2010. Predicting the future with social media. In *Proceedings of the 2010 IEEE/WIC/ACM International Conference on Web Intelligence and Intelligent Agent Technology - Volume 01*, WI-IAT '10, pp. 492–499. Washington, DC, USA: IEEE Computer Society. http://dx.doi.org/10.1109/WI-IAT.2010.63. 133, 134

Auguie, B. 2014. *Package gridExtra: Functions in grid graphics*. Comprehensive R Archive Network. 2014. http://cran.r-project.org/web/packages/gridExtra/gridExtra.pdf.

Baayen, R. H. 2008. *Analyzing Linguistic Data: A Practical Introduction to Statistics with R*. Cambridge, UK: Cambridge University Press. 259

Bache, K. and M. Lichman 2013. UCI machine learning repository. http://archive.ics.uci.edu/ml. 296

Baeza-Yates, R. and B. Ribeiro-Neto 1999. *Modern Information Retrieval*. New York: ACM Press.

Barabási, A.-L. 2003. *Linked: How Everything is Connected to Everything Else*. New York: Penguin/Plume.

Barabási, A.-L. 2010. *Bursts: The Hidden Pattern Behind Everything We Do*. New York: Penguin/Dutton.

Barabási, A.-L. and R. Albert 1999. Emergence of scaling in random networks. *Science* 286 (5439):509–512. 224

Bates, D. and M. Maechler 2014. *Matrix: Sparse and Dense Matrix Classes and Methods*. Comprehensive R Archive Network. 2014. http://cran.r-project.org/web/packages/support.CEs/support.CEs.pdf. 210

Bates, D. M. and D. G. Watts 2007. *Nonlinear Regression Analysis and Its Applications*. New York: Wiley.

Beazley, D. M. 2009. *Python Essential Reference* (fourth ed.). Upper Saddle River, N.J.: Pearson Education.

Beazley, D. M. and B. K. Jones 2013. *Python Cookbook* (third ed.). Sebastopol, Calif.: O'Reilly.

Becker, R. A. and W. S. Cleveland 1996. *S-Plus Trellis^{TM} Graphics User's Manual*. Seattle: MathSoft, Inc. 252

Behnel, S. 2014, September 10. lxml. Supported at http://lxml.de/. Documentation retrieved from the World Wide Web, October 26, 2014, at http://lxml.de/3.4/lxmldoc-3.4.0.pdf.

Belew, R. K. 2000. *Finding Out About: A Cognitive Perspective on Search Engine Technology and the WWW*. Cambridge: Cambridge University Press.

Belsley, D. A., E. Kuh, and R. E. Welsch 1980. *Regression Diagnostics: Identifying Influential Data and Sources of Collinearity*. New York: Wiley.

Berk, R. A. 2008. *Statistical Learning from a Regression Perspective*. New York: Springer.

Berners-Lee, T. 2000. *Weaving the Web: The Original Design and Ultimate Destiny of the World Wide Web*. New York: HarperBusiness.

Berry, M. and M. Browne 2005, October. Email surveillance using non-negative matrix factorization. *Computational and Mathematical Organization Theory* 11(3):249–264. 81

Berry, M. W. and M. Browne 1999. *Understanding Search Engines: Mathematical Modeling and Text Retrieval.* Philadelphia: Society for Industrial and Applied Mathematics.

Berthold, M. R., N. Cebron, F. Dill, T. R. Gabriel, T. Kötter, T. Meinl, P. Ohl, C. Sieb, K. Thiel, and B. Wiswedel 2007. KNIME: The Konstanz Information Miner. In *Studies in Classification, Data Analysis, and Knowledge Organization (GfKL 2007).* New York: Springer. ISSN 1431-8814. ISBN 978-3-540-78239-1.

Bilton, N. 2013. *Hatching Twitter: A True Story of Money, Power, Friendship, and Betrayal.* New York: Portfolio/Penguin.

Bird, S., E. Klein, and E. Loper 2009. *Natural Language Processing with Python: Analyzing Text with the Natural Language Toolkit.* Sebastopol, Calif.: O'Reilly. http://www.nltk.org/book/. 172

Bishop, Y. M. M., S. E. Fienberg, and P. W. Holland 1975. *Discrete Multivariate Analysis: Theory and Practice.* Cambridge: MIT Press.

Bizer, C., J. Lehmann, G. Kobilarov, S. Auer, C. Becker, R. Cyganiak, and S. Hellmann 2009. DBpedia—A crystallization point for the web of data. *Web Semantics* 7:154–165.

Blei, D., A. Ng, and M. Jordan 2003. Latent Dirichlet allocation. *Journal of Machine Learning Research* 3:993–1022. 176, 251

Bogdanov, P. and A. Singh 2013. Accurate and scalable nearest neighbors in large networks based on effective importance. *CIKM '13*:1–10. ACM CIKM '13 conference paper retrieved from the World Wide Web at http://www.cs.ucsb.edu/~petko/papers/bscikm13.pdf.

Bojanowski, M. 2014. *Package intergraph: Coercion Routines for Network Data Objects in R.* Comprehensive R Archive Network. 2014. http://cran.r-project.org/web/packages/intergraph/intergraph.pdf.

Bollen, J., H. Mao, and X. Zeng 2011. Twitter mood predicts the stock market. *Journal of Computational Science* 2:1–8.

Borg, I. and P. J. F. Groenen 2010. *Modern Multidimensional Scaling: Theory and Applications* (second ed.). New York: Springer.

Borgatti, S. P., M. G. Everett, and J. C. Johnson 2013. *Analyzing Social Networks.* Thousand Oaks, Calif.: Sage. 102

Boser, B. E., I. M. Guyon, and V. N. Vapnik 1992. A training algorithm for optimal margin classifiers. In *Proceedings of the Fifth Conference on Computational Learning Theory*, pp. 144–152. Association for Computing Machinery Press. 129

Bourne, W. 2013, February 27. 2 reasons to keep and eye on github. *Inc.com.* Retrieved from the World Wide Web, November 6, 2014, at http://www.inc.com/magazine/201303/will-bourne/2-reasons-to-keep-an-eye-on-github.html. 235

Box, G. E. P. and D. R. Cox 1964. An analysis of transformations. *Journal of the Royal Statistical Society, Series B (Methodological)* 26(2):211–252.

Boztug, Y. and T. Reutterer 2008. A combined approach for segment-specific market basket analysis. *European Journal of Operational Research* 187(1):294–312. 210

Bradley, R. A. 1976, June. Science, statistics, and paired comparisons. *Biometrics* 32(2): 213–239.

Bradley, R. A. and M. E. Terry 1952, December. Rank analysis of incomplete block designs: I. the method of paired comparisons. *Biometrika* 39(3/4):324–345.

Bradley, S. P., A. C. Hax, and T. L. Magnanti 1977. *Applied Mathematical Programming.* Reading, Mass.: Addison-Wesley. 225

Brandes, U. and T. Erlebach (eds.) 2005. *Network Analysis: Methodological Foundations.* New York: Springer.

Brandes, U., L. C. Freeman, and D. Wagner 2014. Social networks. In R. Tamassia (ed.), *Handbook of Graph Drawing and Visualization*, Chapter 26, pp. 805–840. Boca Raton, Fla.: CRC Press/Chapman & Hall.

Breese, J. S., D. Heckerman, and C. M. Kadie 1998, May. Empirical analysis of predictive algorithms for collaborative filtering. Technical Report MSR-TR-98-12, Microsoft Research. 18 pp. http://research.microsoft.com/apps/pubs/default.aspx?id=69656. 210, 296

Breiman, L. 2001a. Random forests. *Machine Learning* 45(1):5–32.

Breiman, L. 2001b. Statistical modeling: The two cultures. *Statistical Science* 16(3):199–215.

Breiman, L., J. H. Friedman, R. A. Olshen, and C. J. Stone 1984. *Classification and Regression Trees.* New York: Chapman & Hall.

Brin, S. and L. Page 1998. The anatomy of a large-scale hypertextual web search engine. *Computer Networks and ISDN Systems* 30:107–117. 46, 103

Brownlee, J. 2011. *Clever Algorithms: Nature-Inspired Programming Recipies.* Melbourne, Australia: Creative Commons. http://www.CleverAlgorithms.com. 251

Bruzzese, D. and C. Davino 2008. Visual mining of association rules. In S. Simoff, M. H. Böhlen, and A. Mazeika (eds.), *Visual Data Mining: Theory, Techniques and Tools for Visual Analytics*, pp. 103–122. New York: Springer.

Buchanan, M. 2002. *Nexus: Small Worlds and the Groundbreaking Science of Networks.* New York: W.W. Norton and Company.

Bühlmann, P. and S. van de Geer 2011. *Statistics for High-Dimensional Data: Methods, Theory and Applications.* New York: Springer. 249

Burt, J. M. and M. B. Garman 1971. Conditional Monte Carlo: A simulation technique for stochastic network analysis. *Management Science* 18(3):207–217. 225

Butts, C. T. 2008a, May 5. network: A package for managing relational data in R. *Journal of Statistical Software* 24(2):1–36. http://www.jstatsoft.org/v24/i02. Updated version available at http://cran.r-project.org/web/packages/network/vignettes/networkVignette.pdf.

Butts, C. T. 2008b, May 8. Social network analysis with sna. *Journal of Statistical Software* 24(6):1–51. http://www.jstatsoft.org/v24/i06.

Butts, C. T. 2014a. *Package network: Classes for Relational Data.* Comprehensive R Archive Network. 2014. http://cran.r-project.org/web/packages/network/network.pdf.

Butts, C. T. 2014b. *Package snad: Statistical Analysis of Network Data with R.* Comprehensive R Archive Network. 2014. http://cran.r-project.org/web/packages/sand/sand.pdf.

Butts, C. T. 2014c. *Package sna: Tools for Social Network Analysis.* Comprehensive R Archive Network. 2014. `http://cran.r-project.org/web/packages/sna/sna.pdf`. Also see the statnet website at `http://www.statnetproject.org/`.

Buvač, V. and P. J. Stone 2001, April 2. The General Inquirer user's guide. Software developed with the support of Harvard University and The Gallup Organization.

Cairo, A. 2013. *The Functional Art: An Introduction to Information Graphics and Visualization.* Berkeley, Calif: New Riders.

Cami, A. and N. Deo 2008. Techniques for analyzing dynamic random graph models of web-like networks: An overview. *Networks* 51(4):211–255.

Campbell, S. and S. Swigart 2014. *Going beyond Google: Gathering Internet Intelligence* (5 ed.). Oregon City, Oreg.: Cascade Insights. `http://www.cascadeinsights.com/gbgdownload`. 59

Cantelon, M., M. Harter, T. J. Holowaychuk, and N. Rajlich 2014. *Node.js in Action.* Shelter Island, N.Y.: Manning. 3

Carley, K. M. and D. Skillicorn 2005, October. Special issue on analyzing large scale networks: The Enron corpus. *Computational and Mathematical Organization Theory* 11(3): 179–181. 291

Carlin, B. P. 1996, February. Improved NCAA basketball tournament modeling via point spread and team strength information. *The American Statistician* 50(1):39–43. 103

Carlin, B. P. and T. A. Louis 1996. *Bayes and Empirical Bayes Methods for Data Analysis.* London: Chapman & Hall. 244

Carr, D., N. Lewin-Koh, and M. Maechler 2014. *hexbin: Hexagonal Binning Routines.* Comprehensive R Archive Network. 2014. `http://cran.r-project.org/web/packages/hexbin/hexbin.pdf`. 252

Carrington, P. J., J. Scott, and S. Wasserman (eds.) 2005. *Models and Methods in Social Network Analysis.* Cambridge, UK: Cambridge University Press.

Casciano, C. 2011. *The CSS Pocket Guide.* Upper Saddle River, N.J.: Pearson/Peachpit.

Cass, S. 2014, July. The top 10 programming languages: Spectrum's 2014 ranking. *IEEE Spectrum* 51(7):68. 298

Chakrabarti, S. 2003. *Mining the Web: Discovering Knowledge from Hypertext Data.* San Francisco: Morgan Kaufmann.

Chambers, J. M. and T. J. Hastie (eds.) 1992. *Statistical Models in S.* Pacific Grove, Calif.: Wadsworth & Brooks/Cole. Champions of S, S-Plus, and R call this "the white book." It introduced statistical modeling syntax using S3 classes.

Chan, E. P. 2009. *Quantitative Trading: How to Build Your Own Algorithmic Trading Business.* New York: Wiley.

Chan, E. P. 2013. *Algorithmic Trading: Winning Strategies and Their Rationale.* New York: Wiley.

Chang, W. 2013. *R Graphics Cookbook.* Sebastopol, Calif.: O'Reilly.

Chapanond, A., M. S. Krishnamoorthy, and B. Yener 2005, October. Graph theoretic and spectral analysis of Enron email data. *Computational and Mathematical Organization Theory* 11(3):265–281. 81

Charmaz, K. 2002. Qualitative interviewing and grounded theory analysis. In J. F. Gubrium and J. A. Holstein (eds.), *Handbook of Interview Research: Context and Method*, Chapter 32, pp. 675–694. Thousand Oaks, Calif.: Sage. 279

Charniak, E. 1993. *Statistical Language Learning*. Cambridge: MIT Press.

Chatterjee, S. and A. S. Hadi 2012. *Regression Analysis by Example* (fifth ed.). New York: Wiley.

Chau, M. and H. Chen 2003. Personalized and focused web spiders. In N. Zhong, J. Liu, and Y. Yao (eds.), *Web Intelligence*, Chapter 10, pp. 198–217. New York: Springer. 59

Cherny, L. 1999. *Conversation and Community: Chat in a Virtual World*. Stanford, Calif.: CSLI Publications. 265

Chodorow, K. 2013. *MongoDB: The Definitive Guide* (second ed.). Sebastopol, Calif.: O'Reilly. 3

Christakis, N. A. and J. H. Fowler 2009. *Connected: The Surprising Power of Our Social Networks and How They Shape Our Lives*. New York: Little Brown and Company.

Christensen, R. 1997. *Log-Linear Models and Logistic-Regression* (second ed.). New York: Springer.

Chun, W. J. 2007. *Core Python Programming* (second ed.). Upper Saddle River, N.J.: Pearson Education.

Cleveland, W. S. 1993. *Visualizing Data*. Murray Hill, N.J.: AT&T Bell Laboratories. Initial documentation for trellis graphics in S-Plus. 252

Clifton, B. 2012. *Advanced Web Metrics with Google Analytics* (third ed.). New York: Wiley. 19

Cohen, J. 1960, April. A coefficient of agreement for nominal data. *Educational and Psychological Measurement* 20(1):37–46. 248

Conway, D. and J. M. White 2012. *Machine Learning for Hackers*. Sebastopol, Calif.: O'Reilly.

Cook, R. D. 1998. *Regression Graphics: Ideas for Studying Regressions through Graphics*. New York: Wiley.

Cook, R. D. 2007. Fisher lecture: Dimension reduction in regression. *Statistical Science* 22: 1–26.

Cook, R. D. and S. Weisberg 1999. *Applied Regression Including Computing and Graphics*. New York: Wiley.

Copeland, R. 2013. *MongoDB Applied Design Patterns* (second ed.). Sebastopol, Calif.: O'Reilly. 3

Cox, T. F. and M. A. A. Cox 1994. *Multidimensional Scaling*. London: Chapman & Hall.

Cranor, L. F. and B. A. LaMacchia 1998. Spam! *Communications of the ACM* 41(8):74–83.

Cristianini, N. and J. Shawe-Taylor 2000. *An Introduction to Support Vector Machines and Other Kernel-Based Learning Methods*. Cambridge: Cambridge University Press.

Crnovrsanin, T., C. W. Mueldera, R. Faris, D. Felmlee, and K.-L. Ma 2014. Visualization techniques for categorical analysis of social networks with multiple edge sets. *Social Networks* 37:56–64.

Crockford, D. 2008. *JavaScript: The Good Parts*. Sebastopol, Calif.: O'Reilly.

Croll, A. and S. Power 2009. *Complete Web Monitoring: Watching Your Visitors, Performance, Communities & Competitors*. Sebastopol, Calif.: O'Reilly.

Csardi, G. 2014a. *Package igraphdata: A Collection of Network Data Sets for the igraph Package*. Comprehensive R Archive Network. 2014. http://cran.r-project.org/web/packages/igraphdata/igraphdata.pdf.

Csardi, G. 2014b. *Package igraph: Network Analysis and Visualization*. Comprehensive R Archive Network. 2014. `http://cran.r-project.org/web/packages/igraph/igraph.pdf`.

Daconta, M. C., L. J. Obrst, and K. T. Smith 2003. *The Semantic Web: A Guide to the Future of XML, Web Services, and Knowledge Management*. New York: Wiley.

Dale, J. and S. Abbott 2014. *Qual-Online The Essential Guide: What Every Researcher Needs to Know about Conducting and Moderating Interviews via the Web*. Ithaca, N.Y.: Paramount Market Publishing.

Datta, A., S. Shulman, B. Zheng, S.-D. Lin, A. Sum, and E.-P. Lim (eds.) 2011. *Social Informatics: Proceedings of the Third International Conference SocInfo 2011*. New York: Springer.

Davenport, T. H. and J. G. Harris 2007. *Competing on Analytics: The New Science of Winning*. Boston: Harvard Business School Press. 239

Davenport, T. H., J. G. Harris, and R. Morison 2010. *Analytics at Work: Smarter Decisions, Better Results*. Boston: Harvard Business School Press. 239

David, H. A. 1963. *The Method of Paired Comparisons*. London: Charles Griffin & Company Limited.

Davidson, R. R. and P. H. Farquhar 1976, June. A bibliography on the method of paired comparisons. *Biometrics* 32(2):241–252.

Davis, B. H. and J. P. Brewer 1997. *Electronic Discourse: Linguistic Individuals in Virtual Space*. Albany, N.Y.: State University of New York Press. 265

Davison, M. L. 1992. *Multidimensional Scaling*. Melbourne, Fla.: Krieger.

de Nooy, W., A. Mrvar, and V. Batagelj 2011. *Exploratory Social Network Analysis with Pajek*. Cambridge, U.K.: Cambridge University Press.

Dean, J. and S. Ghemawat 2004. MapReduce: Simplifed Data Processing on Large Clusters. Retrieved from the World Wide Web at `http://static.usenix.org/event/osdi04/tech/full_papers/dean/dean.pdf`.

Dehuri, S., M. Patra, B. B. Misra, and A. K. Jagadev (eds.) 2012. *Intelligent Techniques in Recommendation Systems*. Hershey, Pa.: IGI Global.

Delen, D., R. Sharda, and P. Kumar 2007. Movie forecast guru: A Web-based DSS for Hollywood managers. *Decision Support Systems* 43(4):1151–1170. 133

Demšar, J. and B. Zupan 2013. Orange: Data mining fruitful and fun—A historical perspective. *Informatica* 37:55–60. 250, 299

Dey, I. 1999. *Grounding Grounded Theory: Guidelines for Qualitative Inquiry*. New York: Wiley. 279

Di Battista, G., P. Eades, R. Tomassia, and I. G. Tollis 1999. *Graph Drawing: Algorithms for the Visualization of Graphs*. Upper Saddle River, N.J.: Prentice Hall.

Di Battista, G. and M. Rimondini 2014. Computer networks. In R. Tamassia (ed.), *Handbook of Graph Drawing and Visualization*, Chapter 25, pp. 763–804. Boca Raton, Fla.: CRC Press/Chapman & Hall.

Diesner, J., T. L. Frantz, and K. M. Carley 2005, October. Communication networks from the Enron email corpus. *Computational and Mathematical Organization Theory* 11(3): 201–228. 81

Dippold, K. and H. Hruschka 2013. Variable selection for market basket analysis. *Computational Statistics* 28(2):519–539. http://dx.doi.org/10.1007/s00180-012-0315-3. ISSN 0943-4062. 210

Dover, D. 2011. *Search Engine Optimization Secrets: Do What You Never Thought Possible with SEO*. New York: Wiley.

Downing, D., M. Covington, M. Covington, C. A. Barrett, and S. Covington 2012. *Dictionary of Computer and Internet Terms* (eleventh ed.). Hauppauge, N.Y.: Barron's Educational Series.

Draper, N. R. and H. Smith 1998. *Applied Regression Analysis* (third ed.). New York: Wiley.

DuCharme, B. 2013. *Learning SPARQL*. Sebastopol, Calif.: O'Reilly. 235

Duda, R. O., P. E. Hart, and D. G. Stork 2001. *Pattern Classification* (second ed.). New York: Wiley.

Dumais, S. T. 2004. Latent semantic analysis. In B. Cronin (ed.), *Annual Review of Information Science and Technology*, Volume 38, Chapter 4, pp. 189–230. Medford, N.J.: Information Today.

Easley, D. and J. Kleinberg 2010. *Networks, Crowds, and Markets: Reasoning about a Highly Connected World*. Cambridge, UK: Cambridge University Press.

Efron, B. 1986. Why isn't everyone a Bayesian (with commentary). *The American Statistician* 40(1):1–11.

Eichenwald, K. 2005. *Conspiracy of Fools: A True Story*. New York: Broadway Books/Random House.

Ellis, D., R. Oldridge, and A. Vasconcelos 2004. Community and virtual community. In B. Cronin (ed.), *Annual Review of Information Science and Technology*, Volume 38, Chapter 3, pp. 145–186. Medford, N.J.: Information Today.

Engelbrecht, A. P. 2007. *Computational Intelligence: An Introduction* (second ed.). New York: Wiley. 251

Erdős, P. and A. Rényi 1959. On random graphs. *Publicationes Mathematicae* 6:290–297.

Erdős, P. and A. Rényi 1960. On the evolution of random graphs. *Publications of the Mathematical Institute of the Hungarian Academy of Sciences* 5:17–61.

Erdős, P. and A. Rényi 1961. On the strength of connectedness of a random graph. *Acta Mathematica Scientia Hungary* 12:261–267.

Erwig, M. 2000. The graph Voronoi diagram with applications. *Networks* 36(3):156–163.

Esipova, N., T. W. Miller, M. D. Zarnecki, J. Elzaurdia, and S. Ponnaiya 2004. Exploring the possibilities of online focus groups. In T. W. Miller and J. Walkowski (eds.), *Qualitative Research Online*, Chapter A, pp. 107–128. Manhattan Beach, Calif.: Research Publishers.

European Parliament 2002. Directive on privacy and electronic communications. Retrieved from the World Wide Web, October 21, 2014, at http://eur-lex.europa.eu/LexUriServ/LexUriServ.do?uri=CELEX:32002L0058:en:HTML. 19

Everitt, B. and G. Dunn 2001. *Applied Multivariate Data Analysis* (second ed.). New York: Wiley. 176

Everitt, B. and S. Rabe-Hesketh 1997. *The Analysis of Proximity Data*. London: Arnold.

Everitt, B. S., S. Landau, M. Leese, and D. Stahl 2011. *Cluster Analysis* (fifth ed.). New York: Wiley.

Fawcett, T. 2003, January 7. ROC graphs: Notes and practical considerations for researchers. http://www.hpl.hp.com/techreports/2003/HPL-2003-4.pdf.

Feinerer, I. 2012. *Introduction to the wordnet Package*. Comprehensive R Archive Network. 2012. http://cran.r-project.org/web/packages/wordnet/vignettes/wordnet.pdf. 259

Feinerer, I. 2014. *Introduction to the tm Package*. Comprehensive R Archive Network. 2014. http://cran.r-project.org/web/packages/tm/vignettes/tm.pdf.

Feinerer, I. and K. Hornik 2014a. *tm: Text Mining Package*. Comprehensive R Archive Network. 2014. http://cran.r-project.org/web/packages/tm/tm.pdf.

Feinerer, I. and K. Hornik 2014b. *wordnet: WordNet Interface*. Comprehensive R Archive Network. 2014. http://cran.r-project.org/web/packages/wordnet/wordnet.pdf. 259

Feinerer, I., K. Hornik, and D. Meyer 2008, 3 31. Text mining infrastructure in R. *Journal of Statistical Software* 25(5):1–54. http://www.jstatsoft.org/v25/i05. ISSN 1548-7660.

Feldman, R. 2002. Text mining. In W. Klösgen and J. M. Żytkow (eds.), *Handbook of Data Mining and Knowledge Discovery*, Chapter 38, pp. 749–757. Oxford: Oxford University Press.

Fellbaum, C. 1998. *WordNet: An Electronic Lexical Database*. Cambridge, Mass.: MIT Press. 259

Fellows, I. 2014a. wordcloud makes words less cloudy. Retrieved from the World Wide Web at http://blog.fellstat.com/. 176, 259

Fellows, I. 2014b. *wordcloud: Word Clouds*. Comprehensive R Archive Network. 2014. http://cran.r-project.org/web/packages/wordcloud/wordcloud.pdf. 176, 259

Fenzel, D., J. Hendler, H. Lieberman, and W. Wahlster (eds.) 2003. *Spinning the Semantic Web: Bringing the World Wide Web to Its Full Potential*. Cambridge: MIT Press.

Ferrucci, D., E. Brown, J. Chu-Carroll, J. Fan, D. Gondek, A. A. Kalyanpur, A. Lally, J. W. Murdock, E. Nyberg, J. Prager, N. Schlaefer, and C. Welty 2010, Fall. Building Watson: An overview of the DeepQA project. *AI Magazine* 31(3):59–79.

Few, S. 2009. *Now You See It: Simple Visualization Techniques and Quantitative Analysis*. Oakland, Calif.: Analytics Press.

Fielding, N. G. and R. M. Lee 1998. *Computer Analysis and Qualitative Research*. London: Sage. 279

Fienberg, S. E. 2007. *Analysis of Cross-Classified Categorical Data* (second ed.). New York: Springer.

Firth, D. 1991. Generalized linear models. In D. Hinkley and E. Snell (eds.), *Statistical Theory and Modeling: In Honour of Sir David Cox, FRS*, Chapter 3, pp. 55–82. London: Chapman and Hall.

Fisher, R. A. 1970. *Statistical Methods for Research Workers* (fourteenth ed.). Edinburgh: Oliver and Boyd. First edition published in 1925. 244

Fisher, R. A. 1971. *Design of Experiments* (ninth ed.). New York: Macmillan. First edition published in 1935. 244

Flam, F. 2014, September 30. The odds, continually updated. *The New York Times*:D1. Retrieved from the World Wide Web at http://www.nytimes.com/2014/09/30/science/the-odds-continually-updated.html?_r=0. 244

Flanagan, D. 2011. *JavaScript: The Definitive Guide* (sixth ed.). Sebastopol, Calif.: O'Reilly. 3

Fox, J. 2002, January. Robust regression: Appendix to an R and S-PLUS companion to applied regression. Retrieved from the World Wide Web at `http://cran.r-project.org/doc/contrib/Fox-Companion/appendix-robust-regression.pdf`. 249

Fox, J. 2014. *car: Companion to Applied Regression*. Comprehensive R Archive Network. 2014. `http://cran.r-project.org/web/packages/car/car.pdf`.

Fox, J. and S. Weisberg 2011. *An R Companion to Applied Regression* (second ed.). Thousand Oaks, Calif.: Sage. 128

Franceschet, M. 2011, June. PageRank: Standing on the shoulders of giants. *Communications of the ACM* 54(6):92–101.

Franks, B. 2012. *Taming the Big Data Tidal Wave: Finding Opportunities in Huge Data Streams with Advanced Analytics*. Hoboken, N.J.: Wiley. 239

Freeman, L. C. 2004. *The Development of Social Network Analysis: A Study in the Sociology of Science*. Vancouver, B.C.: Empirical Press.

Freeman, L. C. 2005. Graphic techniques for exploring social network data. In P. J. Carrington, J. Scott, and S. Wasserman (eds.), *Models and Methods in Social Network Analysis*, Chapter 12, pp. 248–269. Cambridge, UK: Cambridge University Press.

Friedl, J. E. F. 2006. *Mastering Regular Expressions* (third ed.). Sebastopol, Calif.: O'Reilly. 259

Fruchterman, T. M. J. and E. M. Reingold 1991. Graph drawing by force-directed placement. *Software—Practice and Experience* 21(11):1129–1164.

Fry, B. 2008. *Visualizing Data: Exploring and Explaining Data with the Processing Environment*. Sebastopol, Calif.: O'Reilly.

Gabriel, K. R. 1971. The biplot graphical display of matrices with application to principal component analysis. *Biometrika* 58:453–467.

Garcia-Molina, H., J. D. Ullman, and J. Widom 2009. *Database Systems: The Complete Book* (second ed.). Upper Saddle River, N.J.: Prentice-Hall.

Gasston, P. 2011. *The Book of CSS3: A Developer's Guide to the Future of Web Design*. San Francisco: No Starch Press.

Gasston, P. 2013. *The Modern Web: Multi-Device Web Development with HTML5, CSS3, and JavaScript*. San Francisco: No Starch Press.

Geisser, S. 1993. *Predictive Inference: An Introduction*. New York: Chapman & Hall. 244

Gelman, A., J. B. Carlin, H. S. Stern, and D. B. Rubin 1995. *Bayesian Data Analysis*. London: Chapman & Hall. 244

Gelman, A., J. Hill, Y.-S. Su, M. Yajima, and M. G. Pittau 2014. *mi: Missing Data Imputation and Model Checking*. Comprehensive R Archive Network. 2014. `http://cran.r-project.org/web/packages/mi/mi.pdf`.

Georgakopoulou, A. and D. Goutsos 1997. *Discourse Analysis: An Introduction*. Edinburgh, U.K.: Edinburgh University Press. 279

Gibson, H., J. Faith, and P. Vickers 2013, 07. A survey of two-dimensional graph layout techniques for information visualisation. *Information Visualization* 12(3-4):324–357.

Glaser, B. G. and A. L. Strauss 1967. *The Discovery of Grounded Theory: Strategies for Qualitative Research*. Hawthorne, N.Y.: Aldine de Gruyter. 279

Gnanadesikan, R. 1997. *Methods for Statistical Data Analysis of Multivariate Observations* (second ed.). New York: Wiley. 176

Goldenberg, A., A. X. Zheng, S. E. Fienberg, and E. M. Airoldi 2009, January. A survey of statistical network models. *Foundations and Trends in Machine Learning* 2(2):129–233.

Golombisky, K. and R. Hagen 2013. *White Space is Not Your Enemy: A Beginner's Guide to Communicating Visually through Graphic, Web, & Multimedia Design* (second ed.). Burlington, Mass.: Focal Press/Taylor & Francis.

Goodreau, S. M., M. S. Handcock, D. R. Hunter, C. T. Butts, and M. Morris 2008, 5 8. A statnet tutorial. *Journal of Statistical Software* 24(9):1–26. http://www.jstatsoft.org/v24/i09. CODEN JSSOBK. ISSN 1548-7660.

Gower, J. C. and D. J. Hand 1996. *Biplots*. London: Chapman & Hall.

Graybill, F. A. 1961. *Introduction to Linear Statistical Models, Volume 1*. New York: McGraw-Hill.

Graybill, F. A. 2000. *Theory and Application of the Linear Model*. Stamford, Conn.: Cengage Learning.

Gries, S. T. 2009. *Quantitative Corpus Linguistics with R: A Practical Introduction*. New York: Routledge.

Gries, S. T. 2013. *Statistics for Linguistics with R: A Pratical Introduction* (second revised ed.). Berlin: De Gruyter Mouton. 259

Grieve, T. 2003, October 14. The decline and fall of the Enron empire. *Salon*:165–178. Retrieved from the World Wide Web at http://www.salon.com/2003/10/14/enron_22/.

Groneveld, R. A. 1990, November. Ranking teams in a league with two divisions of *t* teams. *The American Statistician* 44(4):277–281. 103

Grolemund, G. and H. Wickham 2014. *lubridate: Make Dealing with Dates a Little Easier*. Comprehensive R Archive Network. 2014. http://cran.r-project.org/web/packages/lubridate/lubridate.pdf. 62

Gross, J. L., J. Yellen, and P. Zhang (eds.) 2014. *Handbook of Graph Theory* (second ed.). Boca Raton, Fla.: CRC Press/Chapman & Hall.

Grothendieck, G. 2014a. *gsubfn: Utilities for Strings and Function Arguments*. Comprehensive R Archive Network. 2014. http://cran.r-project.org/web/packages/gsubfn/gsubfn.pdf.

Grothendieck, G. 2014b. *gsubfn: Utilities for Strings and Function Arguments (Vignette)*. Comprehensive R Archive Network. 2014. http://cran.r-project.org/web/packages/gsubfn/vignettes/gsubfn.pdf.

Groves, R. M., F. J. Fowler, Jr., M. P. Couper, J. M. Lepkowski, E. Singer, and R. Tourangeau 2009. *Survey Methodology* (second ed.). New York: Wiley.

Guilford, J. P. 1936. *Psychometric Methods*. New York: McGraw-Hill.

Gunning, D., V. K. Chaudhri, P. Clark, K. Barker, S.-Y. Chaw, M. Greaves, B. Grosof, A. Leung, D. McDonald, S. Mishra, J. Pacheco, B. Porter, A. Spaulding, D. Tecuci, and J. Tien 2010, Fall. Project Halo update—Progress toward digital Aristotle. *AI Magazine* 31(3):33–58.

Hagbert, A. and D. Schult 2014. *NetworkX: Python Software for Complex Networks*. NetworkX Development Team. 2014. Retrieved from the World Wide Web at https://github.com/networkx/.

Hahsler, M. 2014a. *recommenderlab: A Framework for Developing and Testing Recommendation Algorithms*. Comprehensive R Archive Network. 2014. `http://cran.r-project.org/web/packages/recommenderlab/vignettes/recommenderlab.pdf`.

Hahsler, M. 2014b. *recommenderlab: Lab for Developing and Testing Recommender Algorithms*. Comprehensive R Archive Network. 2014. `http://cran.r-project.org/web/packages/recommenderlab/recommenderlab.pdf`.

Hahsler, M., C. Buchta, B. Grün, and K. Hornik 2014a. *arules: Mining Association Rules and Frequent Itemsets*. Comprehensive R Archive Network. 2014. `http://cran.r-project.org/web/packages/arules/arules.pdf`.

Hahsler, M., C. Buchta, B. Grün, and K. Hornik 2014b. *Introduction to arules: A Computational Environment for Mining Association Rules and Frequent Itemsets*. Comprehensive R Archive Network. 2014. `http://cran.r-project.org/web/packages/arules/vignettes/arules.pdf`.

Hahsler, M., C. Buchta, and K. Hornik 2008. Selective association rule generation. *Computational Statistics* 23:303–315. 210

Hahsler, M. and S. Chelluboina 2014a. *arulesViz: Visualizing Association Rules and Frequent Itemsets*. Comprehensive R Archive Network. 2014. `http://cran.r-project.org/web/packages/arulesViz/arulesViz.pdf`.

Hahsler, M. and S. Chelluboina 2014b. *Visualizing Association Rules: Introduction to the R-extension Package arulesViz*. Comprehensive R Archive Network. 2014. `http://cran.r-project.org/web/packages/arulesViz/vignettes/arulesViz.pdf`.

Hahsler, M., S. Chelluboina, K. Hornik, and C. Buchta 2011. The arules R-package ecosystem: Analyzing interesting patterns from large transaction data sets. *Journal of Machine Learning Research* 12:2021–2025.

Hahsler, M., B. Grün, and K. Hornik 2005, September 29. arules: A computational environment for mining association rules and frequent item sets. *Journal of Statistical Software* 14(15):1–25. `http://www.jstatsoft.org/v14/i15`.

Hale, C. and J. Scanlon 1999. *Wired Style: Principles of English Usage in the Digital Age*. New York: Broadway Books.

Hand, D. J. 1997. *Construction and Assessment of Classification Rules*. New York: Wiley.

Handcock, M. S. and K. Gile 2007, April 28. Modeling social networks with sampled or missing data. `http://www.csss.washington.edu/Papers/wp75.pdf`. 109

Handcock, M. S., D. R. Hunter, C. T. Butts, S. M. Goodreau, and M. Morris 2008, May 5. statnet: Software tools for the representation, visualization, analysis and simulation of network data. *Journal of Statistical Software* 24(1):1–11. `http://www.jstatsoft.org/v24/i01`.

Hanneman, R. A. and M. Riddle 2005. *Introduction to Social Network Methods*. Riverside, Calif.: University of California Riverside. `http://www.faculty.ucr.edu/~hanneman/nettext/`.

Harrell, Jr., F. E. 2001. *Regression Modeling Strategies: With Applications to Linear Models, Logistic Regression, and Survival Analysis*. New York: Springer.

Harrington, P. 2012. *Machine Learning in Action*. Shelter Island, N.Y.: Manning. 210

Hart, R. P. 2000a. *Campaign Talk: Why Elections Are Good for Us*. Princeton, N.J.: Princeton University Press.

Hart, R. P. 2000b. *DICTION 5.0: The Text Analysis Program*. Thousand Oaks, Calif.: Sage.

Hart, R. P. 2001. Redeveloping Diction: theoretical considerations. In M. D. West (ed.), *Theory, Method, and Practice in Computer Content Analysis*, Chapter 3, pp. 43–60. Westport, Conn.: Ablex.

Hastie, T., R. Tibshirani, and J. Friedman 2009. *The Elements of Statistical Learning: Data Mining, Inference, and Prediction* (second ed.). New York: Springer.

Hausser, R. 2001. *Foundations of Computational Linguistics: Human-Computer Communication in Natural Language* (second ed.). New York: Springer-Verlag.

Hearst, M. A. 1997. TextTiling: Segmenting text into multi-paragraph subtopic passages. *Computational Linguistics* 23(1):33–64. 259

Heer, J., M. Bostock, and V. Ogievetsky 2010, May 1. A tour through the visualization zoo: A survey of powerful visualization techniques, from the obvious to the obscure. *acmqueue: Association for Computing Machinery*:1–22. Retrieved from the World Wide Web at http://queue.acm.org/detail.cfm?id=1805128.

Hemenway, K. and T. Calishain 2004. *Spidering Hacks: 100 Industrial-Strength Tips & Tools*. Sabastopol, Calif.: O'Reilly. 59

Herman, A. and T. Swiss (eds.) 2000. *The World Wide Web and Contemporary Cultural Theory*. New York: Routledge. 265

Hinkley, D. V., N. Reid, and E. J. Snell (eds.) 1991. *Statistical Theory and Modeling*. London: Chapman and Hall. 244

Ho, Q. and E. P. Xing 2014. Analyzing time-evolving networks using an evolving cluster mixed membership blockmodel. In E. M. Airoldi, D. Blei, E. A. Erosheva, and S. E. Fienberg (eds.), *Handbook of Mixed Membership Models and Their Applications*, Chapter 23, pp. 489–525. Boca Raton, Fla.: CRC Press.

Hoberman, S. 2014. *Data Modeling for MongoDB: Building Well-Designed and Supportable MongoDB Databases*. Basking Ridge, N.J.: Technics Publications. 3

Hoerl, A. E. and R. W. Kennard 2000. Ridge regression: biased estimation for non-orthogonal problems. *Technometrics* 42(1):80–86. Reprinted from *Technometrics*, volume 12. 249

Hoffman, P. and Scrapy Developers 2014, June 26. Scrapy documentation release 0.24.0. Supported at http://scrapy.org/. Documentation retrieved from the World Wide Web, October 26, 2014, at https://media.readthedocs.org/pdf/scrapy/0.24/scrapy.pdf.

Holden, D. 2006. Hierarchical edge bundles: Visualization of adjacency relations in hierarchical data. *IEEE Transactions on Visualization and Computer Graphics* 12(5):741–748.

Holeton, R. (ed.) 1998. *Composing Cyberspace: Identity, Community, and Knowledge in the Electronic Age*. Boston: McGraw-Hill. 265

Holland, P. W., K. B. Laskey, and S. Leinhardt 1983. Stochastic blockmodels: First steps. *Social Networks* 5:109–138.

Honaker, J., G. King, and M. Blackwell 2014. *Amelia II: A Program for Missing Data*. Comprehensive R Archive Network. 2014. http://cran.r-project.org/web/packages/Amelia/Amelia.pdf.

Hornik, K. 2014a. *RWeka Odds and Ends*. Comprehensive R Archive Network. 2014. http://cran.r-project.org/web/packages/RWeka/vignettes/RWeka.pdf.

Hornik, K. 2014b. *RWeka: R/Weka Interface.* Comprehensive R Archive Network. 2014. `http://cran.r-project.org/web/packages/RWeka/RWeka.pdf`.

Hosmer, D. W., S. Lemeshow, and S. May 2013. *Applied Survival Analysis: Regression Modeling of Time to Event Data* (second ed.). New York: Wiley.

Hosmer, D. W., S. Lemeshow, and R. X. Sturdivant 2013. *Applied Logistic Regression* (third ed.). New York: Wiley. 128

Houston, B., L. Bruzzese, and S. Weinberg 2002. *The Investigative Reporter's Handbook: A Guide to Documents, Databases and Techniques* (fourth ed.). Boston: Bedford/St. Martin's.

Hu, M. and B. Liu 2004, August 22–25. Mining and summarizing customer reviews. *Proceedings of the ACM SIGKDD International Conference on Knowledge Discovery & Data Mining (KDD-2004).* Full paper available from the World Wide Web at `http://www.cs.uic.edu/~liub/publications/kdd04-revSummary.pdf` Original source for opinion and sentiment lexicon, available from the World Wide Web at `http://www.cs.uic.edu/~liub/FBS/sentiment-analysis.html#lexicon`.

Huberman, B. A. 2001. *The Laws of the Web: Patterns in the Ecology of Information.* Cambridge: MIT Press. 265

Huet, S., A. Bouvier, M.-A. Poursat, and E. Jolivet 2004. *Statistical Tools for Nonlinear Regression: A Practical Guide with S-Plus and R Examples* (second ed.). New York: Springer.

Hughes-Croucher, T. and M. Wilson 2012. *Node Up and Running.* Sebastopol, Calif.: O'Reilly. 3

Huisman, M. 2010. Imputation of missing network data: Some simple procedures. *Journal of Social Structure* 10. `http://www.cmu.edu/joss/content/articles/volume10/huisman.pdf`. 109

Hyslop, B. 2010. *The HTML Pocket Guide.* Upper Saddle River, N.J.: Pearson/Peachpit.

Ihaka, R., P. Murrell, K. Hornik, J. C. Fisher, and A. Zeileis 2014. *colorspace: Color Space Manipulation.* Comprehensive R Archive Network. 2014. `http://cran.r-project.org/web/packages/colorspace/colorspace.pdf`.

Indurkhya, N. and F. J. Damerau (eds.) 2010. *Handbook of Natural Language Processing* (second ed.). Boca Raton, Fla.: Chapman and Hall/CRC.

Ingersoll, G. S., T. S. Morton, and A. L. Farris 2013. *Taming Text: How to Find, Organize, and Manipulate It.* Shelter Island, N.Y.: Manning. 176, 251

Izenman, A. J. 2008. *Modern Multivariate Statistical Techniques: Regression, Classification, and Manifold Learning.* New York: Springer. 176

Jackson, M. O. 2008. *Social and Economic Networks.* Princeton, N.J.: Princeton University Press.

James, G., D. Witten, T. Hastie, and R. Tibshirani 2013. *An Introduction to Statistical Learning with Applications in R.* New York: Springer.

Janssen, J., M. Hurshman, and N. Kalyaniwalla 2012. Model selection for social networks using graphlets. *Internet Mathematics* 8(4):338–363.

Johnson, K. 2008. *Quantitative Methods in Linguistics.* Malden, Mass.: Blackwell Publishing. 259

Johnson, R. A. and D. W. Wichern 1998. *Applied Multivariate Statistical Analysis* (fourth ed.). Upper Saddle River, N.J.: Prentice Hall. 176

Johnson, S. 1997. *Interface Culture*. New York: Basic Books. 265

Jones, S. G. (ed.) 1997. *Virtual Culture: Identity & Communication in Cybersociety*. Thousand Oaks, Calif.: Sage. 265

Jones, S. G. (ed.) 1999. *Doing Internet Research: Critical Issues and Methods for Examining the Net*. Thousand Oaks, Calif.: Sage. 265

Jonscher, C. 1999. *The Evolution of Wired Life: From the Alphabet to the Soul-Catcher Chip—How Information Technologies Change Our World*. New York: Wiley. 265

Joula, P. 2008. *Authorship Attribution*. Hanover, Mass.: Now Publishers. 259

Jung, C. G. 1968. *The Archetypes of the Collective Unconscious (Collected Works of C. G. Jung, Vol. 9, Part 1)* (second ed.). Princeton, N.J.: Princeton University Press.

Jünger, M. and P. Mutzel (eds.) 2004. *Graph Drawing Software*. New York: Springer. 82

Jurafsky, D. and J. H. Martin 2009. *Speech and Language Processing: An Introduction to Natural Language Processing, Computational Linguistics, and Speech Recognition* (second ed.). Upper Saddle River, N.J.: Prentice Hall.

Kadushin, C. 2012. *Understanding Social Networks*. New York: Oxford University Press.

Kamada, T. and S. Kawai 1989. An algorithm for drawing general undirected graphs. *Information Processing Letters* 31(1):7–15.

Kaufman, L. and P. J. Rousseeuw 1990. *Finding Groups in Data: An Introduction to Cluster Analysis*. New York: Wiley.

Kaushik, A. 2010. *Web Analytics 2.0: The Art of Online Accountability & Science of Customer Centricity*. New York: Wiley/Sybex.

Kay, M. 2008. *XSLT 2.0 and XPath 2.0 Programmer's Reference* (fourth ed.). New York: Wiley/Wrox. 29

Keener, J. P. 1993, March. The Perron-Frobenius theorem and the ranking of football teams. *SIAM Review* 35(1):80–93.

Keila, P. and D. B. Skillicorn 2005, October. Structure of the Enron email dataset. *Computational and Mathematical Organization Theory* 11(3):183–199. 81

Kelle, U. (ed.) 1995. *Computer-Aided Qualitative Data Analysis: Theory, Methods and Practice*. Thousand Oaks, Calif.: Sage. 279

Kendall, M. G. and B. B. Smith 1940, March. On the method of paired comparisons. *Biometrika* 31(3/4):324–345.

Kirk, R. E. 2013. *Experimental Design: Procedures for the Behavioral Sciences* (fourth ed.). Thousand Oaks, Calif.: Sage.

Klimmt, B. and Y. Yang 2004. Introducing the Enron corpus. `http://ceas.cc/2004/168.pdf`.

Kolaczyk, E. D. 2009. *Statistical Analysis of Network Data: Methods and Models*. New York: Springer. 224

Kolaczyk, E. D. and G. Csárdi 2014. *Statistical Analysis of Network Data with R*. New York: Springer.

Koller, M. 2014. *Simulations for Sharpening Wald-type Inference in Robust Regression for Small Samples*. Comprehensive R Archive Network. 2014. `http://cran.r-project.org/web/packages/robustbase/vignettes/lmrob_simulation.pdf`. 249

Koller, M. and W. A. Stahel 2011. Sharpening Wald-type inference in robust regression for small samples. *Computational Statistics and Data Analysis* 55(8):2504–2515. 249

Kossinets, G. 2006. Effects of missing data in social networks. *Social Networks* 28(3):247–268. 109

Kratochvíl, J. (ed.) 1999. *Graph Drawing: 7th International Symposium / Proceedings GD'99.* New York: Springer.

Krippendorff, K. H. 2012. *Content Analysis: An Introduction to Its Methodology* (third ed.). Thousand Oaks, Calif.: Sage. 133

Krug, S. 2014. *Don't Make Me Think: A Common Sense Approach to Web Usability* (third ed.). Upper Saddle River, N.J.: Pearson/New Riders.

Kuhn, M. 2014. *caret: Classification and Regression Training.* Comprehensive R Archive Network. 2014. `http://cran.r-project.org/web/packages/caret/caret.pdf`.

Kuhn, M. and K. Johnson 2013. *Applied Predictive Modeling.* New York: Springer.

Kutner, M. H., C. J. Nachtsheim, J. Neter, and W. Li 2004. *Applied Linear Statistical Models* (fifth ed.). Boston: McGraw-Hill.

Kvale, S. 1996. *InterViews: An Introduction to Qualitative Research Interviewing.* Thousand Oaks, Calif.: Sage. 275

Lacy, L. W. 2005. *Owl: Representing Information Using the Web Ontology Language.* Victoria, B.C.: Trafford. 235

Lamp, G. 2014. *ggplot for Python.* GitHub. 2014. `https://github.com/yhat/ggplot`. 252

Landauer, T. K., P. W. Foltz, and D. Laham 1998. An introduction to latent semantic analysis. *Discourse Processes* 25:259–284. Retrieved from the World Wide Web, October 31, 2014, at `http://lsa.colorado.edu/papers/dp1.LSAintro.pdf`. 176

Landauer, T. K., D. S. McNamara, S. Dennis, and W. Kintsch (eds.) 2014. *Handbook of Latent Semantic Analysis* (reprint ed.). New York: Psychology Press. 176

Lander, J. P. 2014. *R for Everyone: Advanced Analytics and Graphics.* Upper Saddle River, N.J.: Pearson Education.

Lang, D. T. 2014a. *Package RCurl: General Network (HTTP/FTP/...) Client Interface for R.* Comprehensive R Archive Network. 2014. `http://cran.r-project.org/web/packages/RCurl/RCurl.pdf`. 62

Lang, D. T. 2014b. *Package XML: Tools for Parsing and Generating XML within R and S-Plus.* Comprehensive R Archive Network. 2014. `http://cran.r-project.org/web/packages/XML/XML.pdf`. 62

Langville, A. N. and C. D. Meyer 2006. *Google's Page Rank and Beyond: The Science of Search Engine Rankings.* Princeton, N.J.: Princeton University Press.

Langville, A. N. and C. D. Meyer 2012. *Who's 1?: The Science of Rating and Ranking.* Princeton, N.J.: Princeton University Press.

Lantz, B. 2013. *Machine Learning with R.* Birmingham, U.K.: Packt Publishing.

Laursen, G. H. N. and J. Thorlund 2010. *Business Analytics for Managers: Taking Business Intelligence Beyond Reporting.* Hoboken, N.J.: Wiley. 239

Law, A. M. 2014. *Simulation Modeling and Analysis* (fifth ed.). New York: McGraw-Hill.

Le, C. T. 1997. *Applied Survival Analysis.* New York: Wiley.

Leber, J. 2013. The immortal life of the Enron e-mails. *MIT Technology Review.* `http://www.technologyreview.com/news/515801/the-immortal-life-of-the-enron-e-mails/`. 291

Ledoiter, J. 2013. *Data Mining and Business Analytics with R*. New York: Wiley.

Leetaru, K. 2011. *Data Mining Methods for Content Analysis: An Introduction to the Computational Analysis of Content*. New York: Routledge. 133

Lefkowitz, M. 2014. Twisted documentation. Contributors listed as Twisted Matrix Laboratories at `http://twistedmatrix.com/trac/wiki/TwistedMatrixLaboratories`. Documentation available at `http://twistedmatrix.com/trac/wiki/Documentation`.

Lehmann, J., R. Isele, M. Jakob, A. Jentzsch, D. Kontokostas, P. N. Mendes, S. Hellmann, M. Morsey, P. van Kleef, S. Auer, and C. Bizer 2014. DBpedia—A large-scale, multilingual knowledge base extracted from wikipedia. *Semantic Web Journal*:in press. Retrieved from the World Wide Web at `http://svn.aksw.org/papers/2013/SWJ_DBpedia/public.pdf`.

Leisch, F. and B. Gruen 2014. *CRAN Task View: Cluster Analysis & Finite Mixture Models*. Comprehensive R Archive Network. 2014. `http://cran.r-project.org/web/views/Cluster.html`.

Lenat, D., M. Witbrock, D. Baxter, E. Blackstone, C. Deaton, D. Schneider, J. Scott, and B. Shepard 2010, Fall. Harnessing Cyc to answer clinical researchers' ad hoc queries. *AI Magazine* 31(3):13–32.

Lengauer, T. and R. E. Tarjan 1979. A fast algorithm for finding dominators in a flowgraph. *ACM Transactions on Programming Languages and Systems (TOPLAS)* 1(1):121–141. 82

Leskovec, J. and C. Faloutsos 2006. Sampling from large graphs. Retrieved from the World Wide Web at `http://cs.stanford.edu/people/jure/pubs/sampling-kdd06.pdf`. 109

Leskovec, J., D. Huttenlocher, and J. Kleinbert 2010a. Predicting positive and negative links in online social networks. Retrieved from the World Wide Web at `http://cs.stanford.edu/people/jure/pubs/triads-chi10.pdf`.

Leskovec, J., D. Huttenlocher, and J. Kleinbert 2010b. Signed networks in social media. Retrieved from the World Wide Web at `http://cs.stanford.edu/people/jure/pubs/triads-chi10.pdf`.

Leskovec, J., K. J. Lang, A. Dasgupta, and M. W. Mahoney 2009. Community structure in large networks: Natural cluster sizes and the absence of large well-defined clusters. *Internet Mathematics* 6(1):29–123. `http://www.technologyreview.com/news/515801/the-immortal-life-of-the-enron-e-mails/`.

Lévy, P. 1997. *Collective Intelligence: Mankind's Emerging World in Cyberspace*. Cambridge, Mass.: Perseus Books. 265

Lévy, P. 2001. *Cyberculture*. Minneapolis: University of Minnesota Press. 265

Lewin-Koh, N. 2014. *Hexagon Binning: an Overview*. Comprehensive R Archive Network. 2014. `http://cran.r-project.org/web/packages/hexbin/vignettes/hexagon_binning.pdf`. 252

Lewis, T. G. 2009. *Network Science: Theory and Applications*. New York: Wiley. 224

Liaw, A. and M. Wiener 2014. *randomForest: Breiman and Cutler's Random Forests for Classification and Regression*. Comprehensive R Archive Network. 2014. `http://cran.r-project.org/web/packages/randomForest/randomForest.pdf`.

Lindahl, C. and E. Blount 2003. Weblogs: simplifying web publishing. *Computer* 36(11): 114–116.

Little, R. J. A. and D. B. Rubin 1987. *Statistical Analysis with Missing Data*. New York: Wiley.

Liu, B. 2010. Sentiment analysis and subjectivity. In N. Indurkhya and F. J. Damerau (eds.), *Handbook of Natural Language Processing* (second ed.)., pp. 627–665. Boca Raton, Fla.: Chapman and Hall/CRC.

Liu, B. 2011. *Web Data Mining: Exploring Hyperlinks, Contents, and Usage Data*. New York: Springer. 251

Liu, B. 2012. *Sentiment Analysis and Opinion Mining*. San Rafael, Calif.: Morgan & Claypool.

Lopez, V., C. Unger, P. Cimiano, and E. Motta 2013. Evaluating question answering over linked data. *Web Semantics* 21:3–13.

Lord, F. M. and M. R. Novick 1968. *Statistical Theories of Mental Test Scores*. Reading, Mass.: Addison-Wesley.

Lovink, G. (ed.) 2011. *Networks Withoug a Cause: A Critique of Social Media*. Cambridge, UK: Polity Press. 134

Luangkesorn, L. 2014. *Simulation Programming with Python*. University of Pittsburgh. 2014. Retrieved from the World Wide Web at `http://users.iems.northwestern.edu/~nelsonb/IEMS435/PythonSim.pdf`. Translation of chapter 4 of Nelson (2014).

Luger, G. F. 2008. *Artificial Intelligence: Structures and Strategies for Complex Problem Solving* (sixth ed.). Boston: Addison-Wesley. 251

Lumley, T. 2010. *Complex Surveys: A Guide to Analysis Using R*. New York: Wiley.

Lumley, T. 2014. *mitools: Tools for Multiple Imputation of Missing Data*. Comprehensive R Archive Network. 2014. `http://cran.r-project.org/web/packages/mitools/mitools.pdf`.

Maas, A. L., R. E. Daly, P. T. Pham, D. Huang, A. Y. Ng, and C. Potts 2011, June. Learning word vectors for sentiment analysis. In *Proceedings of the 49th Annual Meeting of the Association for Computational Linguistics: Human Language Technologies*, pp. 142–150. Portland, Ore.: Association for Computational Linguistics. Retrieved from the World Wide Web at `http://ai.stanford.edu/~amaas/papers/wvSent_acl2011.pdf`.

Macal, C. M. and M. J. North 2006. *Introduction to Agent-Based Modeling and Simulation*. `http://www.docstoc.com/docs/36015647/Introduction-to-Agent-based-Modeling-and-Simulation`.

Maechler, M. 2014a. *Package cluster*. Comprehensive R Archive Network. 2014. `http://cran.r-project.org/web/packages/cluster/cluster.pdf`.

Maechler, M. 2014b. *robustbase: Basic Robust Statistics*. Comprehensive R Archive Network. 2014. `http://cran.r-project.org/web/packages/robustbase/robustbase.pdf`. 249

Maisel, L. S. and G. Cokins 2014. *Predictive Business Analytics: Forward-Looking Capabilities to Improve Business Performance*. New York: Wiley. 239

Manly, B. F. J. 1994. *Multivariate Statistical Methods: A Primer* (second ed.). London: Chapman & Hall. 176

Mann, C. and F. Stewart 2000. *Internet Communication and Qualitative Research: A Handbook for Researching Online*. London: Sage. 265

Mann, C. and F. Stewart 2002. Internet interviewing. In J. F. Gubrium and J. A. Holstein (eds.), *Handbook of Interview Research: Context and Method*, Chapter 29, pp. 603–627. Thousand Oaks, Calif.: Sage.

Manning, C. D. and H. Schütze 1999. *Foundations of Statistical Natural Language Processing*. Cambridge: MIT Press.

Markham, A. N. 1998. *Life Online: Researching Real Experience in Virtual Space*. Walnut Creek, Calif.: AltaMira Press. 265

Maronna, R. A., D. R. Martin, and V. J. Yohai 2006. *Robust Statistics Theory and Methods*. New York: Wiley. 249

Matloff, N. 2011. *The Art of R Programming*. San Francisco: no starch press.

Maybury, M. T. (ed.) 1997. *Intelligent Multimedia Information Retrieval*. Menlo Park, Calif./ Cambridge: AAAI Press / MIT Press.

McArthur, R. and P. Bruza 2001. The ABCs of online community. In N. Zhong, Y. Yao, J. Liu, and S. Ohsuga (eds.), *Web Intelligence: Research and Development*, pp. 141–147. New York: Springer.

McCallum, Q. E. (ed.) 2013. *Bad Data Handbook*. Sebastopol, Calif.: O'Reilly.

McCullagh, P. and J. A. Nelder 1989. *Generalized Linear Models* (second ed.). New York: Chapman and Hall.

McGrayne, S. B. 2011. *The Theory that Would Not Die: How Bayes' Rule Cracked the Enigma Code, Hunted Down Russian Submarines and Emerged Triumphant from Two Centuries of Controversy*. New Haven, Conn.: Yale University Press. 244

McKellar, J. and A. Fettig 2013. *Twisted Network Programming Essentials: Event-driven Network Programming with Python* (second ed.). Sebastopol, Calif.: O'Reilly.

McLean, B. and P. Elkind 2003. *The Smartest Guys in the Room: The Amazing Rise and Scandalous Fall of Enron*. New York: Penguin.

McTaggart, R. and G. Daroczi 2014. *Package Quandl: Quandl Data Connection*. Comprehensive R Archive Network. 2014. `http://cran.r-project.org/web/packages/Quandl/Quandl.pdf` with additional online documentation available at `https://www.quandl.com/`. 62

Meadow, C. T., B. R. Boyce, and D. H. Kraft 2000. *Text Information Retrieval Systems* (second ed.). San Diego: Academic Press.

Merkl, D. 2002. Text mining with self-organizing maps. In W. Klösgen and J. M. Żytkow (eds.), *Handbook of Data Mining and Knowledge Discovery*, Chapter 46.9, pp. 903–910. Oxford: Oxford University Press.

Mertz, D. 2002. *Charming Python: SimPy Simplifies Complex Models*. IBM developerWorks. 2002. Retrieved from the World Wide Web at `http://www.ibm.com/developerworks/linux/library/l-simpy/l-simpy-pdf.pdf`.

Meyer, D. 2014a. *Proximity Measures in the proxy Package for R*. Comprehensive R Archive Network. 2014. `http://cran.r-project.org/web/packages/proxy/vignettes/overview.pdf`.

Meyer, D. 2014b. *proxy: Distance and Similarity Measures*. Comprehensive R Archive Network. 2014. `http://cran.r-project.org/web/packages/proxy/proxy.pdf`.

Meyer, D., E. Dimitriadou, K. Hornik, A. Weingessel, and F. Leisch 2014. *e1071: Misc Functions of the Department of Statistics (e1071), TU Wien*. Comprehensive R Archive Network. 2014. `http://cran.r-project.org/web/packages/e1071/e1071.pdf`.

Michalawicz, Z. and D. B. Fogel 2004. *How to Solve It: Modern Heuristics*. New York: Springer. 251

Mikowski, M. S. and J. C. Powell 2014. *Single Page Web Applications JavaScript End-to-End.* Shelter Island, N.Y.: Manning. 3

Milborrow, S. 2014. *rpart.plot: Plot rpart models. An Enhanced Version of plot.rpart.* Comprehensive R Archive Network. 2014. http://cran.r-project.org/web/packages/rpart.plot/rpart.plot.pdf.

Milgram, S. 1967. The small world problem. *Psychology Today* 1(1):60–67. 99

Miller, D. and D. Slater 2000. *The Internet: An Ethnographic Approach*. Oxford: Berg. 265

Miller, G. A. 1995. Wordnet: A lexical database for English. *Communications of the ACM* 38 (11):39–41. 259

Miller, J. H. and S. E. Page 2007. *Complex Adaptive Systems: An Introduction to Computational Models of Social Life*. Princeton, N. J.: Princeton University Press.

Miller, J. P. 2000a. *Millennium Intelligence: Understanding and Conducting Competitive Intelligence in the Digital Age*. Medford, N.J.: Information Today, Inc. 59

Miller, T. W. 2000b. Marketing research and the information industry. *CASRO Journal* 2000:21–26.

Miller, T. W. 2001, Summer. Can we trust the data of online research. *Marketing Research*: 26–32. Reprinted as "Make the call: Online results are a mixed bag," *Marketing News*, September 24, 2001:20–25.

Miller, T. W. 2002. Propensity scoring for multimethod research. *Canadian Journal of Marketing Research* 20(2):46–61.

Miller, T. W. 2005. *Data and Text Mining: A Business Applications Approach*. Upper Saddle River, N.J.: Pearson Prentice Hall.

Miller, T. W. 2015a. *Modeling Techniques in Predictive Analytics with Python and R: A Guide to Data Science*. Upper Saddle River, N.J.: Pearson Education. http://www.ftpress.com/miller.

Miller, T. W. 2015b. *Modeling Techniques in Predictive Analytics: Business Problems and Solutions with R* (revised and expanded ed.). Upper Saddle River, N.J.: Pearson Education. http://www.ftpress.com/miller.

Miller, T. W. and P. R. Dickson 2001. On-line market research. *International Journal of Electronic Commerce* 5(3):139–167.

Miller, T. W., D. Rake, T. Sumimoto, and P. S. Hollman 2001. Reliability and comparability of choice-based measures: Online and paper-and-pencil methods of administration. *2001 Sawtooth Software Conference Proceedings*:123–130. Published by Sawtooth Software, Sequim, WA.

Miller, T. W. and J. Walkowski (eds.) 2004. *Qualitative Research Online*. Manhattan Beach, Calif.: Research Publishers LLC. 261, 313

Miniwatts Marketing Group 2014. Internet users of the world. Retrieved from the World Wide Web, October 21, 2014, at http://www.internetworldstats.com/stats.htm. 45

Mitchell, M. 1996. *An Introduction to Genetic Algorithms*. Cambridge: MIT Press. 251

Mitchell, M. 2009. *Complexity: A Guided Tour*. Oxford, U.K.: Oxford University Press. 227

Moran, M. and B. Hunt 2009. *Search Engine Marketing, Inc.: Driving Search Traffic to Your Company's Web Site* (second ed.). Boston: IBM Press/Pearson Education.

Moreno, J. L. 1934. Who shall survive?: Foundations of sociometry, group psychotherapy, and sociodrama. Reprinted in 1953 (second edition) and in 1978 (third edition) by Beacon House, Inc., Beacon, N.Y. 96

Mosteller, F. and D. L. Wallace 1984. *Applied Bayesian and Classical Inference: The Case of "The Federalist" Papers* (second ed.). New York: Springer. Earlier edition published in 1964 by Addison-Wesley, Reading, Mass. The previous title was *Inference and Disputed Authorship: The Federalist*.

Müller, K., T. Vignaux, and C. Chui 2014. *SimPy: Discrete Event Simulation for Python.* SimPy Development Team. 2014. Retrieved from the World Wide Web at `https://simpy.readthedocs.org/en/latest/` with code available at `https://bitbucket.org/simpy/simpy/`.

Murphy, K. P. 2012. *Machine Learning: A Probabilistic Perspective.* Cambridge, Mass.: MIT Press. 176, 244, 251

Murrell, P. 2011. *R Graphics* (second ed.). Boca Raton, Fla.: CRC Press.

Nair, V. G. 2014. *Getting Started with Beautiful Soup: Build Your Own Web Scraper and Learn All About Web Scraping with Beautiful Soup.* Birmingham, UK: PACKT Publishing.

Nelson, B. L. 2013. *Foundations and Methods of Stochastic Simulation: A First Course.* New York: Springer. Supporting materials available from the World Wide Web at `http://users.iems.northwestern.edu/~nelsonb/IEMS435/`.

Nelson, W. B. 2003. *Recurrent Events Data Analysis for Product Repairs, Disease Recurrences, and Other Applications.* Series on Statistics and Applied Probability. Philadelphia and Alexandria, Va.: ASA-SIAM.

Neuendorf, K. A. 2002. *The Content Analysis Guidebook.* Thousand Oaks, Calif.: Sage.

Neuwirth, E. 2014. *Package RColorBrewer: ColorBrewer palettes.* Comprehensive R Archive Network. 2014. `http://cran.r-project.org/web/packages/RColorBrewer/RColorBrewer.pdf`.

Newman, M. E. J. 2010. *Networks: An Introduction.* Oxford, UK: Oxford University Press. 224

Nolan, D. and D. T. Lang 2014. *XML and Web Technologies for Data Sciences with R.* New York: Springer. 29

North, M. J. and C. M. Macal 2007. *Managing Business Complexity: Discovering Strategic Solutions with Agent-Based Modeling and Simulation.* Oxford, U.K.: Oxford University Press. 226

Nunnally, J. C. 1967. *Psychometric Theory.* New York: McGraw-Hill.

O'Hagan, A. 2010. *Kendall's Advanced Theory of Statistics: Bayesian Inference*, Volume 2B. New York: Wiley. 244

Osborne, J. W. 2013. *Best Practices in Data Cleaning: A Complete Guide to Everything You Need to Do Before and After Collecting Your Data.* Los Angeles: Sage.

Osgood, C. 1962. Studies in the generality of affective meaning systems. *American Psychologist* 17:10–28. 133

Osgood, C., G. Suci, and P. Tannenbaum (eds.) 1957. *The Measurement of Meaning.* Urbana, Ill.: University of Illinois Press. 133

Pang, B. and L. Lee 2008. Opinion mining and sentiment analysis. *Foundations and Trends in Information Retrieval* 2(1–2):1–135.

Pedregosa, F., G. Varoquaux, A. Gramfort, V. Michel, B. Thirion, O. Grisel, M. Blondel, P. Prettenhofer, R. Weiss, V. Dubourg, J. Vanderplas, A. Passos, D. Cournapeau, M. Brucher, M. Perrot, and E. Duchesnay 2011. Scikit-learn: Machine learning in Python. *Journal of Machine Learning Research* 12:2825–2830.

Penenberg, A. L. and M. Barry 2000. *Spooked: Espionage in Corporate America.* Cambridge, Mass.: Perseus Publishing. 56

Pentland, A. 2014. *Social Physics: How Good Ideas Spread—The Lessons from a New Science.* New York: Penguin.

Peters, G. and J. T. Woolley 2014. State of the Union Addresses and Messages. Online papers available from The American Presidency Project. Retrieved from the World Wide Web, November 1–14, 2014 at `http://www.presidency.ucsb.edu`. Documentation provided at `http://www.presidency.ucsb.edu/sou.php`. 295

Pinker, S. 1994. *The Language Instinct.* New York: W. Morrow and Co.

Pinker, S. 1997. *How the Mind Works.* New York: W.W. Norton & Company.

Pinker, S. 1999. *Words and Rules: The Ingredients of Language.* New York: HarperCollins.

Popping, R. 2000. *Computer-Assisted Text Analysis.* Thousand Oaks, Calif.: Sage. 133, 279

Porter, D. (ed.) 1997. *Internet Culture.* New York: Routledge. 265

Potts, C. 2011. On the negativity of negation. In *Proceedings of Semantics and Linguistic Theory 20*, pp. 636–659. CLC Publications. Retrieved from the World Wide Web at `http://elanguage.net/journals/salt/article/view/20.636/1414`.

Powazak, D. M. 2001. *Design for Community: The Art of Connecting Real People in Virtual Places.* Indianapolis: New Riders.

Powers, S. 2003. *Practical RDF.* Sebastopol, Calif.: O'Reilly. 235

Preece, J. 2000. *Online Communities: Designing Usability, Supporting Sociability.* New York: Wiley.

Priebe, C. E., J. M. Conroy, D. J. Marchette, and Y. Park 2005, October. Scan statistics on Enron graphs. *Computational and Mathematical Organization Theory* 11(3):229–247.

Priebe, J. 2009, April 29. A study of Internet users' cookie and JavaScript settings. Retrieved from the World Wide Web, October 21, 2014, at `http://smorgasbork.com/component/content/article/84-a-study-of-internet-users-cookie-and-javascript-settings`. 16

Provost, F. and T. Fawcett 2014. *Data Science for Business: What You Need to Know About Data Mining and Data-Analytic Thinking.* Sebastopol, Calif.: O'Reilly. 239

Pustejovsky, J. and A. Stubbs 2013. *Natural Language Annotation for Machine Learning.* Sebastopol, Calif.: O'Reilly.

Putler, D. S. and R. E. Krider 2012. *Customer and Business Analytics: Applied Data Mining for Business Decision Making Using R.* Boca Raton, Fla: Chapman & Hall/CRC.

Radcliffe-Brown, A. R. 1940. On social structure. *Journal of the Royal Anthropological Society of Great Britain and Ireland* 70:1–12. 96

Rajaraman, A. and J. D. Ullman 2012. *Mining of Massive Datasets.* Cambridge, UK: Cambridge University Press. 210

Reingold, E. M. and J. S. Tilford 1981. A fast algorithm for finding dominators in a flow-graph. *IEEE Transactions on Software Engineering* 7:223–228. 82

Reitz, K. 2014a, October 15. Python guide documentation (release 0.0.1). Retrieved from the World Wide Web, October 23, 2014, at `https://media.readthedocs.org/pdf/python-guide/latest/python-guide.pdf`.

Reitz, K. 2014b. Requests: HTTP for humans. Documentation available from the World Wide Web at `http://docs.python-requests.org/en/latest/`.

Rencher, A. C. and G. B. Schaalje 2008. *Linear Models in Statistics* (second ed.). New York: Wiley.

Resig, J. and BearBibeault 2013. *Secrets of the JavaScript Ninja.* Shelter Island, N.Y.: Manning. 3

Resnick, M. 1998. *Turtles, Termites, and Traffic Jams: Explorations in Massively Parallel Microworlds.* Cambridge, Mass.: MIT Press. 227

Rheingold, H. 2000. *The Virtual Community: Homesteading on the Electronic Frontier* (revised ed.). Cambridge: The MIT Press. 265

Ricci, F., L. Rokach, B. Shapira, and P. B. Kantor (eds.) 2011. *Recommender Systems Handbook.* New York: Springer.

Richardson, L. 2014. Beautiful soup documentation. Available from the World Wide Web at `http://www.crummy.com/software/BeautifulSoup/bs4/doc/`.

Robbins, J. N. 2003. *Learning Web Design: A Beginners Guide to HTML, CSS, JavaScript, and Web Graphics* (fourth ed.). Sebastopol, Calif.: O'Reilly.

Robert, C. P. 2007. *The Bayesian Choice: From Decision Theoretic Foundations to Computational Implementation* (second ed.). New York: Springer. 244

Robert, C. P. and G. Casella 2009. *Introducing Monte Carlo Methods with R.* New York: Springer. 244

Roberts, C. W. (ed.) 1997. *Text Analysis for the Social Sciences: Methods for Drawing Statistical Inferences from Texts and Transcripts.* Mahwah, N.J.: Lawrence Erlbaum Associates. 133, 279

Robinson, I., J. Webber, and E. Eifrem 2013. *Graph Databases.* Sebastopol, Calif.: O'Reilly.

Rosen, J. 2001. *The Unwanted Gaze: The Destruction of Privacy in America.* New York: Vintage. 134

Rossant, C. 2014. *IPython Interactive Computing and Visualization Cookbook.* Birmingham, UK: Packt Publishing.

Rousseeuw, P. J. 1987. Silhouettes: A graphical aid to the interpretation and validation of cluster analysis. *Journal of Computational and Applied Mathematics* 20:53–65.

Rubin, D. B. 1987. *Multiple Imputation for Nonresponse in Surveys.* New York: Wiley.

Russell, M. A. 2014. *Mining the Social Web: Data Mining Facebook, Twitter, LinkedIn, Google+, GitHub, and More.* Sebastopol, Calif.: O'Reilly. 61

Russell, S. and P. Norvig 2009. *Artificial Intelligence: A Modern Approach* (third ed.). Upper Saddle River, N.J.: Prentice Hall. 251

Rusu, A. 2014. Tree drawing algorithms. In R. Tamassia (ed.), *Handbook of Graph Drawing and Visualization*, Chapter 5, pp. 155–192. Boca Raton, Fla.: CRC Press/Chapman & Hall. 82

Ryan, G. W. and H. R. Bernard 2000. Data management and analysis methods. In N. K. Denzin and Y. S. Lincoln (eds.), *Handbook of Qualitative Research: Context and Method* (second ed.), Chapter 29, pp. 769–802. Thousand Oaks, Calif.: Sage. 279

Ryan, J. A. 2014. *Package quantmod: Quantitative Financial Modelling Framework.* Comprehensive R Archive Network. 2014. `http://cran.r-project.org/web/packages/quantmod/quantmod.pdf`. 62

Ryan, J. A. and J. M. Ulrich 2014. *Package xts: eXtensible Time Series.* Comprehensive R Archive Network. 2014. `http://cran.r-project.org/web/packages/xts/xts.pdf`. 62

Ryan, T. P. 2008. *Modern Regression Methods* (second ed.). New York: Wiley. 128

Šalamon, T. 2011. *Design of Agent-Based Models: Developing Computer Simulations for a Better Understanding of Social Processes.* Czech Republic: Tomáš Bruckner, Řepín-Živonín.

Samara, T. 2007. *Design Elements: A Graphic Style Manual.* Gloucester, Mass.: Rockport Publishers.

Sarkar, D. 2008. *Lattice: Multivariate Data Visualization with R.* New York: Springer.

Sarkar, D. 2014. *lattice: Lattice Graphics.* Comprehensive R Archive Network. 2014. `http://cran.r-project.org/web/packages/lattice/lattice.pdf`.

Sarkar, D. and F. Andrews 2014. *latticeExtra: Extra Graphical Utilities Based on Lattice.* Comprehensive R Archive Network. 2014. `http://cran.r-project.org/web/packages/latticeExtra/latticeExtra.pdf`.

Savage, D., X. Zhang, X. Yu, P. Chou, and Q. Wang 2014. Anomaly detection in online social networks. *Social Networks* 39:62–70.

Schafer, J. L. 1997. *Analysis of Incomplete Multivariate Data.* London: Chapman & Hall.

Schauerhuber, M., A. Zeileis, D. Meyer, and K. Hornik 2008. Benchmarking open-source tree learners in R/RWeka. In C. Preisach, H. Burkhardt, L. Schmidt-Thieme, and R. Decker (eds.), *Data Analysis, Machine Learning, and Applications*, pp. 389–396. New York: Springer.

Schaul, J. 2014. *ComplexNetworkSim Package Documentation.* pypi.python.org. 2014. Retrieved from the World Wide Web at `https://pythonhosted.org/ComplexNetworkSim/` with code available at `https://github.com/jschaul/ComplexNetworkSim`.

Schnettler, S. 2009. A structured overview of 50 years of small-world research. *Social Networks* 31:165–178. 99

Schrott, P. R. and D. J. Lanoue 1994. Trends and perspectives in content analysis. In I. Borg and P. Mohler (eds.), *Trends and Perspectives in Empirical Social Research*, pp. 327–345. Berlin: Walter de Gruyter.

Scott:, J. 2013. *Social Network Analysis* (third ed.). Thousand Oaks, Calif.: Sage.

Seale, C. 2002. Computer-assisted analysis of qualitative interview data. In J. F. Gubrium and J. A. Holstein (eds.), *Handbook of Interview Research: Context and Method*, Chapter 31, pp. 651–670. Thousand Oaks, Calif.: Sage. 279

Sebastiani, F. 2002. Machine learning in automated text categorization. *ACM Computing Surveys* 34(1):1–47.

Seber, G. A. F. 2000. *Multivariate Observations.* New York: Wiley. Originally published in 1984. 176

Segaran, T., C. Evans, and J. Taylor 2009. *Programming the Semantic Web*. Sebastopol, Calif.: O'Reilly. 235

Sen, R. and M. H. Hansen 2003, March. Predicting web user's next access based on log data. *Journal of Computational and Graphical Statistics* 12(1):143–155.

Senkul, P. and S. Salin 2012, March. Improving pattern quality in web usage mining by using semantic information. *Knowledge and Information Systems* 30(3):527–541.

SEOMoz, Inc. 2013. Search engine ranking factors: survey and correlation data. Retrieved from the World Wide Web, November 1, 2014, at `http://moz.com/search-ranking-factors`.

Shakhnarovich, G., T. Darrell, and P. Indyk (eds.) 2006. *Nearest-Neighbor Methods in Learning and Vision: Theory and Practice*. Cambridge, Mass.: MIT Press.

Sharda, R. and D. Delen 2006. Predicting box office success of motion pictures with neural networks. *Expert Systems with Applications* 30:243–254. 133

Sharma, S. 1996. *Applied Multivariate Techniques*. New York: Wiley. 176

Sherry, Jr., J. F. and R. V. Kozinets 2001. Qualitative inquiry in marketing and consumer research. In D. Iacobucci (ed.), *Kellogg on Marketing*, Chapter 8, pp. 165–194. New York: Wiley.

Shmueli, G. 2010. To explain or predict? *Statistical Science* 25(3):289–310.

Shroff, G. 2013. *The Intelligent Web: Search, Smart Algorithms, and Big Data*. Oxford, U.K.: Oxford University Press.

Siegel, E. 2013. *Predictive Analytics: The Power to Predict Who Will Click, Buy, Lie, or Die*. Hoboken, N.J.: Wiley. 239

Silverman, D. 1993. *Interpreting Qualitative Data: Methods for Analysing Talk, Text, and Interaction*. London: Sage. 279

Silverman, D. 2000a. Analyzing talk and text. In N. K. Denzin and Y. S. Lincoln (eds.), *Handbook of Qualitative Research: Context and Method* (second ed.), Chapter 31, pp. 821–834. Thousand Oaks, Calif.: Sage. 279

Silverman, D. 2000b. *Doing Qualitative Research: A Practical Approach*. Thousand Oaks, Calif.: Sage.

Silverman, D. 2001. *Interpreting Qualitative Data: Methods for Analysing Talk, Text, and Interaction* (second ed.). Thousand Oaks, Calif.: Sage.

Simpson, J. 2002. *XPath and XPointer: Locating Content in XML Documents*. Sebastopol, Calif.: O'Reilly. 29

Sing, T., O. Sander, N. Beerenwinkel, and T. Lengauer 2005. ROCR: Visualizing classifier performance in R. *Bioinformatics* 21(20):3940–3941.

Slater, J. B., P. V. M. Broekman, M. Corris, A. Iles, B. Seymour, and S. Worthington (eds.) 2013. *Proud to Be Flesh—A Mute Magazine Anthology of Cultural Politics After the Net*. London: Mute Publishing.

Smith, J. A. and J. Moody 2013. Structural effects of network sampling coverage I: Nodes missing at random. *Social Networks* 35:652–668. 109

Smith, M. A. and P. Kollock (eds.) 1999. *Communities in Cyberspace*. New York: Routledge. 265

Snedecor, G. W. and W. G. Cochran 1989. *Statistical Methods* (eighth ed.). Ames, Iowa: Iowa State University Press. First edition published by Snedecor in 1937. 244

Socher, R., J. Pennington, E. H. Huang, A. Y. Ng, and C. D. Manning 2011. Semi-Supervised Recursive Autoencoders for Predicting Sentiment Distributions. In *Proceedings of the 2011 Conference on Empirical Methods in Natural Language Processing (EMNLP)*.

Srivastava, A. N. and M. Sahami (eds.) 2009. *Text Mining: Classification, Clustering, and Applications*. Boca Raton, Fla.: CRC Press.

StatCounter 2014. StatCounter global stats. Retrieved from the World Wide Web, October 21, 2014, at `http://gs.statcounter.com/#browser-ww-monthly-200807-201410`.

Stefanov, S. 2010. *JavaScript Patterns: Building Better Applications with Coding and Design Patterns*. Sebastopol, Calif.: O'Reilly. 3

Stone, P. J. 1997. Thematic text analysis: New agendas for analyzing text content. In C. W. Roberts (ed.), *Text Analysis for the Social Sciences: Methods for Drawing Statistical Inferences from Texts and Transcripts*, Chapter 2, pp. 35–54. Mahwah, N.J.: Lawrence Erlbaum Associates. 279

Stone, P. J., D. C. Dunphy, M. S. Smith, and D. M. Ogilvie 1966. *The General Inquirer: A Computer Approach to Content Analysis*. Cambridge: MIT Press.

Strauss, A. and J. Corbin 1998. *Basics of Qualitative Research: Techniques and Procedures for Developing Grounded Theory* (second ed.). Thousand Oaks, Calif.: Sage. 279

Stuart, A., K. Ord, and S. Arnold 2010. *Kendall's Advanced Theory of Statistics: Classical Inference and the Linear Model*, Volume 2A. New York: Wiley. 244

Suess, E. A. and B. E. Trumbo 2010. *Introduction to Probability Simulation and Gibbs Sampling with R*. New York: Springer. 244

Sullivan, D. 2001. *Document Warehousing and Text Mining: Techniques for Improving Business Operations, Marketing, and Sales*. New York: Wiley.

Supowit, K. J. and E. M. Reingold 1983. The complexity of drawing trees nicely. *Acta Informatica* 18:177–392. 82

Taddy, M. 2013a. Measuring political sentiment on Twitter: factor-optimal design for multinomial inverse regression. Retrieved from the World Wide Web at `http://arxiv.org/pdf/1206.3776v5.pdf`.

Taddy, M. 2013b. Multinomial inverse regression for text analysis. Retrieved from the World Wide Web at `http://arxiv.org/pdf/1012.2098v6.pdf`.

Taddy, M. 2014. *textir: Inverse Regression for Text Analysis*. 2014. `http://cran.r-project.org/web/packages/textir/textir.pdf`.

Tamassia, R. (ed.) 2014. *Handbook of Graph Drawing and Visualization*. Boca Raton, Fla.: CRC Press/Chapman & Hall.

Tan, P.-N., M. Steinbach, and V. Kumar 2006. *Introduction to Data Mining*. Boston: Addison-Wesley. 210

Tang, W., H. He, and X. M. Tu 2012. *Applied Categorical and Count Data Analysis*. Boca Raton, Fla.: Chapman & Hall/CRC.

Tannenbaum, A. S. and D. J. Wetherall 2010. *Computer Networks* (fifth ed.). Upper Saddle River, N.J.: Pearson/Prentice Hall.

Tanner, M. A. 1996. *Tools for Statistical Inference: Methods for the Exploration of Posterior Distributions and Likelihood Functions* (third ed.). New York: Springer. 244

Tennison, J. 2001. *XSLT and XPath On The Edge*. New York: Wiley. 29

Therneau, T. 2014. *survival: Survival Analysis.* Comprehensive R Archive Network. 2014. `http://cran.r-project.org/web/packages/survival/survival.pdf`.

Therneau, T., B. Atkinson, and B. Ripley 2014. *rpart: Recursive Partitioning.* Comprehensive R Archive Network. 2014. `http://cran.r-project.org/web/packages/rpart/rpart.pdf`.

Therneau, T. and C. Crowson 2014. *Using Time Dependent Covariates and Time Dependent Coefficients in the Cox Model.* Comprehensive R Archive Network. 2014. `http://cran.r-project.org/web/packages/survival/vignettes/timedep.pdf`.

Therneau, T. M. and P. M. Grambsch 2000. *Modeling Survival Data: Extending the Cox Model.* New York: Springer.

Thiele, J. C. 2014. *Package RNetLogo: Provides an Interface to the Agent-Based Modelling Platform NetLogo.* Comprehensive R Archive Network. 2014. User manual: `http://cran.r-project.org/web/packages/RNetLogo/RNetLogo.pdf`. Additional documentation available at `http://rnetlogo.r-forge.r-project.org/`.

Thompson, M. 1975. On any given Sunday: Fair competitor orderings with maximum likelihood methods. *Journal of the American Statistical Association* 70(351):536–541. 103

Thompson, T. (ed.) 2000. *Writing about Business* (second ed.). New York: Columbia University Press.

T.Hothorn, F. Leisch, A. Zeileis, and K. Hornik 2005, September. The design and analysis of benchmark experiments. *Journal of Computational and Graphical Statistics* 14(3):675–699.

Thurstone, L. L. 1927. A law of comparative judgment. *Psychological Review* 34:273–286.

Tibshirani, R. 1996. Regression shrinkage and selection via the lasso. *Journal of the Royal Statistical Society, Series B* 58:267–288. 249

Tidwell, J. 2011. *Designing Interfaces: Patterns for Effective Interactive Design* (second ed.). Sebastopol, Calif.: O'Reilly.

Toivonen, R., L. Kovanen, J.-P. O. Mikko Kivelä, J. SaramŁki, and K. Kaski 2009. A comparative study of social network models: Network evolution models and nodal attribute models. *Social Networks* 31:240–254.

Tong, S. and D. Koller 2001. Support vector machine active learning with applications to text classification. *Journal of Machine Learning Research* 2:45–66.

Torgerson, W. S. 1958. *Theory and Methods of Scaling.* New York: Wiley.

Travers, J. and S. Milgram 1969, December. Experimental study of small world problem. *Sociometryy* 32(4):425–443. 99

Trybula, W. J. 1999. Text mining. In M. E. Williams (ed.), *Annual Review of Information Science and Technology,* Volume 34, Chapter 7, pp. 385–420. Medford, N.J.: Information Today, Inc.

Tsvetovat, M. and A. Kouznetsov 2011. *Social Network Analysis for Startups: Finding Connections on the Social Web.* Sebastopol, Calif.: O'Reilly.

Tufte, E. R. 1990. *Envisioning Information.* Cheshire, Conn.: Graphics Press.

Tufte, E. R. 1997. *Visual Explanations: Images and Quantities, Evidence and Narrative.* Cheshire, Conn.: Graphics Press.

Tufte, E. R. 2004. *The Visual Display of Quantitative Information* (second ed.). Cheshire, Conn.: Graphics Press.

Tufte, E. R. 2006. *Beautiful Evidence*. Cheshire, Conn.: Graphics Press.

Tukey, J. W. 1977. *Exploratory Data Analysis*. Reading, Mass.: Addison-Wesley.

Tukey, J. W. and F. Mosteller 1977. *Data Analysis and Regression: A Second Course in Statistics*. Reading, Mass.: Addison-Wesley.

Turney, P. D. 2002, July 8–10. Thumbs up or thumbs down? Semantic orientation applied to unsupervised classification of reviews. *Proceedings of the 40th Annual Meeting of the Association for Computational Linguistics (ACL '02)*:417–424. Available from the National Research Council Canada publications archive. 134

Turow, J. 2013. *The Daily You: How the New Advertising Industry Is Defining Your Identity and Your Worth*. New Haven, Conn.: Yale University Press. 134

Uddin, S., J. Hamra, and L. Hossain 2013. Exploring communication networks to understand organizational crisis using exponential random graph models. *Computational and Mathematical Organization Theory* 19:25–41.

Uddin, S., A. Khan, L. Hossain, M. Piraveenan, and S. Carlsson 2014. A topological framework to explore longitudinal social networks. *Computational and Mathematical Organization Theory*:21 pages. Manuscript published online by Springer Science+Business Media, New York.

Unwin, A., M. Theus, and H. Hofmann (eds.) 2006. *Graphics of Large Datasets: Visualizing a Million*. New York: Springer. 252

Vanderbei, R. J. 2014. *Linear Programming: Foundations and Extensions* (fourth ed.). New York: Springer. 225

Vapnik, V. N. 1998. *Statistical Learning Theory*. New York: Wiley. 129

Vapnik, V. N. 2000. *The Nature of Statistical Learning Theory* (second ed.). New York: Springer. 129

Venables, W. N. and B. D. Ripley 2002. *Modern Applied Statistics with S* (fourth ed.). New York: Springer-Verlag. Champions of S, S-Plus, and R call this *the mustard book*.

Vine, D. 2000. *Internet Business Intelligence: How to Build a Big Company System on a Small Company Budget*. Medford, N.J.: Information Today, Inc. 59

W3Techs 2014. Usage of traffic analysis tools for websites. Retrieved from the World Wide Web, October 21, 2014, at http://w3techs.com/technologies/overview/traffic_analysis/all. 3, 14

Walther, J. B. 1996. Computer-mediated communication: impersonal, interpersonal, and hyperpersonal interaction. *Communication Research* 23(1):3–41. 265

Wanderschneider, M. 2013. *Learning Node.js: A Hands-On Guide to Building Web Applications in JavaScript*. Upper Saddle River, N.J.: Pearson Education/Addison-Wesley. 3

Wang, X., H. Tao, Z. Xie, and D. Yi 2013. Mining social networks using wave propogation. *Computational and Mathematical Organization Theory* 19:569–579.

Wasserman, L. 2010. *All of Statistics: A Concise Course in Statistical Inference*. New York: Springer. 244

Wasserman, S. and C. Anderson 1987. Stochastic *a posteriori* blockmodels: Construction and assessment. *Social Networks* 9:1–36.

Wasserman, S. and K. Faust 1994. *Social Network Analysis: Methods and Applications*. Cambridge, UK: Cambridge University Press. Chapter 4: Graphs and Matrices contributed by D. Iacobucci.

Wasserman, S. and D. Iacobucci 1986. Statistical analysis of discrete relational data. *British Journal of Mathematical and Statistical Psychology* 39:41–64.

Wasserman, S. and P. Pattison 1996, September. Logit models and logistic regression for social networks: I An introduction to markov graphs and p^*. *Psychometrika* 61(3):401–425.

Watts, D. J. 1999. *Small Worlds: The Dynamics of Networks between Order and Randomness*. Princeton, N.J.: Princeton University Press. 99

Watts, D. J. 2003. *Six Degrees: The Science of a Connected Age*. New York: W.W. Norton.

Watts, D. J. and S. H. Strogatz 1998. Collective dynamics of 'small-world' networks. *Nature* 393(6684):440–442. 99, 224

Wei, T. H. 1952. The algebraic foundations of ranking theory. Ph.D. thesis, Cambridge University, Cambridge, UK.

Weiner, J. 2014. *Package riverplot: Sankey or ribbon plots*. Comprehensive R Archive Network. 2014. http://cran.r-project.org/web/packages/riverplot/riverplot.pdf.

Weiss, S. M., N. Indurkhay, and T. Zhang 2010. *Fundamentals of Predictive Text Mining*. New York: Springer.

Weiss, S. M., N. Indurkhya, and T. Zhang 2010. *Fundamentals of Predictive Text Mining*. New York: Springer.

Weitzman, E. A. 2000. Software and qualitative research. In N. K. Denzin and Y. S. Lincoln (eds.), *Handbook of Qualitative Research: Context and Method* (second ed.)., Chapter 30, pp. 803–820. Thousand Oaks, Calif.: Sage. 279

Weitzman, E. A. and M. B. Miles 1995. *Computer Programs for Qualitative Data Analysis*. Thousand Oaks, Calif.: Sage. 279

Werry, C. and M. Mowbray (eds.) 2001. *Online Communities: Commerce, Community Action, and the Virtual University*. Upper Saddle River, N.J.: Prentice-Hall.

West, B. T. 2006. A simple and flexible rating method for predicting success in the NCAA basketball tournament. *Journal of Quantitative Analysis in Sports* 2(3):1–14. 103

West, M. D. (ed.) 2001. *Theory, Method, and Practice in Computer Content Analysis*. Westport, Conn.: Ablex. 133, 279

White, T. 2011. *Hadoop: The Definitive Guide* (second ed.). Sebastopol, Calif.: O'Reilly.

Wickham, H. 2010. stringr: Modern, consistent string processing. *The R Journal* 2(2):38–40.

Wickham, H. 2014a. *Advanced R*. Boca Raton, Fla.: Chapman & Hall/CRC.

Wickham, H. 2014b. *stringr: Make It Easier to Work with Strings*. Comprehensive R Archive Network. 2014. http://cran.r-project.org/web/packages/stringr/stringr.pdf. 259

Wickham, H. and W. Chang 2014. *ggplot2: An Implementation of the Grammar of Graphics*. Comprehensive R Archive Network. 2014. http://cran.r-project.org/web/packages/ggplot2/ggplot2.pdf. 62, 252

Wikipedia 2014a. 2011 Virginia earthquake—Wikipedia, the free encyclopedia. Retrieved from the World Wide Web, September 18, 2014, at url = http://en.wikipedia.org/w/index.php?title=2011_Virginia_earthquake&oldid=625563344.

Wikipedia 2014b. Directory on privacy and electronic communications. Retrieved from the World Wide Web, October 21, 2014, at `http://en.wikipedia.org/wiki/Directive_on_Privacy_and_Electronic_Communication`. 19

Wikipedia 2014c. History of Wikipedia—Wikipedia, the free encyclopedia. Retrieved from the World Wide Web, September 23, 2014, at `http://en.wikipedia.org/wiki/History_of_Wikipedia`.

Wilensky, U. 1999. Netlogo. NetLogo 5.1.0 User Manual and documentation available from the Center for Connected Learning and Computer-Based Modeling, Northwestern University, Evanston, IL at `http://ccl.northwestern.edu/netlogo/`.

Wilkinson, L. 2005. *The Grammar of Graphics* (second ed.). New York: Springer.

Williams, H. P. 2013. *Model Building in Mathematical Programming* (fifth ed.). New York: Wiley. 225

Witten, I. H., E. Frank, and M. A. Hall 2011. *Data Mining: Practical Machine Learning Tools and Techniques*. Burlington, Mass.: Morgan Kaufmann. 210, 250

Witten, I. H., A. Moffat, and T. C. Bell 1999. *Managing Gigabytes: Compressing and Indexing Documents and Images* (second ed.). San Francisco: Morgan Kaufmann.

Wolcott, H. F. 1994. *Transforming Qualitative Data: Description, Analysis, and Interpretation.* Thousand Oaks, Calif.: Sage.

Wolcott, H. F. 1999. *Ethnography: A Way of Seeing.* Walnut Creek, Calif.: AltaMira.

Wolcott, H. F. 2001. *The Art of Fieldwork.* Walnut Creek, Calif.: AltaMira.

Wood, D., M. Zaidman, and L. Ruth 2014. *Linked Data: Structured Data on the Web.* Shelter Island, N.Y.: Manning.

Yau, N. 2011. *Visualize This: The FlowingData Guide to Design, Visualization, and Statistics.* New York: Wiley.

Yau, N. 2013. *Data Points: Visualization That Means Something.* New York: Wiley.

Youmans, G. 1990. Measuring lexical style and competence: The type-token vocabulary curve. *Style* 24(4):584–599. 259

Youmans, G. 1991. A new tool for discourse analysis: The vocabulary management profile. *Language* 67(4):763–789. 259

Zeileis, A., G. Grothendieck, and J. A. Ryan 2014. *Package zoo: S3 Infrastructure for Regular and Irregular Time Series (Z's ordered observations).* Comprehensive R Archive Network. 2014. `http://cran.r-project.org/web/packages/zoo/zoo.pdf`. 62

Zeileis, A., K. Hornik, and P. Murrell 2009, July. Escaping RGBland: Selecting colors for statistical graphics. *Computational Statistics and Data Analysis* 53(9):3259–3270.

Zeileis, A., K. Hornik, and P. Murrell 2014. *HCL-Based Color Palettes in R.* Comprehensive R Archive Network. 2014. `http://cran.r-project.org/web/packages/colorspace/vignettes/hcl-colors.pdf`.

Zubcsek, P. P., I. Chowdhury, and Z. Katona 2014. Information communities: The network structure of communication. *Social Networks* 38:50–62.

Index

351